GPRS in Practice

GPRS in Practice
A Companion to the Specifications

Peter McGuiggan

PMCG Consultancy, UK

John Wiley & Sons, Ltd

Other Wiley Editorial Offices

John Wiley & Sons Inc., 111 River Street, Hoboken, NJ 07030, USA

Jossey-Bass, 989 Market Street, San Francisco, CA 94103-1741, USA

Wiley-VCH Verlag GmbH, Boschstr. 12, D-69469 Weinheim, Germany

John Wiley & Sons Australia Ltd, 33 Park Road, Milton, Queensland 4064, Australia

John Wiley & Sons (Asia) Pte Ltd, 2 Clementi Loop #02-01, Jin Xing Distripark, Singapore 129809

John Wiley & Sons Canada Ltd, 22 Worcester Road, Etobicoke, Ontario, Canada M9W 1L1

British Library Cataloguing in Publication Data

A catalogue record for this book is available from the British Library

ISBN 0-470-09507-5

Typeset in 10/12pt Times by Integra Software Services Pvt. Ltd, Pondicherry, India

This book is printed on acid-free paper responsibly manufactured from sustainable forestry
in which at least two trees are planted for each one used for paper production.

To Angela

and Maximillian

Contents

Preface

The GSM system has now been in operation for 12 years and is a phenomenal success.

This beautifully engineered system has passed all its tests with flying colours and is probably the most successful self-contained technology ever!

As they say, 1000 million (purported) customers cannot all be wrong!

Is the GPRS system 'bolted on' to the GSM system as elegant in design and will it become equally successful?

The answer to the first part of this question is probably no, it is not such an elegant design as it has to be adapted to a system which is designed for telephone conversations. The answer to the second part of the question is wait and see. It is becoming popular, but whether it will become so popular as to extend the life of GSM remains to be seen. EGPRS will assist in this direction, but only time will tell.

This book is constructed from four years of presenting the course 'GPRS Operations' around the world. The course has been improved over the years as the GPRS system has become operational and answers have been provided to '. . . what will happen if' and also in no small part to the intense scrutiny to which the course notes have been subjected and the ensuing lively discussions with engineers expert in their fields.

The book tries to present the subject matter in a simple way using simple English, whilst at the same time attempting to be thorough and rigorous.

Any engineer who has used the specifications to learn and understand the GSM and GPRS systems will know that the specifications '. . . do what it says on the tin'; they give specifications and not explanations. In general it is left to the reader to construct the underlying concepts and puzzle out how the specifications will work in practice. There are some exceptions to this rule, but they are unfortunately thin on the ground. Learning from the specifications is very hard work.

This book attempts to give the engineer the concepts of GPRS and also to explain in detail how the GPRS air interface works. It is hoped that it will make a useful companion to the specifications, with the specifications giving the word and this book casting some light upon those words.

There are many explanatory diagrams in the book. This is because I believe we try to think in pictures as the best way of conceptualising quite complex ideas. Many of the diagrams are 'busy', but if the reader takes the time to examine the diagrams in conjunction with the text, then I believe they will gain a good understanding of the GPRS concepts.

I would like to thank the many engineers around the world who have (in many cases unwittingly) helped me to write (and re-write!) this book. Your contributions to my understanding of the GPRS system have been invaluable. In particular I would like to thank Eric Mrzynski of Agere, Germany, a remarkable engineer who has contributed much through his discussions and correspondence. Also Dr Philip Williams of Sharp, UK who has always been open-minded in his discussions and constructive in his criticisms.

Finally I must thank my wife Angela; as she well knows, the writing of a book is a most solipsist, insufferable occupation, demanding patience, forbearance and tact from those closest to the author.

Peter McGuigann
Tamworth England
April 2004

1

Introduction

1.1 The purpose of GPRS

Because of the Internet, data communications over the fixed line telephone network now exceed voice telephone calls in minutes and revenue. GPRS may achieve the same end for mobile customers and, it could be said, has the business aim to provide wireless Internet access to a mass market.

Wireless Internet access is also commonly available using standard GSM data circuits and this capability is enhanced with the advent of HSCSD – high speed circuit switched data, which allows suitably equipped mobile stations to access multiple physical channels, raising the access speed to multiples of 9.6 kb/s, the maximum rate for a single physical channel. In fact GPRS can do nothing that cannot already be done by HSCSD. CS means circuit switched, and HSCSD is high speed circuit switched data which uses multiple physical channels to attain higher speeds.

1.2 So why GPRS?

A mass market probably emerges if the service provided is desirable and affordable. HSCSD is desirable, but whether it is affordable to a mass market remains to be seen. GPRS should be both desirable and affordable, as illustrated in Figure 1.1.

The available GSM radio bandwidth, the communication medium affected, is used efficiently with packet transmission over the air interface, and this results in savings to the customer. In addition, a customer uses the radio resources only when there is data to be sent on the downlink or uplink and this should result in the customer being charged only for the packets which are transmitted or received. Compare this with HSCSD, where the radio resources are provided continuously until the call is closed down, and the cost to the customer using GPRS should be reduced significantly.

GPRS in Practice: A Companion to the Specifications   Peter McGuiggan
© 2004 John Wiley & Sons, Ltd   ISBN: 0-470-09507-5

(a)   In circuit switched communications a full duplex circuit is provided between the
       DTEs. As full duplex is rarely used and one direction, when in use is not used for
       100 % of the time, CS communication wastes resources and the user pays for the
       hire of the circuit irrespective of the percentage of the hired time that is used. This
       makes CS communications expensive, especially in Internet applications where
       there are long 'idle' periods in both directions.

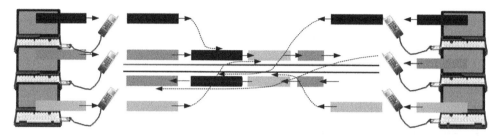

(b)   Packet switched data communications uses the link much more efficiently and
       should be cheaper to the user.

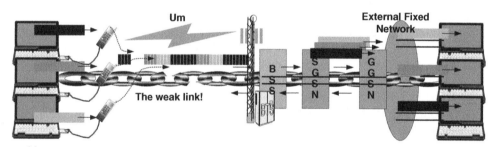

(c)   The sole purpose of GPRS is to condition standard IP 'packets' for transmission
       across the 'weak link' – the air interface Um, thus providing cheap mobile Internet
       access. It also provides an opaque vocabulary!

Figure 1.1   (a) Circuit switched data communications link; (b) Packet switched data communications link; (c) The purpose of GPRS

A GPRS subscriber may also stay connected to the Internet all day, being charged only when the radio link is actually used to transfer data. This is largely true but oversimplified, as we shall see. (What this really means is the mobile station, once communication is established with the Internet, is then in a position to receive data from the Internet – this condition can indeed last all day – or days if the GPRS sub-network operator tolerates it for so long!) The same capability should be possible with standard GSM data circuits but is currently not provided.

This feature compares very favourably with current fixed network practices where Internet data can only be sent to a customer after the customer has initiated a call to the Internet, and the communication is generally then closed down. However, at this time (January 2004) some Internet service providers (ISPs) and ISDN network providers are offering all-day Internet connection at very reasonable fixed monthly charges – in some cases even free connection – and of course broadband connections over landlines already do provide 'all-day' online services.

If GPRS over GSM does not become a mass-market service, then it could face a decline as has happened with other telecommunications services. If GPRS over GSM does become a mass-market service, its capabilities will be enhanced through the application of EDGE – enhanced data rates for GSM evolution – which in theory will raise the data rate of a GSM physical channel to 60 kb/s. These data rates could even delay the introduction of UMTS (the cellular technology that will eventually replace GSM), but in Europe enormous sums are committed already to UMTS air-space, so it could be, because of this, that GSM GPRS services have a short lifespan.

Technically, GPRS over GSM is designed for one purpose – to adapt standard data packet communications (TCP/IP, and other Internet standards) to the communication medium of the GSM air interface.

1.3   Internet communication

Figure 1.2 shows the GPRS sub-network used for Internet communication. It is called a sub-network because it is a conduit to another network, the Internet.

Note from this diagram that the SGSN (serving GPRS support node) has a split personality. Facing downlink to the radio side it is a sub-network (the GPRS network is called the sub-network in the specifications), and facing uplink towards the Internet it is looking at the backbone network. The backbone network is almost identical to the standard Internet transmission network protocol.

There are then three networks to consider:

1. The true GPRS network which is called the sub-network.
2. The backbone network linking the SGSN with GGSN (gateway GPRS support nodes).
3. The network, which is the external network (most commonly the Internet) to which the GPRS sub-network provides connection services to its customers.

It is interesting to note that the GPRS mobile station includes both network and sub-network layers. This book will take the liberty of referring to the mobile station network layer interchangeably as the network, user or application layer.

The *SGSN* has the responsibility of setting up data calls when requested by a mobile station. This is called *PDP context activation*. The SGSN checks with the HLR that the requested service is valid, and then asks the GGSN to set up the data call to the Internet. A part of this call set-up includes the allocation of an IP address to the call – if the call set-up is successful, the mobile station is informed of the allocated IP address. This is dynamic IP address allocation.

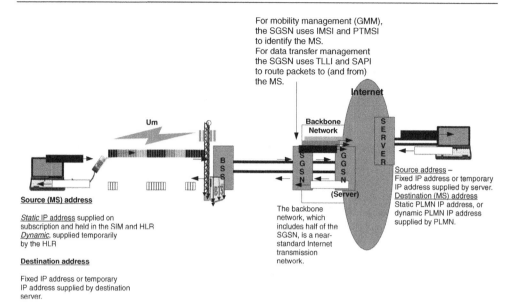

For mobility management (GMM),
the SGSN uses IMSI and PTMSI
to identify the MS.
For data transfer management
the SGSN uses TLLI and SAPI
to route packets to (and from)
the MS.

Source (MS) address

Static IP address supplied on
subscription and held in the SIM and HLR
Dynamic, supplied temporarily
by the HLR

Destination address

Fixed IP address or temporary
IP address supplied by destination
server.

The backbone
network, which
includes half of the
SGSN, is a near-
standard Internet
transmission
network.

Source address –
Fixed IP address or temporary
IP address supplied by server.
Destination (MS) address
Static PLMN IP address, or
dynamic PLMN IP address
supplied by PLMN.

Figure 1.2   The GPRS network for Internet communications

The SGSN also manages the attachment of mobiles to the GPRS sub-network
(exactly analogous to GSM IMSI attach), and the tracking of the location, (cell or
routeing area) for each attached mobile station.

Each *GGSN* has a specific IP address of the form gatewayxx. (mnc). (mcc). gprs and
the Internet regards the GGSN as a server or router in the same way as any other
Internet server or router.

The GPRS IP addresses allocated to mobile stations may be dynamic or static. The
GPRS sub-network allocates dynamic IP addresses to a GPRS subscriber when a call is
set up (PDP context activation). The mobile station keeps this allocated dynamic IP
address when a call is completed and the Internet may use this address to send data to
the mobile station. However, dynamic IP addresses are only allocated when a mobile
station initiates a call and the mobile station cannot receive incoming Internet initiated
calls – the Internet cannot initiate calls to an unknown address! Static addresses are
permanent IP addresses. These can receive Internet initiated calls, but there are com-
plications, as we shall see.

Figure 1.2 shows three subscribers time division multiplexed onto one physical
channel (not, it must be added, in standard GPRS form!). In fact GPRS allows up
to eight subscribers to be multiplexed on to one physical channel if dynamic
allocation is used. There are three methods of allocating radio resources to a mobile
station:

- **Dynamic allocation**, where the GPRS sub-network gives a series of allocations during
  a TBF (temporary block flow);
- **Fixed allocation**, where the GPRS sub-network gives just one allocation for a TBF; and

- **Extended dynamic allocation**, where, for a mobile station that is allocated more than one physical channel, only one instruction is given to allow the mobile station to use all of the allocated channels.

1.4   Current Internet protocol – static addresses

With current Internet protocol (IP), the IP address may be allocated permanently. It is then a static IP address; these are not in favour in IPv4 as the Internet addresses are nearing exhaustion. It is common for the ISP or corporate Intranet administrator to allocate IP addresses to the user dynamically on a per-communication basis (dynamic IP addresses).

For security reasons, many corporate Intranet users do have static IP addresses. The Internet will use the server as the primary address and the server will then distribute messages to the individual addresses on its network. This is shown in Figure 1.3.

However, if a user with a static IP address moves networks, then that user cannot receive Internet communications.

It is evident that if the home server has the forwarding address of the server on the foreign network, then communications to the static IP address can be forwarded to that foreign server. This forms the basis of *mobile IP*, which allows mobility of static IP addresses between various networks. We will call it *portable IP* to distinguish it from cellular radio mobility aspects.

1.5   Current Internet protocol – dynamic addresses

Figure 1.4 shows a home network which has allocated a dynamic address to a user for a communication. The dynamic address is unknown to other users until they are informed of it when the owner of the dynamic address initiates communication and all initial communication must be initiated by the dynamic address owner. This is commonplace for home users of the Internet, where communication from the Internet cannot be received until the home user sets up a link to the Internet service provider; similarly when surfing the Internet, Internet sites are informed of the dynamic address which is accessing them.

When the user moves to a foreign network, a new dynamic address may be allocated to allow Internet communication to be initiated if the foreign network allows this.

1.6   GPRS Internet addresses

The mobile station on the left in Figure 1.5 is allocated a dynamic address (which is given when a mobile originated call is set up – PDP context activation), and can receive communications from the external IP address that it has accessed. If this accessed address is, for example, the mobile station ISP then the mobile station and GPRS sub-network can leave the PDP context open all day (or a longer or shorter period contingent upon the QoS and the GPRS operator policy), and the ISP,

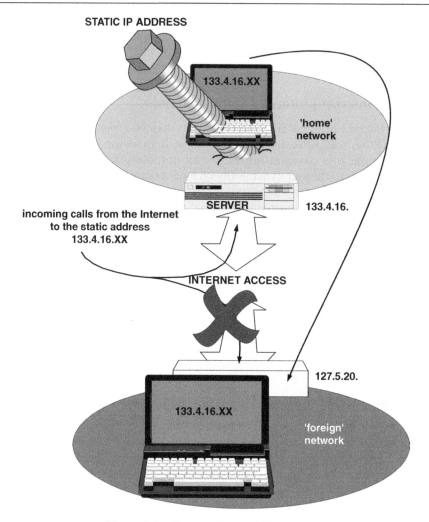

Figure 1.3   Current IP – static addresses

knowing the mobile station dynamic address, can forward communication to the subscriber.

For roaming, the IMSI is authenticated, as is normally the case for GSM operations (extracting the subscriber data and the authentication and ciphering triplets from the home HLR). Chapter 10 covers roaming procedures in more detail.

Figure 1.6 further illustrates the working of dynamic IP addressing in GPRS sub-networks. The upper diagram shows a mobile station establishing a PDP context activation to its ISP. The GPRS sub-network supplies the IP address of the mobile station dynamically. Establishing the call to the ISP, the network informs the ISP of the mobile station's dynamic address.

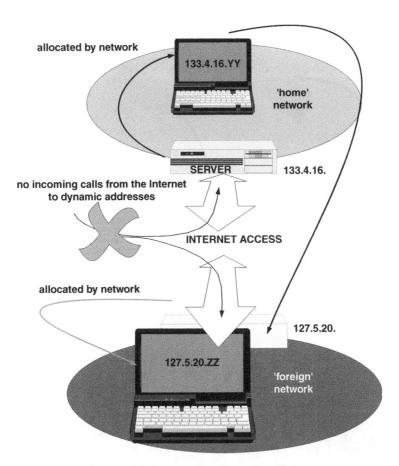

Figure 1.4 Current IP – dynamic addresses

Upon completion of the communication the mobile station will go to GMM standby mode (a period exists after the TBF is completed, when the mobile station stays in GMM ready mode, before it changes to GMM standby), but the PDP context – in particular the dynamic address – remains allocated to the mobile station. If the ISP subsequently has a communication for that mobile station, this will be forwarded to that GPRS IP dynamic address, and the GPRS sub-network will page the mobile station which owns that address. The Internet message are then downloaded to the mobile station. The lower diagram in Figure 1.6 shows this in operation.

1.7 Portable IP

Portable IP is a development of IP designed to allow a node or host with a fixed IP address to move from the home to a foreign network and yet receive communications from the Internet. A simplified principle of operation is shown in Figure 1.7.

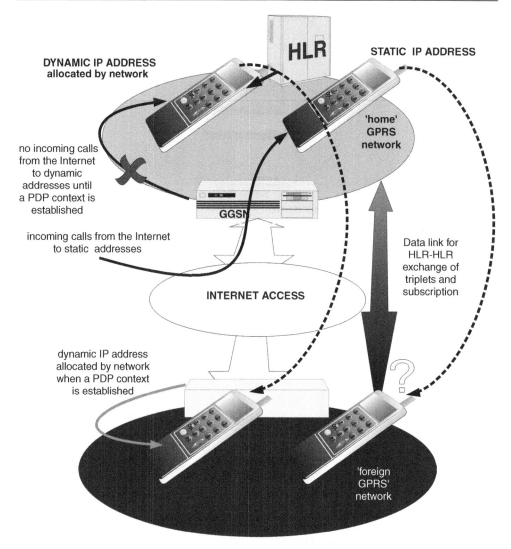

DYNAMIC IP ADDRESS
allocated by network

STATIC IP ADDRESS

'home'
GPRS
network

no incoming calls
from the Internet
to dynamic
addresses until
a PDP context is
established

incoming calls from the Internet
to static addresses

Data link for
HLR-HLR
exchange of
triplets and
subscription

INTERNET ACCESS

dynamic IP address
allocated by network
when a PDP context
is established

'foreign
GPRS'
network

Figure 1.5   GPRS Internet addresses

The host with address 132.6.16.XX moves from its home network (server address 132.6.16) to a foreign network (server address 126.7.13). The foreign network router broadcasts an *agent advertisement* and the visiting host receiving this, knows portable IP is used and registers with the foreign network. From that it gets the *care-of address* of the foreign network, and communicates that to the home network server.

Incoming data to the home network addressed to the absent host will be encapsulated by the home server within the care-of address supplied and *tunnelled* down to the foreign network. The foreign network receives this packet and removes the IP data with its 'absent host' address from the tunnelling encapsulation and delivers it to the visiting host.

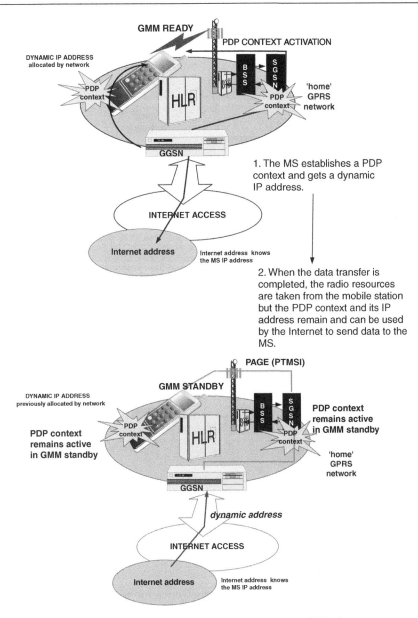

Figure 1.6 'Hanging on' to the Internet all day?

When the visiting host establishes a communication with the Internet, the forward routeing from the visiting host can be very different from the return routeing which will always be via the home network server. If the visiting mobile station is in a foreign country and the ISP (e.g. Intranet) is in the mobile's home country, then all communications to the Internet will be via the home country server.

Portable IP is not yet in general use.

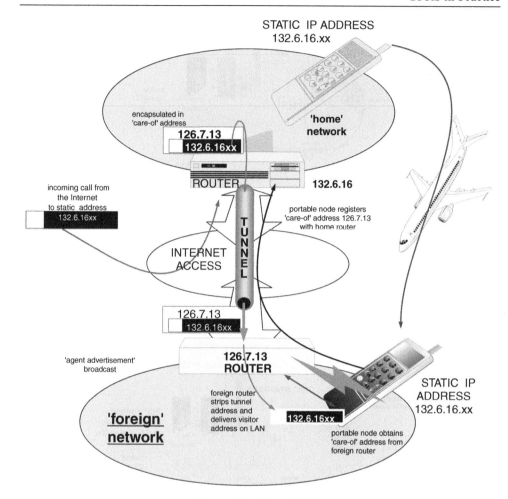

Figure 1.7   Portable IP

1.8   The GPRS sub-network

This book looks at the GPRS services primarily from a mobile station viewpoint, whilst considering the flow of messages across the GPRS sub-network. This section briefly introduces the GPRS sub-network. Figure 1.8 shows the components of a combined GSM/GPRS network.

The GPRS sub-network is subsidiary to and dependent upon the GSM circuit switched network and cannot exist without it. In particular, the BCCH (broadcast control channel) carrier of the GSM network must be accessed so that the mobile station knows that GPRS services are provided.

The additional components required for a GPRS sub-network are the SGSN (serving GPRS support node) which is the hierarchical equivalent of the GSM MSC (mobile

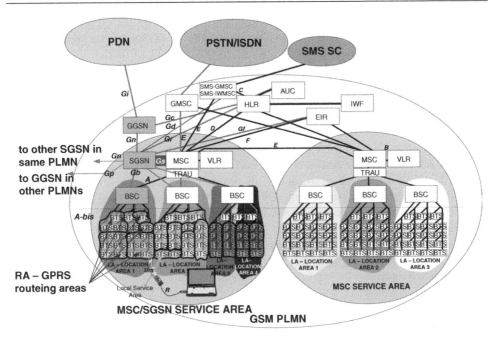

Figure 1.8   Phase 2+ network configuration

switching centre) but in this case applied to packet switched services, and the GGSN (gateway GPRS support node) which has an hierarchical equivalence to the GMSC, but in this case applied to Internet working. The GGSN, from the Internet's point of view, is the server providing access to a private network.

In the configuration shown in Figure 1.8, there is no connection between the SGSN and the VLR; this is currently the norm and means that the SGSN must incorporate the mobility management functions which, for GSM, are performed through the VLR. However, for NMO1 (network mode of operation), the Gs interface provides a connection between the SGSN and MSC, allowing simultaneous IMSI and GPRS attach and authentication. Paging for circuit switched operations via the GPRS paging channel is also available if the Gs interface and a packet broadcast control channel (PBCCH) are functional.

An additional feature seen in Figure 1.8 is the *routeing area*. This is used for location identification when the mobile station is in GMM (GPRS mobility management) states standby and ready. The figure shows a multitude of routeing areas for one SGSN; whether the GPRS sub-network operators will follow this or ascribe just one routeing area to one SGSN is left to the operators' discretion, however as the paging load for packet switched operations is far greater than that for circuit switched operations it is necessary to provide more routeing areas than location areas in order to reduce the GPRS sub-network paging loading.

The names of the various interfaces are shown, for example to the left of Figure 1.8, the Gn interface is the link connecting all SGSNs within a PLMN together. This is necessary for handovers (or cell change orders as they are called for GPRS operations) from a cell in one SGSN service area to a cell in a different SGSN service area. It is also necessary for identification of a mobile station, as these use TLLI to identify themselves and TLLI is derived from a PTMSI allocated by a particular SGSN. Should an SGSN receive a TLLI which is not from within the group of PTMSIs belonging to that SGSN, then it will contact the SGSN to which the TLLI does belong and the details of the mobile station and its current data transfer status are transferred across the Gn interface.

The Gp interface is used for PLMN roaming – it is expected that this will connect to an international switching centre so that each PLMN has only one roaming contact point rather than a connection to each of its roaming partners.

Finally, for this brief introduction, GPRS QoS (Quality of Service) parameters are measured between the R interface to the bottom of the diagram and the Gi interface to the Internet. The interactions between the elements of the GPRS sub-network and mobile stations are covered in later chapters.

Table 1.1 summarises some of the interfaces, abbreviations and other characteristics of the GPRS system.

Table 1.1

External protocols used	IP
Multiplexing	FDD radio channels
	TDM physical channels
	TDM logical channels
Location management	Routeing area and SoLSA
Interfaces	A-bis BTS-BSC
	Gb BSC-SGSN
	Gc GGSN-HLR
	Gd SGSN-SMS-GMSC
	Gf SGSN-EIR
	Gl GGSN-PDN
	Gn SGSN-GGSN/SGSN
	Gp SGSN-GGSN & other PLMN
	Gr SGSN-HLR
	Gs MSC-SGSN
Internal protocols	
GPRS tunnelling protocol (GTP)	SGSN-GGSN/SGSN
Transmission control protocol (TCP)	SGSN-GGSN/SGSN
User datagram protocol (UDP)	SGSN-GGSN/SGSN
Sub-network dependent convergence protocol (SNDCP)	MS-SGSN
Logical link control (LLC)	MS-SGSN
Radio link control/medium access control (RLC/MAC)	BSS-MS
Relay	Transfers from BSS Um side to BSS NSS side
Network service	SGSN-BSS

Network mode of operation

NMO1	Gs interface operational and PBCCH
	CS Paging on both physical channels CCCH and PCCCH
NMO2	No PBCCH
	All paging on CCCHs
NMO3	PBCCH operational, no Gs interface
	Paging separately on PCCCH and CCCHs

MS types

A	Simultaneous CS and packet traffic operation
B	Packet traffic suspended to take CS calls
C	Only CS or packet traffic
Type 1	Non-simultaneous Tx and Rx
Type 2	Simultaneous Tx and Rx
Multislot class	Determines uplink and downlink physical channels

Air Interface Logical Channels	Packet broadcast control channel (PBCCH)
	Packet common control channel (PCCCH)
	Packet access grant channel (DL PAGCH)
	Packet paging channel (DL PPCH)
	Packet random access channel (UL PRACH)
	Packet associated control channel (PACCH)
	Packet data traffic channel (PDTCH)
Air Interface Physical Channels	Packet data channel (PDCH)

Maximum data rates per physical channel

GPRS	9.05 kb/s	GMSK
GPRS	13.4 kb/s	GMSK
GPRS	15.6 kb/s	GMSK
GPRS	21.4 kb/s	GMSK
EGPRS	8.8 kb/s	GMSK
EGPRS	11.2 kb/s	GMSK
EGPRS	14.8 kb/s	GMSK
EGPRS	17.6 kb/s	GMSK
EGPRS	22.4 kb/s	8-PSK
EGPRS	29.6 kb/s	8-PSK
EGPRS	44.8 kb/s	8-PSK
EGPRS	54.4 kb/s	8-PSK
EGPRS	59.2 kb/s	8-PSK

1.9 Abbreviations used in this chapter

BCCH	Broadcast control channel
CS	Circuit switched
EDGE	Enhanced data rate for GSM evolution
GGSN	Gateway GPRS support node
GMM	GPRS mobility management
GPRS	General packet radio service
GSM	Global system for mobile communication

HLR   Home location register
HSCSD   High speed circuit switched data
IMSI   International mobile subscriber identity
IP   Internet protocol
ISDN   Integrated services digital network
ISP   Internet service provider
MSC   Mobile switching centre
NMO   Network mode of operation
PBCCH   Packet broadcast control channel
PDP   Packet data protocol
PLMN   Public land mobile network
PTMSI   Packet temporary mobile subscriber identity
QoS   Quality of service
SGSN   Serving GPRS support node
TBF   Temporary block flow
TCP/IP   Transmission control protocol/Internet protocol
TLLI   Temporary logical link identifier
UMTS   Universal mobile telecommunications system
VLR   Visitor location register

2

Radio Channels, Physical Channels and Logical Channels – the GSM/GPRS Air Interface

The air interface is the troublesome part of GPRS and of GSM. The GPRS protocols, procedures, encoding and modulation are designed to overcome this potential weak link in the communications system.

Quite surprisingly, in view of its complexity, GSM works extremely well, so well in fact that it is often difficult to know whether a caller is on the GSM mobile network or the fixed ISDN network. This is due to the remarkable engineering effort that has been put into GSM. Undoubtedly, GPRS will work just as well.

This chapter is about the radio channels, physical channels and logical channels involved in establishing communication on the GPRS air interface. It is the part which most people new to the subject find difficulty with at first. As familiarity with the plethora of acronyms, abbreviations and especially the new vocabulary of GPRS grows, things become easier.

As GPRS uses the GSM air interface, then the GSM air interface in general is covered in this chapter.

2.1 The radio channels (GSM 45.001)

The air interface, Um, provides the radio channel connection between a cell belonging to a base transceiver station (BTS) and the mobile station.

GPRS in Practice: A Companion to the Specifications Peter McGuiggan
© 2004 John Wiley & Sons, Ltd ISBN: 0-470-09507-5

The characteristics of the radio channel are the strengths of the GSM system, which have led to its overwhelming worldwide popularity; unfortunately embedded within these same characteristics is a shortcoming which will lead to its demise in the near future. These characteristics are:

- A bandwidth allocation of 200 kHz per radio channel;
- A maximum aggregate bit rate, or more accurately, symbol rate of 271 kb/s per radio channel;
- A maximum voice data rate of 13 kb/s (up to eight of these per radio channel);
- A maximum data rate of 22.8 kb/s for data (up to eight of these per radio channel).

There are:

- 124 radio channels in GSM900;
- 374 radio channels in GSM1800; and
- 299 radio channels available in PCS1900.

A base transceiver station (BTS) which controls a cell or many cells may have just one radio channel or many radio channels in each cell.

A transceiver (TRX) is the fundamental radio channel communication unit in a cell. A TRX is normally allocated one of the available radio channels (except when synthesiser frequency hopping is in use, when the TRX hops through many radio channels).

The traffic capacity of a cell can be increased by allocating more TRXs – additional radio channels – to the cell. Using Gaussian minimum shift keying (GMSK), it is possible to modulate a 200 kHz bandwidth carrier at a bit rate of 271 kb/s; this is the aggregate bit rate of GSM (and GPRS) radio channels. Using regular pulse excitation, long-term prediction (RPE-LTP) or enhanced full rate (EFR) encoding, it is possible to encode a speech signal at a rate of 14 kb/s with a quality very close to that of the ISDN. However, as this speech signal must communicate through an extremely harsh radio environment, protection is necessary and this is provided by forward error protection encoding, the added redundancy raising the speech channel bit rate to 22.8 kb/s. Further error protection consisting of a TSC – training sequence code (to reduce the effects of time dispersion) and bit periods for ramping up and down the transmitted power within the air interface burst structure raises the speech channel to a rate of about 33.8 kb/s.

This rate leads directly to eight physical channels of 33.8 kb/s in a 271 kb/s radio channel. The physical channels are organised as timeslot numbers TN. Eight of these TNs form a frame with a period of 4.615 ms.

The organization of the GSM radio channel spectrum is given in Figure 2.1. This shows the frequency bands originally agreed upon by the European member countries (GSM900 and 1800) and the United States (PCS). The two bands, uplink and downlink, are divided into paired radio channels each separated by frequency duplexing and each radio channel has an allocated bandwidth of 200 kHz.

Figure 2.2 shows a GSM900 cell transmitting four carriers with absolute radio frequency channel numbers (ARFCNs) 1, 4, 10 and 16 out of the possible 124. Each carrier occupies a bandwidth of 200 kHz.

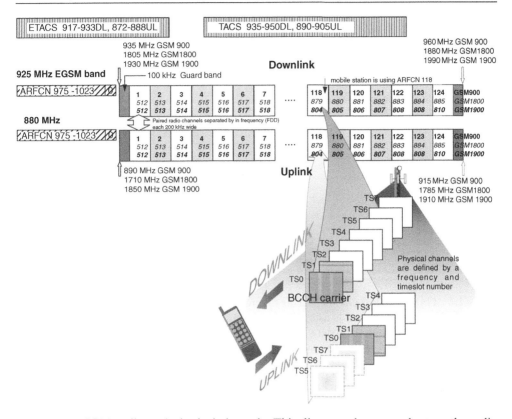

Figure 2.1 GSM radio and physical channels. This diagram shows, at the top, the radio channels with their ARFCN (absolute radio channel frequency number) for the GSM900, 1800 and 1900 systems. To the left is shown an extra allocation of frequencies for the 900 band, called the EGSM (extended GSM band). To the right is shown a cell transmitting and receiving on ARFCN 118. This radio channel is allocated, in this example, as a BCCH radio carrier. Timeslots 0 and 1 are reserved as control channels. It is seen that there is a three-burst offset between the MS receiving on Tn0 and transmitting on Tn0

2.2 Physical channels (GSM 45.001)

A GSM physical channel is defined by a timeslot number (TN) and a frequency or set of frequencies (called MA, mobile allocation in the specification) if the physical channel is frequency hopping.

The physical channels in GSM are derived from the basic characteristics of the radio channel and the original voice data rate of 13 kb/s is raised to 22.8 kb/s with added protective redundancy. These characteristics enable a radio channel to carry eight physical channels.

A physical channel is a burst of radio energy. There are eight bursts of radio energy in 4.615 ms and each burst is labelled, from TN0 to TN7. Figure 2.2 shows the structure of the air interface physical channels as a series of bursts forming a timeslot number,

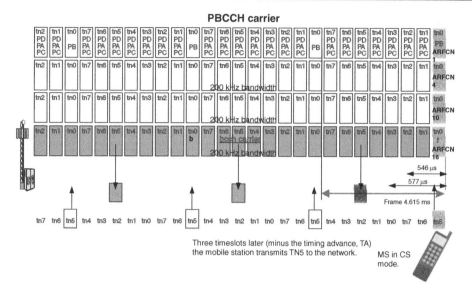

Figure 2.2 GSM physical channels. This diagram shows a BTS using four ARFCNs, 1, 4, 10 and 16. The mobile station is using TN5 in CS mode on ARFCN 16, which is the BCCH (broadcast control channel) carrier. TN0 of the BCCH carrier is carrying the CCHs (common channels) F = FCCH (the frequency correction channel), S = SCH (the synchronisation channel), B = BCCH (the broadcast control channel) and, not shown, C = CCCH (common control channels). The mobile station transmits TN5 after a three burst delay, allowing the use of a simple switch within the handset. The mobile's TN5 is in fact transmitted a little earlier than the three TN delay to allow the transmitted burst to arrive at the BTS at the correct time, accounting for the two-way transmission delay BTS – MS and MS – BTS. The BTS, when it transmits a TN, simultaneously opens the receiver and expects to receive, at that time, a TN which is three TNs lower in order than the one transmitted. Hence, when it transmits TN5 it expects to receive TN2, and will only expect to receive TN5 when it transmits TN0. ARFCN 1 is completely designated as a packet switched service radio, as all the physical channels are providing GPRS services

each burst lasting for 546 microseconds, with an added guard period giving a total of 577 microseconds for each radio burst.

The eight time slots or physical channels in a radio channel may be allocated as traffic channels (TCH), or some of the physical channels may be used as control channels. The purpose of the control channels is to inform the mobile stations using the cell of the cell operating parameters (BCCH and SCH) and to allow the mobile station to access the network services (CCCHs).

Figure 2.2 also shows an offset in the TN time base between a cell transmitting and receiving. When the cell is transmitting a burst corresponding to TN0, it is receiving a burst corresponding to TN5; there is a three-burst shift between transmitting and receiving the same TN. This is built-in to ease the design of the mobile station. When allocated a physical channel it simply has to switch between its receiver and transmitter after two bursts.

For GPRS mobile stations this is made more demanding, with the mobile station sometimes having only one burst between transmitting and receiving; GPRS Class 'A' mobile stations must be able to transmit and receive simultaneously.

Figure 2.2 shows a mobile station working in GSM circuit switched mode, allocated physical channel TN5. The mobile station receives a burst corresponding to TN5 and, after a delay of two radio bursts, transmits a radio burst TN5. (Note that, with the mobile station transmitting only one burst out of eight, the burst train is effectively an amplitude modulated waveform with an envelope frequency of about 200 Hz. This sometimes causes interference problems in hearing aids and other electronic equipment.)

For GPRS the physical channel is called a packet data channel (PDCH). The PDCH is limited to a maximum data rate of 21.4 kb/s, somewhat less than the full capability of 22.8 kb/s. To achieve this rate, no redundancy is added to the data in the form of FEC; this is the CS4 – coding system 4 for GPRS.

There are three other coding systems: CS1, CS2 and CS3 which, starting from CS1, add progressively less redundancy, giving physical channel data information rates of 9.05 kb/s, 13.4 kb/s and 15.6 kb/s respectively.

Unfortunately, these higher data rates will require progressively higher C/I ratios than the standard GSM 9 dB, and this will effectively reduce the working radius of cells offering the higher rates.

2.2.1   The characteristics of the GSM/GPRS physical channels (GSM 45.001)

- There are eight physical channels to one radio channel;
- The eight physical channels are labelled TN0 to TN7;
- A physical channel is transmitted as one burst of radio energy every 4.615 ms period;
- A burst of radio energy in GSM/GPRS lasts for 546 μs, including a guard period this increases to 577μs;
- The 4.615 ms period in which eight bursts belonging to eight physical channels are transmitted by a cell is the basic GSM frame;
- A physical channel for GSM circuit switched operations is a timeslot number, TN, at a particular frequency pair (uplink and downlink). This pair may be frequency hopping;
- A physical channel for GPRS packet switched operations is called a PDCH, it consists of a TN and frequency (or set of frequencies for hopping) reserved solely for GPRS use. As the GPRS service is dynamic, the GPRS physical channel(s) is (are) activated or deactivated at any time by the GPRS sub-network operator;
- A physical channel has a data rate of 22.8 kb/s with GMSK. Generally this includes FEC redundancy, but for GPRS CS4 no redundancy is added to the data and the customer data rate is very close to this.

Time division multiplexed on to a physical channel are *logical channels*, which can be control channels, signalling channels or traffic channels.

Generally, logical channels are information channels, and the type of information carried on a logical channel defines the logical channel type. For example, customer-generated data is carried on the GSM logical channel TCH (traffic channel), and GPRS customer-generated data is carried on the GPRS logical channel PDTCH (packet data traffic channel).

2.3   Logical channels (GSM 45.001, 45.002, 43.064)

Logical channels are sub-channels time division multiplexed onto a physical channel. They can be either traffic channels, signalling channels or control channels. Figure 2.2 shows a number of logical channels carried upon the physical channels. This section is restricted to listing the GSM and GPRS logical channels and giving a simple explanation of their function.

2.3.1   GSM logical channels

Logical channel	Abbreviation used in this book	Functional description
FCCH (DL)	F	The frequency correction channel. When a mobile station finds this channel it knows it has found a BCCH carrier and knows it has found TN0 of the BCCH carrier (and all other carriers in the cell). The FCCH is modulated with zeroes, which GMSK causes to raise the carrier centre frequency by about 67 kHz. The mobile station uses this to synchronise its synthesiser to the network frequency.
SCH (DL)	S	The synchronisation channel. This carries the BSIC (base station identity code) and FN (frame sequence number) for all the carriers in the cell. On decoding this the mobile station is fully synchronised to the cell.
BCCH (DL)	B	The broadcast control channel, one of the many logical channels on the eponymous BCCH carrier. The broadcast control channel carries system information telling the mobile station of the cell operating parameters. It also tells the mobile station if the cell offers GPRS services.
CCCH (DL and UL)	CCCH	Common control channel – common in that the use of this logical channel is common to all mobile stations using the cell. This channel provides access to the GSM network services. The CCCH may also provide access to the GPRS services. There are three sub-channels inside the CCCH. These are listed below.
AGCH (DL)	AGCH	CCCH-AGCH (the access grant channel) provides 'immediate assignment' of dedicated channels to a mobile requesting network services with a *'channel request'*. If providing GPRS access the channel becomes the PAGCH (packet access grant channel) and the downlink message is *packet immediate assignment*.

RACH (UL)	RACH	CCCH-RACH (the random access channel) is used on the uplink by the mobile station to request network services by sending a *channel request*. A response is received on the AGCH (see above). If used to request GPRS services, the channel becomes the PRACH and the request *packet channel request*.
PCH (DL)	PCH	Paging channel; mobile stations assign themselves to a paging channel and when incoming calls from an external network arrive for a mobile station, the mobile station's IMSI or TMSI is placed on the paging channel to alert the mobile station. When used for GPRS paging, it is called the PPCH.
EBCCH (DL)	EBCCH	Extended broadcast control channel, an extra BCCH, which may be used to speed up the transmission of system information.
NCH (DL)	NCH	The channel used to inform mobile stations of the existence and location of a voice broadcast channel or a group of mobile stations of the existence of a VGCS (voice group calling service).
SACCH (DL and UL)	A	Slow associated control channel – this channel is always associated or paired with a dedicated channel – TCH or SDCCH. The purpose of the channel is to look after the radio resources so that the dedicated channel is always on a satisfactory physical channel. Its presence or absence is also an indicator of the integrity of the radio link. This channel is also used for delivery of SMS messages if a dedicated TCH channel is in use.
SDCCH (DL and UL)	SDCCH, Dm	Standalone dedicated control channel, a dedicated signalling channel used for call control and mobility management signalling. The equivalent of the 'D' channel in ISDN it is sometimes abbreviated to Dm. This channel is also used to deliver SMS messages if a TCH is not already in use.
FACCH (DL and UL)	FACCH	Fast associated control channel, probably misnamed, definitely not a fast version of SACCH! This channel is used for signalling in the same way as the SDCCH (but faster) and 'steals' TCH frames when a dedicated channel is in use and signalling is required.
IDLE	I	An idle channel, used with a fixed pattern of bits (dummy bursts) where it is necessary to transmit a channel when no information is to be sent, e.g. a physical channel (apart from TN0) on the BCCH carrier, which is not carrying any other logical channel, transmits idle channels. (The BCCH carrier must always transmit on all physical channels). On non-BCCH carriers the idle logical channel is not transmitted.

(Continued)

Logical channel	Abbreviation used in this book	Functional description
TCH/FR	TCH/FR, Bm	A full rate voice traffic channel at 13 kb/s.
TCH/HR	TCH/H	A half rate traffic channel at 5.6 kb/s.
TCH/D	TCH/D	A traffic channel assigned for data communications at 9.6 kb/s, 4.8 kb/s, 2.4 kb/s.
TCH/EFR	TCH/EFR	Traffic channel enhanced full rate.

2.3.2   GPRS channels which are used with or without a PBCCH

Logical channel	Abbreviation used in this book	Functional description
PACCH	PACCH	The packet associated control channel is always associated with a GPRS traffic channel, however it displaces the GPRS traffic channel when RLC/MAC signalling is required. RLC/MAC signalling is concerned with management of the radio resources, so the PACCH can be regarded in the same light as the SACCH.
PDTCH	PDTCH	The packet data traffic channel, used for transfer of traffic or higher layer signalling RLC/MAC blocks during a TBF.
PTCCH	PTCCH	Packet timing (advance) control channel. There are sixteen of these channels in the positions which would be occupied by SACCH in circuit switched operations – the 13th frame of a 26-frame multiframe (more correctly for GPRS the 13th frame of a 52-frame multiframe). Mobile stations in a downlink TBF send short bursts to this channel to allow the GPRS sub-network to calculate TA requirements.

2.3.3   GPRS logical channels which are used only in conjunction with PBCCH

Logical channel	Abbreviations used in this book	Functional description
PBCCH (DL)	PBCCH	The packet broadcast control channel. This broadcasts packet system information. This is always in the position of block 0 of a 52-frame multiframe. The BCCH channel tells mobile stations which physical channel it occupies.

PCCCH (DL and UL)	PCCCH	The packet common control channel used to provide access to the GPRS sub-network services. It includes the three sub-channels below.
PRACH	PRACH, P	Packet random access channel on which the mobile station requests GPRS sub-network services with a *packet channel request*.
PAGCH	PAGCH	Packet access grant channel which the GPRS sub-network uses to respond to a channel request with the message *packet uplink assignment*.
PPCH	PPCH	The packet paging channel which is used to alert mobile stations to an incoming packet.

2.3.4   GPRS logical channels which are used in the absence of a PBCCH

Logical channel	Abbreviations used in this book	Functional description
BCCH (DL)	B	The broadcast control channel, one of the many logical channels on the eponymous BCCH carrier. The broadcast control channel carries system information telling the mobile station of the cell operating parameters. It also tells the mobile station if the cell offers GPRS services and, in the absence of a PBCCH, carries the GPRS operating parameters.
CCCH (DL and UL)	CCCH	Common control channel – common in that the use of this logical channel is common to all mobile stations using the cell. This channel provides access to the GSM network services. The CCCH will also provide access to the GPRS services in the absence of a PBCCH. There are three sub-channels inside the CCCH. These are listed below.
AGCH (DL)	AGCH	CCCH-AGCH (the access grant channel) provides 'immediate assignment' of dedicated channels to a mobile requesting network services with a '*channel request*'. If providing GPRS access the channel becomes a PAGCH (packet access grant channel) and the downlink message is *packet immediate assignment*.

(Continued)

Logical channel	Abbreviations used in this book	Functional description
RACH	RACH, R	CCCH-RACH, (the random access channel) used on the uplink by the mobile station to request network services by sending *channel request*. The response is received on the AGCH (see above). If used to request GPRS services, the channel becomes the PRACH and the request *packet channel request*.
PCH (DL)	PCH	Paging channel, mobile stations assign themselves to a paging channel and when incoming calls from an external network arrive for a mobile station, the mobile station's IMSI or TMSI is placed on the paging channel to alert the mobile station. When used for GPRS paging, it is called the PPCH.
EBCCH (DL)	EBCCH	Extended broadcast control channel, an extra BCCH which may be used to speed up the transmission of system information and may include GPRS parameters in the absence of a PBCCH.

Finally in this chapter, two important channels are considered.

2.4 The BCCH radio carrier

The most important radio channel in any cell is that which carries the BCCH. The BCCH radio channel *defines* a cell.

Figure 2.2 shows that one of the transmitted carriers, or radio channels, is labelled the BCCH carrier. This is the broadcast control channel carrier, carrying mandatory instructions for mobile stations wishing to access the cell. All cells in the GSM system must transmit a BCCH carrier and if a cell transmits only one carrier then that must be a BCCH carrier. Generally, common control channels occupy only TN0, leaving TN1–7 free for use as traffic channels. A network operator may put common control channels on additional timeslots if they wish.

TN0 of the BCCH carrier always carries control channels which include the BCCH channel. Common channels (CCH) are carried on TN0, as are common control channels (CCCH) another type of common channel. Figure 2.2 is an example of the invariant structure of a non-combined BCCH carrier, with mandatory common channels (FCCH, the frequency correction channel; SCH, synchronisation channel; BCCH, broadcast control channel; CCCH, common control channels) on TN0, and SDCCH – standalone dedicated control channels (used for signalling to set up a service and sometimes referred to as the Dm channel, as in the D signalling channel for ISDN) on TN1.TN0 on the uplink (mobile to network) is always used as a RACH (random access channel) for a mobile station to request access to the network services. (Again RACH channels may also be used on other TNs if these other TNs are in use as CCCHs.) There is a three burst delay between transmissions on the downlink and uplink – this greatly simplifies the design of mobile station equipment as it does not need to transmit and receive simultaneously and a simple switch is all that is necessary to change from transmit to receive modes, eliminating the necessity for rapid frequency changing synthesisers (or even two synthesisers, one for the transmitter and one for the receiver), and duplexers.

2.5 The PBCCH

Radio channel ARFCN 1 of Figure 2.2 has a PBCCH (packet broadcast control channel) carried on TN0. The physical channel carrying the PBCCH is specified in the BCCH system information. In the same way as the BCCH carries essential network information for circuit switched mode communications (and also for GPRS operations if the cell offers GPRS services without a PBCCH), the PBCCH carries essential information for mobiles wishing to use the packet switched services on GSM. Whether a PBCCH is used in GPRS operations is operator dependent. In the absence of a PBCCH, packet system information is carried on the BCCH channel.

2.6 Abbreviations used in this chapter

AGCH   Access grant channel
ARFCN   Absolute radio frequency channel number
BCCH   Broadcast control channel
BTS   Base transceiver station
CCCH   Common control channel
C/I   Carrier to interference ratio
CS   Coding System
DL   Downlink
EBCCH   Extended broadcast control channel
EFR   Enhanced full rate
FACCH   Fast associated control channel
FCCH   Frequency correction channel
FEC   Forward error correction
FR   Full rate
GMSK   Gaussian minimum shift keying
GPRS   General packet radio service
GSM   Global system for mobile communication
HR   Half rate
IMSI   International mobile subscriber identity
ISDN   Integrated services digital network
LTP   Long term prediction
MA   Mobile allocation
MAC   Medium access control
PACCH   Packet associated control channel
PAGCH   Packet access grant channel
PBCCH   Packet broadcast control channel
PCH   Paging channel

PCCCH	Packet common control channel
PDCH	Packet data channel
PDTCH	Packet data traffic channel
PPCH	Packet paging channel
PRACH	Packet random access channel
PTCCH	Packet timing control channel
RACH	Random access channel
RLC	Radio link control
RPE	Regular pulse excitation
SACCH	Slow associated control channel
SCH	Synchronisation channel
SDCCH	Standalone dedicated control channel
SMS	Short message service
TBF	Temporary block flow
TCH	Traffic channel
TN	Timeslot number
TRX	Transceiver
TSC	Training sequence code
UL	Uplink
Um	Air interface
VGCS	Voice group calling service

3

Air Interface Frame and Multiframe Structures

(GSM 45.002, 43.064)

Before starting upon a detailed examination of the air interface framing structure for packet switched services it is helpful to explain how the structures are represented as this can sometimes lead to misunderstanding.

Figure 3.1 shows a cell of a BTS transmitting its physical layer radio bursts, frame after frame, frame 0 to frame 12, consecutively and linearly in time. Each frame has a period of 4.615 ms.

If we could see radio waves, the cell transmissions would look something like Figure 3.1. However, this method of representing frame transmission is inconvenient as the time axis is linear and it becomes difficult to represent a large number of frames on a single page.

In order to overcome this limitation of representation, each frame is normally shown lying side by side with its immediately preceding and following frame. This is shown in Figure 3.2. In this representation all the timeslot bursts of a particular TN number are contiguous. The bursts of timeslot number '0', for example, all lie alongside one another on the first row.

The period between repetitions of bursts of TN0 is still 4.615 ms – a frame period, but the time axis follows a saw tooth route as shown, with the 4.615 ms period lying behind each burst of TN0. In this way, the frames and the logical channels they carry can be compressed into a small area, allowing ease of representation.

This is the graphical representation we will use for the air interface frames and multiframes and it is also the representation that is used in the GSM specifications, although generally only one physical channel is shown, leaving the reader to imagine the full eight physical channels!

GPRS in Practice: A Companion to the Specifications Peter McGuiggan
© 2004 John Wiley & Sons, Ltd ISBN: 0-470-09507-5

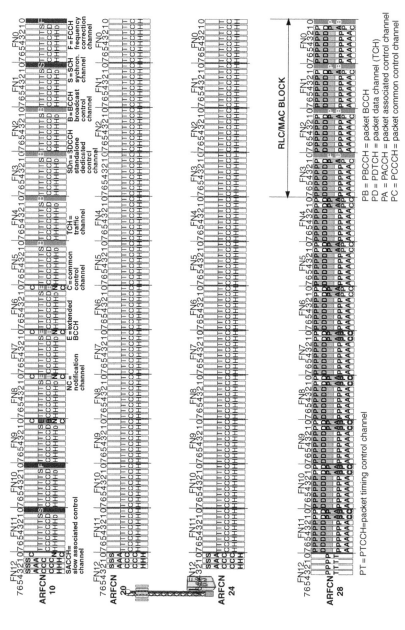

Figure 3.1 A possible phase 2+ configuration of radio channels in a cell. This figure shows a cell transmitting four carriers, the top one of these is the BCCH carrier which always defines a cell; two carriers are devoted entirely to circuit switched traffic channels, and the fourth carrier is entirely devoted to GPRS packet data channels. This carrier includes a PBCCH channel, which is not entirely necessary as the BCCH and CCCHs of the BCCH carrier could handle the control of GPRS traffic channels (PDTCHs). Where a physical channel burst is shown carrying more than one logical channel, this implies that the burst can be carrying just one (but one of the many shown) of these logical channels at that time

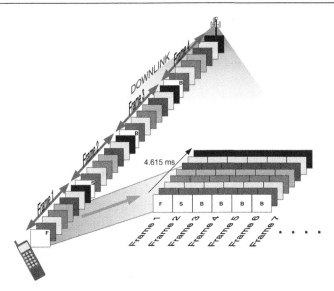

Figure 3.2 A more convenient way of representing air interface bursts and frames. Each frame is laid alongside the next so that similar TNs are adjacent to each other. The saw tooth time axis is a little convoluted but this way of representing the frames makes it much easier to get a lot of information on one page!

3.1 The basic frame

A basic frame has eight bursts constituting the eight physical channels, the timeslots shown in Figure 3.1. The timeslots are numbered TN0 to TN7 and each burst of a timeslot lasts for 577 microseconds.

A *normal* burst contains 156.25 bits during this period. (A normal burst is the common information-carrying burst such as a traffic channel – the timeslots labelled TCH in Figure 3.1).

A series of bursts forming eight TNs have a combined bit rate of 156.25/577 microseconds, or approximately 271 kb/s. A physical channel, occupying one burst out of eight, has a full bit rate of one eighth of this, 33.8 kb/s.

The protected information content of a normal burst for a GSM physical channel is 114 bits (plus two bits which are used as flags adjacent to the training sequence), or 114/156.25 × 33.8 kb/s = 24.66 kb/s, but as we shall see, only 24 frames out of 26 carry user information (or traffic), therefore the traffic channel rate is 22.8 kb/s.

The 'raw information' rate for a full rate traffic channel is 13 kb/s; the higher rate of 22.8 kb/s is the result of FEC encoding adding redundancy to the raw rate. The eight timeslots in a frame give a frame period of 8 × 577 microseconds = 4.615 ms. The basic frame is a part of the RLC/MAC block in GPRS as four frames make up an RLC/MAC block. As there are eight physical channels in each frame, then four frames may accommodate eight RLC/MAC blocks.

It takes 4 × 4.615 ms = 18.46 ms to send an RLC/MAC block. As the data rate for a packet with encoding system 1 FEC is 9.05 kb/s, then each RLC/MAC block must carry 168 information bits. We shall examine the validity of this statement in a later chapter.

3.2   The GPRS 52-frame multiframe and logical channel structures

The GPRS logical channel structure is different in type from circuit switched logical channel structures in that the logical channels share physical channels dynamically as required.

Figure 3.1 shows a cell transmitting four carriers. One of these carriers is the BCCH carrier, which must be transmitted by all cells.

At the bottom of Figure 3.1 is shown a radio channel carrying a GPRS PBCCH (packet broadcast control channel). This is a possible configuration for a PBCCH carrier, but the PBCCH carrier is not necessary to GPRS working, as the BCCH carrier control channels can accommodate both CS and packet switched information and control channels. However, this may lead to overloaded control channels on the BCCH carrier. The fine control of GPRS operations when a PBCCH is used is also lost in the absence of a PBCCH.

Figure 3.1 shows the difficulty of the unmodified physical representation of logical channel structures. Thirteen frames are shown, not enough to show all the combinations of logical channels on the physical channels.

Figure 3.2 demonstrates how the specification and many textbooks represent a radio channel in such a way that all the logical channels can be easily shown. Instead of each frame or radio burst following one another linearly in time, each frame is placed alongside its predecessor, so that time is following a saw tooth path. Although this may appear puzzling at first, it is in fact a very good representation of the physical and logical channel structures in a radio channel.

The representation in Figure 3.2 is repeated in Figure 3.3 for a radio channel carrying a PBCCH. The representation shown in the demonstration Figures 3.2 and 3.3 is extended in Figure 3.4, which shows most of the logical channels used for CS and packet operations.

Examining Figure 3.4, TN0 is carrying the logical control channels, which identify this radio channel as the BCCH carrier. Two complete 51-frame multiframes of the BCCH carrier are shown, the second directly beneath the first. The circuit switched signalling channels – standalone dedicated control channels (SDCCH) occupy TN1, and this operator has chosen to put more common control channels (CCCH) on TN2.

Focusing on TN3 which is a packet data channel (PDCH), a GPRS physical channel, we see that the PDCH is arranged in blocks – RLC/MAC blocks naturally – and twelve of these blocks (with some extra frames) form the GPRS 52-frame multiframe.

The first block in this 52-frame multiframe is 'block 0'. This block always starts in frame '0' and has a 52-frame repetition period. This is followed directly by blocks 1 and 2 and then frame 12, which is not part of the RLC/MAC block structure, but is a packet timing control channel (PTCCH).

Blocks 3–5 follow this, and then frame 25 is an idle channel. This pattern repeats itself in the next 26 frames which include blocks 6–11, giving a total of 52 frames, the GPRS multiframe. Now the 52-frame multiframe is a multiple of two circuit switched 26-frame traffic channel (TCH) multiframes, with the idle channels falling in the same positions.

The 52-frame multiframe is designed to interact with the 51-frame multiframe of neighbour cells allowing the GPRS mobile station to capture the SCH of neighbour cells within one second when taking neighbour cell measurements whilst in a TBF. This 52-frame multiframe applies to all GPRS PDCHs, as can be seen for TN7 in Figure 3.4.

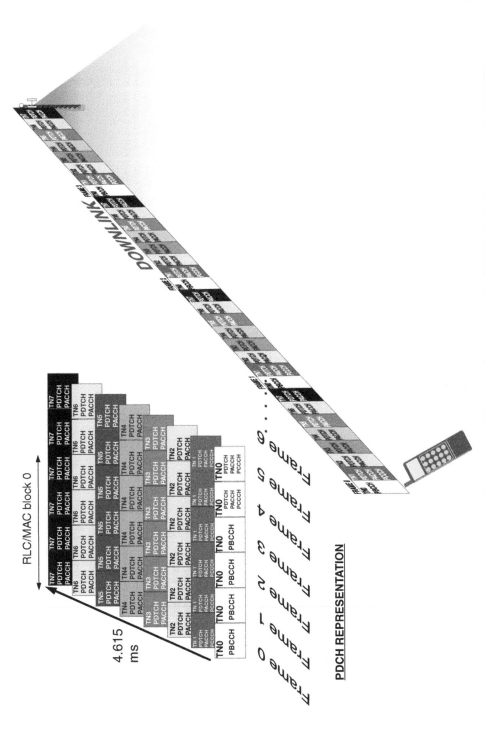

Figure 3.3 Figure 3.2 repeated to show a radio channel carrying eight GPRS physical channels, called PDCH

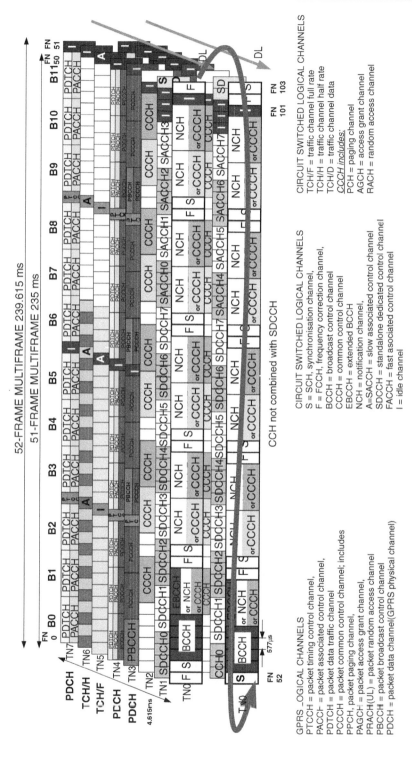

Figure 3.4 The full development of Figures 3.2 and 3.3 showing a non-combined BCCH carrier. CCCHs are shown on TN0 and TN2; SDCCHs are shown on TN1; GPRS services are provided with a PBCCH on TN3; TN3 shows all possible options for the PBCCH physical channel; TN7 is activated as a combined PDTCH/PACCH; TN4 shows all possible combinations of a GPRS PDCH; TN5 is a CS TCH full rate and TN6 shows two CS half rate channels

Examining the logical channel structure within the 52-frame multiframe of TN3 PDCH, 'block 0' is carrying the PBCCH. The PBCCH is always in this 'block 0' position (more accurately, the PBCCH is always in this position if the GPRS services are using a PBCCH). The mobile station 'finds' the PBCCH because the BCCH contains a *PBCCH assignment* message giving the physical channel that is carrying the PBCCH.

The mobile station switches to the assigned physical channel, performs 'modulo 52' on the frame numbers of that physical channel, and when the result is zero, has found the PBCCH.

Now blocks 3, 6 and 9 on TN3 of of Figure 3.4, may also carry PBCCH blocks. This is an option available to the GPRS sub-network operators. If they decide to have extra blocks for the PBCCH in this way, then the 'block 0' PBCCH will inform the mobile station of their existence.

'Blocks 1–11' of TN3, the PBCCH physical channel, always carry PCCCHs (unless blocks 3, 6 and 9 are assigned to carry additional PBCCH channels). The packet common control channel's prime purpose is to allow mobile stations to access the GPRS services. The sub-logical channels of the PCCCH are the PPCH (packet paging channel) on the downlink, the PRACH – the uplink packet random access channel which requests access to the GPRS services, and the downlink PAGCH (packet access grant channel) which responds to the request on the PRACH channel by giving the mobile station uplink radio resources.

However, the PCCCH may be time-shared with PDTCH (packet data traffic channel) and PACCH (packet associated control channels). This may seem odd at first sight, but may prove useful in handling GMM or SM procedures without sacrificing another physical channel. If there is no PDTCH/PACCH on the physical channel carrying PCCCH then the mobile station must be assigned another physical channel where these channels can be accessed.

Frames 12 and 38 (and 52-frame multiples) on TN3, the PBCCH physical channel, contain the PTCCH (packet timing (advance) control channel) in the position which would be taken by SACCH in circuit switched operation. These channels occupy the same positions in all PDCHs as is seen from TN4 and TN7.

Frames 25 and 51 (and 52-frame multiples) of TN3, the PBCCH physical channel, contain the idle channel in the same position as the idle channel (or SACCH for half-rate TCH) in circuit switched operations.

Now let us examine the logical channel structure within the 52-frame multiframe of TN4 PDCH. This GPRS PDCH will rarely be encountered. It is assigned as an additional GPRS physical channel carrying extra packet common control channels (PCCCH). It is unlikely in practice that additional PCCCHs will be needed, but the specification does allow for it.

The mobile station is told of the position of these extra control channels by the PBCCH. All blocks are PCCCH in this configuration of this physical channel, but as the illustration shows, the PCCCH may be time-shared with PDTCH/PACCH logical channels.

Examining the logical channel structure within the 52-frame multiframe of TN7 PDCH, we see that this channel is active and is assigned as a PDTCH, changing to PACCH as required. It is active because the PDTCH does not exist unless it is active! In other words, the allocation of GPRS traffic channels is dynamic to meet customer demand. All blocks on this physical channel are PDTCH/PACCH and the other frames are used for the purposes discussed above.

Figure 3.5 shows a simpler configuration of a radio channel for GPRS operations.

And that just about covers all the possible configurations of GPRS logical channels! At least on the 52-frame multiframe.

The PBCCH can be on any physical channel (except TN0 of the BCCH carrier!) that the operator requires – in the case of Figure 3.4 it is on a physical channel of the BCCH carrier, but that is just for convenience of explanation.

Some time periods of interest are:

- the frame period of 4.615 ms;
- the RLC/MAC block period of 18.46 ms; and
- the 52-frame multiframe period of 239.98 ms. (240 ms)

3.3   The 52-frame multiframe uplink PRACH channel (GSM 45.002, 43.064)

This section briefly considers the PRACH channel. PRACH is not limited to the PCCCH, but can take advantage of any unused uplink block on the physical channel carrying PCCCHs.

Unidirectional logical channels such as the downlink only PBCCH on TN3 of Figure 3.4 have a corresponding block on the uplink which is not used, and mobile stations may transmit *packet channel request* on this uplink block.

Moving to the next block, block 1 on TN3 of Figure 3.4, the downlink PCCCH has two sub-channels, PPCH and PAGCH on the downlink side and PRACH on the uplink side; mobile stations may always use the uplink PCCCH to transmit service requests.

Frames 12 and 38 (and 52-frame multiples of these) contain the PTCCH, which is a two-way communications channel, and mobile stations cannot transmit requests on these uplink frames.

Frames 25 and 51 (and 52-frame multiples of these) are idle and mobile stations could use the corresponding uplink block for service requests.

Figure 3.6 illustrates the uplink PRACH channels available in a 52-frame multiframe. A *packet channel request* on the PRACH requires only one burst and therefore there are four PRACH channels in a block.

The PRACH channel uses the 'short access bursts' in the same way as the GSM random access channel (RACH). This short access burst, about half the size of a normal, full length burst, guarantees that the short burst will be received within the correct timeslot at the BTS receiver without spilling over into the next TN if the mobile station is within 35 km of the cell.

As we have seen, the PCCCHs may be time-shared with PDTCH. If that is the case, it means that *packet channel request* must be controlled by the GPRS sub-network in case a PCCCH block is in fact in use as a PDTCH or PACCH (these two logical channels are bidirectional channels).

This is achieved by including an uplink status flag (USF) in the headers of each downlink RLC/MAC block on the physical channel carrying PCCCHs. The mobile

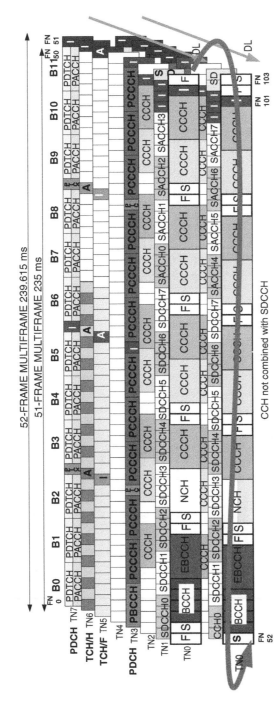

Figure 3.5 A simpler arrangement than Figure 3.4 for PDCHs. This shows a single PBCCH block on block 0 with the PCCCHs devoted entirely to control channels on TN3. TN7 is configured as a PDCH carrying packet data traffic channels and PACCH signalling channels. Other physical channels may be configured as PDCH-PDTCH/PACCH on demand. The BCCH physical channel is configured with one NCH and an extended BCCH channel

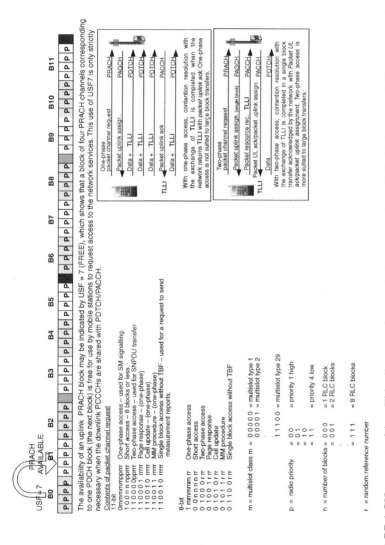

USF=7 AVAILABLE   PRACH

B0 B1 B2 B3 B4 B5 B6 B7 B8 B9 B10 B11

The availability of an uplink PRACH block may be indicated by USF = 7 (FREE), which shows that a block of four PRACH channels corresponding to one PDCH block (the next block) is free for use by mobile stations to request access to the network services. This use of USF7 is only strictly necessary when the downlink PCCCHs are shared with PDTCH/PACCH.

Contents of packet channel request

1-bit

0mmmmmpprrr	One-phase access – used for SM signalling.
100nn npprrr	Short access – 8 blocks or less.
11000 0pprrr	Two-phase access – used for SNPDU transfer
110001 rrrr	Page response – (one-phase)
110010 rrrr	Cell update – (one-phase)
110011 rrrr	MM procedure – (one-phase)
110010 rrrr	Single block access without TBF – used for a request to send measurement reports.

8-bit

1 mmmmmn rr	One-phase access
0 0nnnnn rr	Short access
0 100 0rr	Two-phase access
0 100 1rr	Page response
0 101 0rr	Cell update
0 101 1rr	MM procedure
0 110 0rr	Single block access without TBF

m = multislot class m = 0 0 0 0 0 = multislot type 1
                        0 0 0 0 1 = multislot type 2

                        1 1 1 0 0 = multislot type 29

p = radio priority   = 0 0 = priority 1 high
                       0 1 = priority
                       1 0
                       1 1 = priority 4 low

n = number of blocks = 0 0 0 = 1 RLC block
                       0 0 1 = 2 RLC blocks
                       . . .
                       1 1 1 = 8 RLC blocks

r = random reference number

One-phase

packet channel request	PRACH
Packet uplink assign.	PAGCH
Data + TLLI	PDTCH
Data + TLLI	PDTCH
Data + TLLI	PDTCH
Packet uplink ack.	PAGCH
Data + TLLI	PDTCH
TLLI	PDTCH

With one-phase access, contention resolution with the exchange of TLLI is completed when the network returns TLLI with packet uplink ack. One-phase access is not suited to large block transfers.

Two-phase

packet channel request	PRACH
Packet uplink assign. (single block)	PAGCH
Packet resource rec. TLLI	PACCH
Packet UL ack/packet uplink assign.	PAGCH
Data	PDTCH

With two-phase access, contention resolution with the exchange of TLLI is completed in a single block transfer acknowledged by the network with Packet UL ack/packet uplink assignment. Two-phase access is more suited to large block transfers.

Figure 3.6 MS PRACH opportunities. This diagram shows, at the top, a downlink 52-frame multiframe carrying packet common control channels, PCCCH. If these PCCCHs are shared with GPRS traffic and signalling channels (PDTCH/PACCH) then it is necessary to limit uplink usage of RLC/MAC blocks to that time when they are carrying PCCCHs. This is indicated by setting the downlink USF header = 7 in the prior block, indicating that the next uplink RLC/MAC block is free for use as four PRACH channels. The **P** in the top 52-frame multiframe indicates the possible availability of uplink PRACH channels and their actual availability is indicated by USF = 7 in the prior block. The table on the left shows the messages sent on the uplink PRACH channel. The network may allow 8 or 11-bit messages. A mobile station will request one- or two-phase access dependent upon the type of LLC PDU it wants to transfer. One- and two-phase access is illustrated on the right, and the full details of these are included in later chapters

stations read these headers, and if the value is '7', then the next uplink RLC/MAC block is free for use as four PRACH channels carrying *packet channel request*.

Figure 3.6 shows the message content of UL PRACH 8-bit and 11-bit messages and defines whether each of these message types should request one-phase or two-phase access. The elements of one- and two-phase access are illustrated in Figure 3.6 and these will be covered in detail in later chapters.

To avoid future confusion, a USF is also used on PDTCHs when the allocation of a PDTCH to a mobile station is dynamic. The range of USFs available is 0–7; hence, on a PDTCH, eight mobile stations may be time division multiplexed on the uplink.

It is only on physical channels carrying PCCCHs that USF 7 is reserved for controlling access to the PRACH channel.

The MS may also be instructed which uplink blocks to use for PRACH by the network packet system information (see Appendix 2 page 340).

3.4   The GSM 51-frame multiframe logical channel structures (non-combined configuration)

Figures 3.4 and 3.5 illustrate the 51-frame multiframe structure for GSM. TN0 in Figure 3.4 has a specific configuration, the *BCCH non-combined configuration*. 'Non-combined' means that the CCCHs (common control channels) are not combined with SDCCHs (standalone dedicated control channels). In that way the SDCCH channels are always on a separate physical channel, TN1 in Figure 3.4, leaving a full complement of nine CCCHs on TN0.

Figure 3.7 shows the non-combined configuration again. This figure also shows a carrier in the *BCCH combined configuration*, where there are only three CCCHs on TN0, the remainder being replaced by standalone dedicated control channels (SDCCH) and slow associated control channels (SACCH). This gives a *combination* of CCCHs and SDCCH/SACCH on TN0.

The *non-combined* structure of the BCCH carrier TN0 in Figure 3.7 shows that frame number 0 of TN0 carries the FCCH (frequency correction channel). All logical control channels on TN0 repeat within a 51-frame period (the FCCH and SCH repeat every ten frames within the 51-frame multiframe).

One frame after the FCCH burst the logical channel is the SCH (synchronisation channel). The eponymous BCCH channel, similar to an RLC/MAC block in that a message requires a radio burst in each of four frames to convey a message, follows the SCH.

Under most circumstances, the next four frames would carry the first of the CCCH (common control channels). However, the first CCCH can be usurped by the EBCCH (extended broadcast control channel) or the NCH (notification channel). This is indicated in Figure 3.4 where all possibilities are shown. (The BCCH channel will inform the mobile station which 'blocks' are used for EBCCH or NCH).

In the next two frames the FCCH and SCH channels repeat. The remaining 'blocks' are all CCCHs interspersed every tenth frame with the FCCH and SCH bursts.

The non-combined structure on the BCCH carrier TN1 of Figures 3.4 and 3.7 shows a commonly used configuration with SDCCHs assigned to this physical channel. There

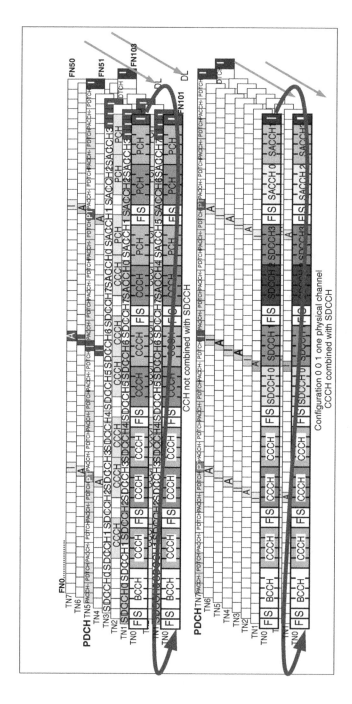

Figure 3.7 CCCH (common control channel) and PDCH (packet data channel) configuration when a cell has no PBCCH. The top diagram shows a cell using a non-combined BCCH carrier. This configuration offers a maximum of nine CCCHs in a 51-frame multiframe (on TN0 and perhaps other TNs of the BCCH carrier). This configuration is usual in large-capacity cells. The lower diagram shows a cell using a combined BCCH carrier. This configuration offers a maximum of three CCCHs on TN0. The remaining spaces are taken up by higher layer service signalling channels, SDCCH and SACCH. This configuration is usual in low-capacity cells. The GPRS MS selects a physical channel carrying CCCHs based upon its IMSI. It then selects its PPCHs on that physical channel, again using its IMSI. For GPRS the repetition period of packet paging channels is 64 51-frame multiframes. It uses the RACH to send *packet channel request*, and then listens to the AGCH for *packet immediate assignment*, that instructs the MS to go to a 52-frame multiframe packet data physical channel PDCH (TN5 or TN7 above)

are eight signalling channels, SDCCHs on TN1, numbered SDCCH0 to SDCCH7. Each of these channels requires a 'block' of four frames.

As the SDCCH is a 'dedicated' channel (the TCH and FACCH are other dedicated channels), it will have a paired associated channel, the SACCH. The paired channel SACCH0, coupled to SDCCH0, immediately follows the 'block' containing SDCCH7.

It is immediately evident that within the 51-frame multiframe only four SDCCHs have a paired SACCH. This is rectified in the *next* 51-frame multiframe where TN1 is identical to TN1 of the first 51-frame multiframe except for the SACCH channels, which were previously SACCH0–3 and now become SACCH4–7.

SACCH channels paired to SDCCH channels have a repetition period of 102 frames.

The BCCH carrier TN2 of Figure 3.4 shows an uncommon configuration with extra common control channels (CCCHs) assigned to this physical channel. It is uncommon because, in general, network operators do not find it necessary to include extra CCCHs.

There are nine CCCHs on TN2. The BCCH channel may also be copied to this physical channel. However, as the FCCH and SCH *define* TN0 of the BCCH carrier, these logical channels are not copied to any other TN.

The BCCH carrier TN5 of Figure 3.4 is assigned as a traffic channel full rate (TCH/ FR). Traffic channels have a 26-frame multiframe structure. FN0–11 within the 26-frame multiframe carry customer data. Frame 12 is one burst of the SACCH, which is paired with the TN5 TCH.

As it takes four bursts to convey a SACCH message, then four 26-frame multiframes are required, that is 104 frames. This combination of four 26-frame multiframes carrying a SACCH message is sometimes called a *SACCH block*.

A SACCH block of 104 frames takes $104 \times 4.615\,ms$ to send. This is approximately 0.5 s, and is the maximum rate of measurement reports.

Frames 13–24 (and their multiples) carry customer data and frame 25 (and its multiples) carry the idle channel. The 26-frame multiframe then repeats.

Two 26-frame multiframes are one frame longer than the 51-frame multiframe. This means there is a one-frame slippage between them for every 51-frame multiframe. This guarantees that the SCH channel of neighbour cells is 'captured' within a maximum time of about 1s when measurements of neighbour cells are taken whilst a TCH is in use. These measurements are always made on the BCCH radio carrier of the neighbour cell. The SCH on TN0 of this radio carrier must be decoded. TN0 has a 51-frame multiframe structure.

The BCCH radio carrier TN6 of Figure 3.4 is assigned as a TCH/H, traffic channel half rate. Two subscribers are shown time division multiplexed onto this physical channel, both using alternate frames. Frame 12 is once again a quarter of a SACCH channel, but in this case, the SACCH belongs to the TCH/H, which starts on FN0 (the dark coloured subscriber). The SACCH channel for the TCH/H starting on FN1 is stolen from what was the idle frame.

The logical channel, which is not illustrated in Figure 3.4, is the fast associated control channel (FACCH). This signalling channel is only evoked when a dedicated TCH is in use, and L3 signalling is required whilst the TCH is in use. If signalling is required during transfer of customer data, half frames of the TCH are 'stolen' and used for signalling over the logical channel now named FACCH.

3.5   The GSM 51-frame multiframe and logical channel (combined configuration)

The lower half of Figure 3.7 shows that there are only three common control channels (CCCHs) on TN0 of the BCCH radio carrier with four standalone dedicated control channels (SDCCHs) and two associated control channels (SACCHs) replacing the CCCHs.

This configuration is commonly used by network operators for small capacity cells such as micro cells or pico cells. It is economical with physical channels as one physical channel carries both CCCHs and SDCCHs and another physical channel that is used in the non-combined mode for SDCCHs, in the combined mode is freed for use as a revenue-earning TCH.

As the cell is small capacity, it does not require as many control and signalling channels as a higher capacity cell (with a non-combined configuration). This combined configuration does not (of course) allow more CCCHs to be placed on other physical channels as is the case for the non-combined configuration.

3.6   The GPRS 51-frame multiframe logical channel structures (GSM 45.002, 43.064)

Figure 3.7 shows the GPRS services using a 51-frame multiframe structure. This structure is used by a cell which provides GPRS services without a PBCCH. In this case, the MS obtains its GPRS related system information from the BCCH SI13 messages. It uses the circuit switched C1 and C2 criteria for cell selection and reselection. The GPRS mobile station uses the BCCH radio carrier CCCH provision of PCH which, in addition to carrying *circuit switched paging* messages, carries *packet paging* messages. The circuit switched access grant channel (AGCH) is also used as the packet access grant channel (PAGCH) and the circuit switched random access channel (RACH) is also used as the packet random access channel. A PDCH (packet data channel) is provided as necessary and this follows the GPRS 52-frame multiframe structure, with PDTCH and PACCH sharing use of the blocks. (PDTCH is the packet traffic channel and PACCH is the packet associated signalling channel). The GPRS MS calculates which physical channel containing CCCHs it must use, as discussed in a chapter 7.

3.7   Using the 51- and 52-frame logical channels

This section is an introduction to how the logical channels are used. Figures 3.8 and 3.9 show the logical channels used for single-phase access to a TBF. Single-phase access is used for GMM signalling, SM signalling or customer data transfer when the data may be transferred using eight RLC/MAC blocks or less.

Not shown in these diagrams are the temporary flow identities (TFIs) used for identifying the block flow.

Figures 3.10 and 3.11 show the logical channels used for two-phase access to a TBF. The interchange of TLLI between the GPRS sub-network and the mobile station is

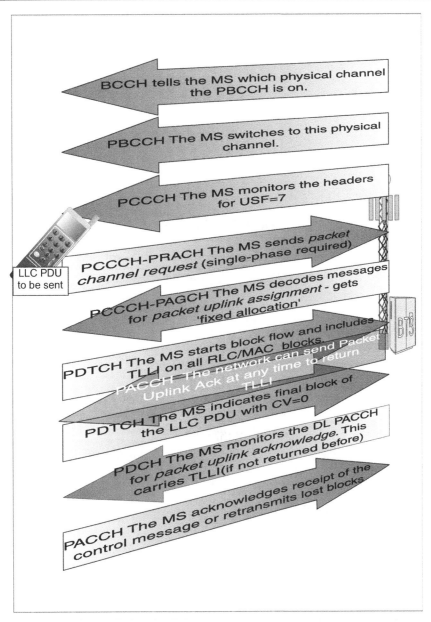

Figure 3.8 Single-phase access in a cell with PBCCH

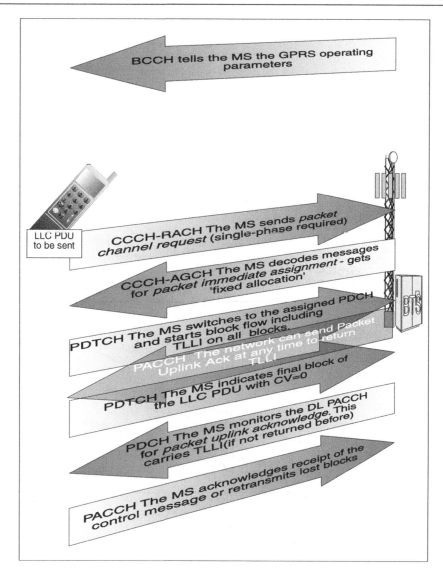

Figure 3.9   Single-phase access in a cell without PBCCH

more immediate and two-phase access is used for TBFs with a large number of RLC/ MAC blocks (as we might expect from customer data).

Two-phase access is used only for customer data transfer (where customer data is an SN PDU, the result of SNDCP layer processing of a network PDU (N PDU).

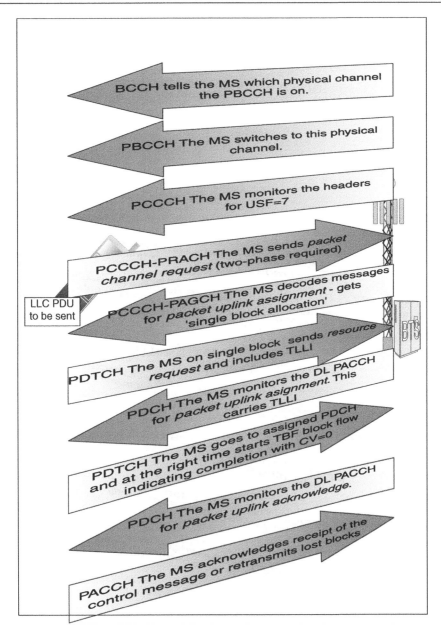

Figure 3.10   Two-phase access in a cell with PBCCH

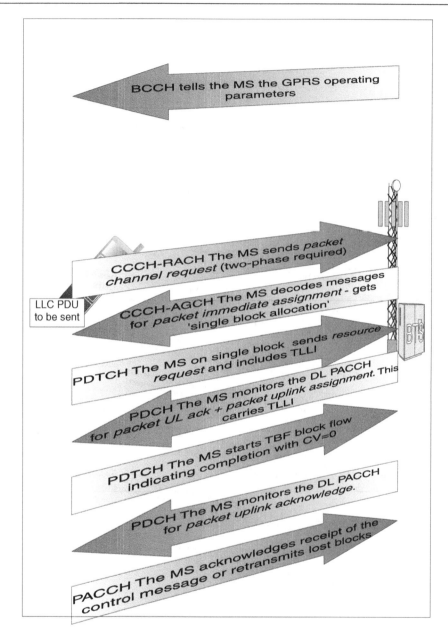

Figure 3.11   Two-phase access in a cell without PBCCH

3.8 Abbreviations used in this chapter

AGCH	Access grant channel
BCCH	Broadcast control channel
BTS	Base transceiver station
CCCH	Common control channel
CS	Circuit switched
DL	Downlink
EBCCH	Extended broadcast control channel
FCCH	Frequency correction channel
FEC	Forward error correction
GMM	GPRS mobility management
GPRS	General packet radio service
GSM	Global system for mobile communication
MAC	Medium access control
MS	Mobile station
NCH	Notification channel
N PDU	Network protocol data unit
PACCH	Packet associated control channel
PAGCH	Packet access grant channel
PBCCH	Packet broadcast control channel
PCH	Paging channel
PCCCH	Packet common control channel
PDCH	Packet data channel
PDTCH	Packet data traffic channel
PPCH	Packet paging channel
PRACH	Packet random access channel
PTCCH	Packet timing control channel
RACH	Random access channel
RLC	Radio link control
SACCH	Slow associated control channel
SCH	Synchronisation channel
SDCCH	Standalone dedicated control channel
SM	Session management
SNDCP	Sub-network dependent convergence protocol
SN PDU	Sub-network protocol data unit
TBF	Temporary block flow
TCH	Traffic channel
TFI	Temporary flow identity
TLLI	Temporary logical link identifier
TN	Timeslot number
UL	Uplink
USF	Uplink status flag

4

The TBF and the MAC Layer

(GSM 44.060, sections 5, 7, 8)

This chapter looks at the operation of the MAC (medium access control) layer which has the task of getting the right information onto the correct logical channel on the allocated physical channel, and all at the right time. However, before the MAC layer is covered, we look at the process at the heart of GPRS, the temporary block flow, TBF.

4.1 What is a TBF? An introduction to the temporary block flow

4.1.1 The radio link control/medium access control (RLC/MAC) block

The whole purpose of the conditioning of customer generated TCP/IP (or other Internet protocols) packets by the GPRS sub-network is to adapt them to the GSM air interface. In order to transmit customer data across the air interface, we segment them into so-called *RLC/MAC blocks*. The orderly transfer across the air interface of a number of blocks constituting a customer data packet is called a TBF.

RLC stands for radio link control and this is the layer where customer data packets are segmented into blocks conveniently transferred across the air interface as a TBF. The customer data packets coming into the RLC layer are called logical link control protocol data units (LLC PDUs).

The customer TCP/IP packet (carried inside an LLC PDU) has been pre-conditioned for reasons which we shall consider in later chapters.

A temporary block flow (TBF) is always initiated by the reception at an RLC layer (in either the GPRS sub-network or mobile station) of an LLC PDU. The RLC layer segments the incoming LLC PDUs into blocks of data (RLC/MAC blocks) for transfer over the air interface. One RLC/MAC block of data occupies one physical channel over four frames on the air interface. Figure 4.1 illustrates this, showing a cell transmitting

GPRS in Practice: A Companion to the Specifications   Peter McGuiggan
© 2004 John Wiley & Sons, Ltd   ISBN: 0-470-09507-5

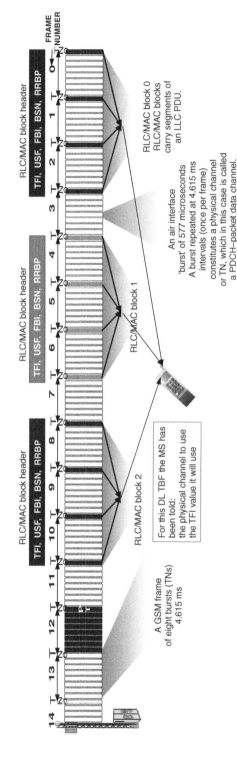

Figure 4.1 The structure of RLC/MAC blocks on the air interface. A three block downlink transfer is shown that does not show allocation procedures or acknowledgments. The MS decodes each DL block of four radio bursts in four radio frames and examines the RLC/MAC block headers. **TFI = temporary flow identity**, the value of this parameter tells the MS that has been given this value that this DL block is addressed to it; **USF = uplink status flag**, the value of this parameter tells the MS that has been given this value that it can transmit on the next UL block; **FBI = final block identity**, this has two values, 0 and 1. The MS which owns the TFI in the block header will look at the FBI and if it has the value 1, recognise that this is the final block of the DL TBF; **BSN = block sequence number**, the sequence number of the current block; **RRBP = relative reserved block period**, this indicates that the MS must send an acknowledgment in the uplink block indicated by the value of this parameter

a single radio carrier which is a stream of 577 microsecond 'bursts' of radio energy (the 577 microsecond period includes a guard period between bursts). A group of eight radio bursts form a 'frame' (4.615 ms). The first burst in each frame is called timeslot number TN0.

In Figure 4.1 TN0 has been allocated to a mobile station for a downlink GPRS data transfer; the allocated physical channel is called a packet data channel (PDCH). The incoming LLC PDU to the RLC layer is segmented and each segment requires four bursts within four frames to transmit over the air interface.

Three RLC/MAC blocks are required to transfer our particular LLC PDU across the air interface as the downlink TBF illustrated, but the access procedures, resource allocation of the physical channel and acknowledgment for the completion of the TBF are not shown. These will be shown later.

If *all* the eight physical channels are allocated to the mobile station as packet data channels (PDCHs) then eight RLC/MAC blocks can be transferred within the four frames. In our example, only one physical channel is assigned to the mobile station and the data rate for one physical channel is about 9 kb/s if coding system 1 (CS1) is used for channel encoding in the physical layer. This rises per physical channel to 13.4 kb/s for CS2, 15.6 kb/s for CS3 and 21.4 kb/s for CS4. Coding system 1 applies the maximum redundancy to data within the RLC/MAC blocks but this redundancy is gradually reduced for CS2 and CS3 until there is no redundancy at all for CS4. This decreases the communication reliability for CS2–4 and effectively reduces the operating radius of the cell, as a higher C/I (carrier to interference ratio) is required to compensate for the loss of reliability.

The RLC function operating in acknowledged mode also provides a layer-two point-to-point data communications link which guarantees delivery of the LLC PDUs, otherwise it delivers unitdata packets in unacknowledged mode.

The RLC layer always acknowledges receipt of a TBF, regardless of whether the RLC data link is acting in acknowledged mode or unacknowledged mode. If it is acting in acknowledged mode, then indication of unreceived blocks causes retransmission of those blocks; in unacknowledged mode, the status of unacknowledged blocks is disregarded. The reason that acknowledgments are required in unitdata (unacknowledged) mode is that the radio link status is also confirmed (or suspect if acknowledgments are not received).

That is the concept of the RLC/MAC block, straightforward and uncomplicated. However, the RLC/MAC block is only a part of the picture of that which we call a TBF.

TBFs are always transferred on the logical channel *packet data traffic channel (PDTCH)* but the PDTCH RLC/MAC blocks can, at any time, be 'stolen' and used for RLC signalling (more specifically radio resource signalling). Then the PDTCH logical channel becomes a PACCH logical channel, the packet associated control channel.

The PACCH has a similar function to the SACCH in GSM circuit switched operations, in that it is a radio resources signalling channel, but unlike the SACCH, which is always present with a dedicated channel, the PACCH is only present when RLC signalling is required and hence cannot be used like the SACCH to verify the 'goodness' of the radio link.

It is appropriate to clarify GPRS signalling channels of which there are two types – the PACCH, already considered and used for radio resource signalling, and the PDTCH, which, in addition to being used as a GPRS 'traffic channel', is also used for 'higher layer service signalling', which is the signalling for the GPRS GMM layer and the GPRS SM layer. These are covered in detail in later chapters.

The RLC/MAC blocks also include the following information in the block headers or information fields:

- **Temporary flow identifier** (*TFI*). This identifies each RLC/MAC block of a TBF on both the uplink and downlink. The mobile station is told the TFI, which identifies the TBF when a radio resource is given, and uses this identity to recognise RLC/MAC blocks addressed to it. On the downlink it is then possible to time division multiplex on one physical channel many TBFs to many mobile stations. It also identifies the TBF on the uplink.
- **Temporary logical link identifier** (*TLLI*). The mobile station must be attached to the GPRS sub-network for a TBF to take place. The GPRS sub-network allocates a packet temporary mobile subscriber identity (PTMSI) as part of the attach process and from this is derived the TLLI. The TLLI is sent by the mobile station on all the RLC/MAC blocks of a single-phase access TBF for identification purposes (it is sent on only one block for two-phase access). TLLI is also used to identify a mobile station for some categories of downlink message to that mobile station.
- **The block sequence number** (*BSN*). Of each block of a TBF. These are used for flow control. The RLC function includes a point-to-point data communications link and the BSN, in acknowledged mode, is used to check that the distant end has received all the transmitted blocks (the distant end indicates which BSNs have been received); if any are missing the sending end will resend the missing BSN RLC/MAC blocks. Acknowledged mode at the RLC level is only used for customer data packets, and then only if the QoS given to the customer justifies RLC acknowledged mode.
- **Countdown value** (*CV*). This uplink parameter counts down for the last sixteen RLC/MAC blocks (or less, sixteen is the default value) and when it reaches zero, the GPRS sub-network knows that this is the last RLC/MAC block of the uplink TBF and will send an acknowledgment.
- **Final block indicator** (*FBI*). This downlink parameter tells the mobile station that the final block of the downlink TBF is received. The FBI is always paired with the relative reserved block period (RRBP), which tells the mobile station which uplink PACCH block it must use (relative to the downlink block in which FBI/RRBP is received) to send an acknowledgment of reception of the TBF.
- **Uplink status flag** (*USF*). This downlink parameter, given to a mobile station with an uplink dynamic allocation, indicates by its presence on a downlink RLC/MAC block that the mobile station owning this USF is allowed to transmit in the next uplink RLC/MAC block. In this way the GPRS sub-network can multiplex up to eight mobile stations onto one uplink physical channel.

  Each of the eight mobile stations is given a value of the USF flag when the radio resources are allocated. The USF allows the GPRS sub-network to control access on the uplink dynamically. On PDCHs carrying PCCCHs the USF value 7 may be reserved for controlling access to the PRACH.
- **USF granularity**. This is allocated to a mobile station in conjunction with a USF value. It tells the mobile station how many RLC/MAC blocks may be transmitted on the uplink when its allocated USF value appears on the downlink.

4.1.2 Introduction to the MAC function (GSSM 44.060 sections 5, 7, 8)

We have referred to RLC/MAC blocks without mentioning the meaning or function of the MAC (medium access control) function.

The medium that is accessed is the GSM air interface and the MAC function is one of the more involved of the GPRS functions. It is responsible, on the mobile station side, for getting access to the GPRS sub-network using the PRACH on the uplink (sending *packet channel requests*), and the PCCCH on the downlink (*packet uplink assignment*). It is then responsible for accessing the assigned uplink PDCH, using the correct logical channel,[1] on the correct physical channels (TN, frequency or hopping sequence) at the correct frame numbers.

The GPRS sub-network through the *packet uplink assignment* gives the mobile station RR layer the control parameters necessary and the RR layer instructs the MAC layer.

For an uplink TBF these instructions include physical channel(s) assignment and dynamic, extended dynamic or fixed allocation of uplink RLC/MAC blocks:

- **Dynamic allocation.** This tells the mobile station that it may only access the uplink RLC/MAC blocks when its allocated USF appears on a downlink RLC/MAC block header of the allocated downlink physical channel.

  The mobile station is also told the USF granularity, which simply means how many RLC/MAC blocks the mobile station is allowed to send upon the appearance of its allocated USF; the values are currently one or four RLC/MAC blocks.

  Many mobile stations will be multiplexed onto the same downlink physical channel, and a particular mobile station must be able to recognise that a downlink message is addressed to itself. The TFI, given by the GPRS sub-network to the mobile station with the uplink assignment accomplishes this. The mobile station is also given an uplink TFI which must be included on all uplink RLC/MAC blocks to allow the GPRS sub-network to positively identify that it is the correct mobile station using the uplink block.

- **Extended dynamic allocation** gives the mobile station the same information as in dynamic allocation with the difference that, if more than one physical channel (or timeslot number) is allocated, then the mobile station has only to examine the USF headers on the first timeslot number (TN) and this implicitly gives access to uplink RLC/MAC blocks on all the uplink TNs of the allocation.

  Comparing this with dynamic allocation, when the mobile station is given more than one physical channel, it is given a USF for each of these channels and must examine each downlink physical channel USF header for permission to use the uplink RLC/MAC blocks on that physical channel.

- **Fixed allocation** gives the mobile station a fixed number of uplink RLC/MAC blocks and specifies the frame number of the start of the first block to be transmitted.

For a downlink TBF the mobile station is alerted to the presence of a downlink TBF by receiving either a paging message on its paging channel (carrying its PTMSI) or by a *packet downlink assignment* on its paging channel, or any PCCCH that the GPRS

---

[1] The correct logical channel on the uplink is determined by the type of packet the MAC layer is asked to send. For customer data, MM and SM signalling it is the PDTCH, and for RLC signalling it is the PACCH.

sub-network knows the mobile station is listening to (TLLI is used to identify the mobile station).

If the mobile station is in an uplink TBF, the downlink PACCH (identified by the TFI) carries the downlink assignment which includes the downlink TFI for each physical channel. The mobile station then examines the RLC/MAC block headers on the allocated downlink physical channels and, receiving its TFI, passes the content of the RLC/MAC blocks to the LLC layer.

The downlink block carrying the TFI will also indicate whether the logical channel is PACCH (carrying RR signalling) or PDTCH (carrying customer data or higher layer service signalling).

4.1.3 *Combining the components of a TBF into a complete TBF*

We are now in a position to construct a TBF that combines the various components considered above.

Figure 4.2 illustrates how a TBF is initiated – the RLC/MAC layer receives a data packet from the LLC layer directly above. This data packet is called, appropriately an LLC PDU and upon receiving one of these the RLC/MAC layer springs into action to send it as a TBF across the air interface.

In this example, one LLC PDU stimulates the TBF, but the RLC/MAC layer can receive one after another, as many LLC PDUs as its buffer can store. The RLC/MAC layer then initiates a TBF with enough RLC/MAC blocks to send all the buffered LLC PDUs across the air interface. The LLC PDUs are segmented into RLC/MAC block size, and some of the RLC/MAC blocks will contain the last part of one LLC PDU and the first part of the next LLC PDU.

Figure 4.3 expands upon a mobile station's actions when the RLC/MAC layer receives an LLC PDU. The RLC/MAC layer segments the LLC PDU into RLC/MAC blocks. The RLC layer now knows how many RLC/MAC blocks are required in the TBF. The RLC/MAC layer requests radio resources from the GPRS sub-network by sending a short access burst on the PRACH channel containing a *packet channel request*. The only identity sent with this access request is a short random number generated by the mobile station. The mobile station RLC/MAC layer may indicate that it requires one-phase access, in which case the *packet channel request* may indicate how many uplink blocks are required to send the LLC PDU in a TBF. It may indicate two-phase access, in which case the mobile station will wait until the next message it sends, only then specifying how many uplink RLC/MAC blocks are required.

The mobile station now keeps its receiver open, decoding all messages on the downlink PCCCH; it is looking for the downlink message *packet uplink assignment* returning the random number generated in the *packet channel request*. With the *packet uplink assignment* the mobile station receives the physical channel assignment and the parameters controlling access to that physical channel. If the request was for single-phase access it delivers the first uplink RLC/MAC block carrying a segment of a L3 PDUs of the TBF.

The RLC/MAC blocks delivered on the assigned uplink physical channel will carry TLLI so that the GPRS sub-network can identify the mobile station. The TBF continues, with all RLC/MAC blocks carrying the TLLI. At some stage, the GPRS

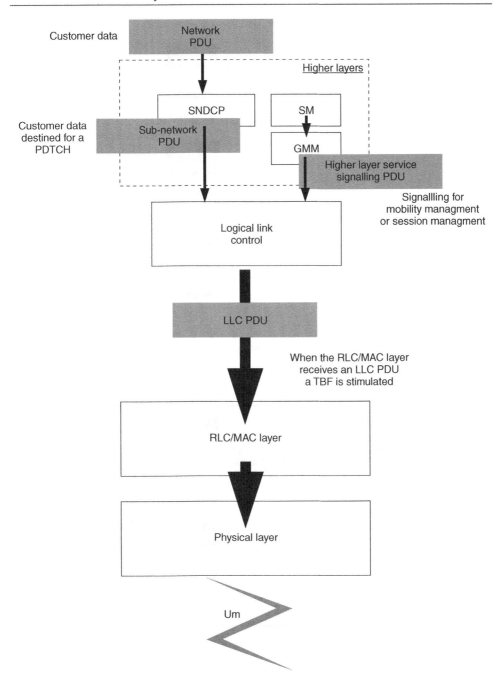

Figure 4.2 The impetus for a TBF

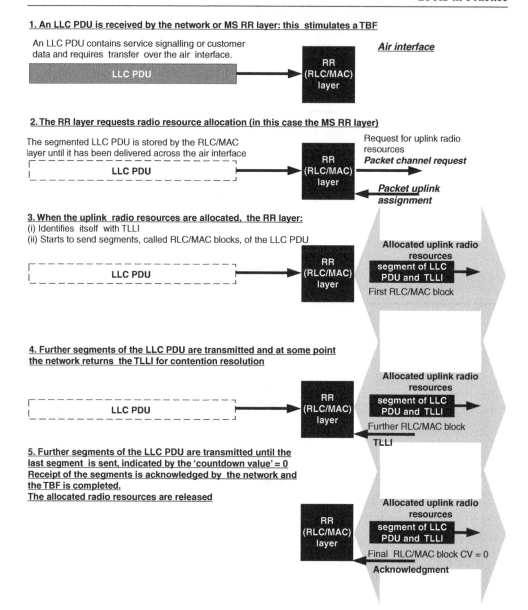

1. An LLC PDU is received by the network or MS RR layer: this stimulates a TBF

An LLC PDU contains service signalling or customer data and requires transfer over the air interface.

LLC PDU

Air interface

RR (RLC/MAC) layer

2. The RR layer requests radio resource allocation (in this case the MS RR layer)

The segmented LLC PDU is stored by the RLC/MAC layer until it has been delivered across the air interface

LLC PDU

RR (RLC/MAC) layer

Request for uplink radio resources
Packet channel request

Packet uplink assignment

3. When the uplink radio resources are allocated, the RR layer:
(i) Identifies itself with TLLI
(ii) Starts to send segments, called RLC/MAC blocks, of the LLC PDU

LLC PDU

RR (RLC/MAC) layer

Allocated uplink radio resources

segment of LLC PDU and TLLI

First RLC/MAC block

4. Further segments of the LLC PDU are transmitted and at some point the network returns the TLLI for contention resolution

LLC PDU

RR (RLC/MAC) layer

Allocated uplink radio resources

segment of LLC PDU and TLLI

Further RLC/MAC block

TLLI

5. Further segments of the LLC PDU are transmitted until the last segment is sent, indicated by the 'countdown value' = 0. Receipt of the segments is acknowledged by the network and the TBF is completed. The allocated radio resources are released

RR (RLC/MAC) layer

Allocated uplink radio resources

segment of LLC PDU and TLLI

Final RLC/MAC block CV = 0

Acknowledgment

Figure 4.3   What is a TBF? An example

sub-network will respond to the mobile station with a *packet uplink acknowledgement*, which returns the TLLI.

When the final RLC/MAC block of the TBF is transmitted by the mobile station, it contains the countdown value $CV = 0$. The mobile station now stops transmitting and monitors the downlink PACCH (which was initially allocated in the *packet uplink assignment*). It is looking for the message *final packet uplink acknowledge* carrying its TFI, which also was allocated in the *packet uplink assignment*. If the *final packet uplink acknowledgment* indicates that the GPRS sub-network has received all blocks, the TBF is complete and the radio resources are released.

If the RLC link is operating in unacknowledged mode, then the status of the received blocks sent in the *packet uplink acknowledge* message is disregarded as irrelevant and the TBF is complete.

If the RLC link is operating in acknowledged mode, and the GPRS sub-network has not received all the numbered blocks, the missing blocks are indicated in the *packet uplink acknowledge/negative acknowledge* message and the mobile station is simultaneously given additional uplink radio resources to retransmit the missing blocks. The blocks which were not received are retransmitted and the GPRS sub-network will acknowledge the retransmitted blocks; the TBF is complete and the radio resources are released.

The LLC PDU (or PDUs) transmitted in this uplink TBF may constitute only part of a customer's session; further TBFs will be stimulated in the same manner when further LLC PDUs are received by the RLC/MAC layer.

4.1.4   TBF arrow diagrams (GSM 44.060 sections 5, 7, 8, 9)

We are now in a position to examine arrow diagrams showing the flow of messages across the air interface before, during and after a TBF.

Figure 4.4 illustrates the basic elements of a TBF; in this case, customer data is transferred on an uplink TBF which uses two-phase access. Two-phase access is used to speed up contention resolution, which may arise if two or more mobile stations in the same cell transmit identical *packet channel requests* in the same PRACH with the same random identity number. (In certain circumstances the GPRS sub-network may be able to decode one message out of two received on the same channel). Both mobile stations will then decode all PCCCH messages looking for *packet uplink assignment*. One of these channels will return *packet uplink assignment*, returning the random number generated by both of the mobile stations. Both mobile stations then regard the assignment as belonging to them and access the assigned uplink PDCH. Two-phase access tries to overcome rapidly this undesirable situation.

The *packet uplink assignment* assigns only one RLC/MAC block and both mobile stations will use that to send *a packet resource request*. Included with this single block message the mobile stations send their identity, TLLI. The GPRS sub-network correlates this to the mobile station IMSI (and IP address, if it has one at this stage).

Normally when two mobile stations clash on the PACCH sending *packet resource request*, the GPRS sub-network cannot decode either message and cannot respond. This causes the contention resolution timer in each mobile station to time out, disconnecting them from the radio resources. Both mobile stations then repeat the access attempt from scratch. However, situations may arise where the GPRS sub-network can decode one of the mobile stations using the uplink PACCH. If the GPRS sub-network can decode one of the messages from the two mobile stations, then the TLLI that is decoded is sent back on the downlink PACCH along with the message *packet uplink acknowledge + packet*

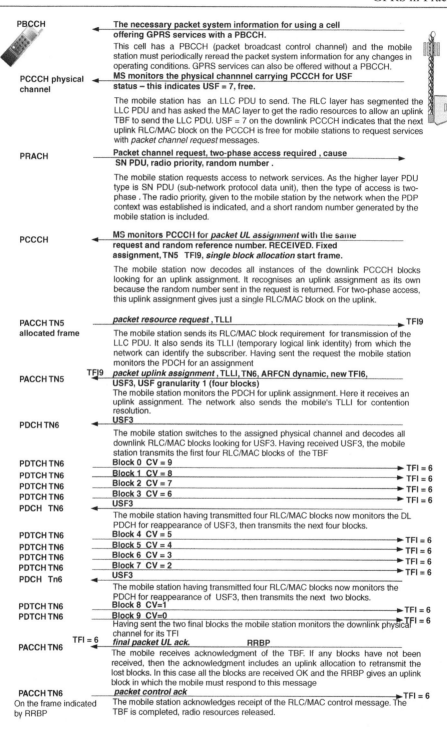

Figure 4.4   An example of the basic elements of a TBF UL two-phase access

uplink assignment. Both mobile stations receive this message and the one which has received the wrong TLLI must immediately disconnect from the radio connection.

The sooner a mobile station receives its returned TLLI, the more economical is the use of radio resources. In two-phase access, TLLI is returned within the exchange of three RLC/MAC blocks. With one-phase access, the TLLI may not, in the worst case, be returned to the mobile station until the TBF is completed, so if the radio conditions are poor, one-phase access may waste radio resources. If the radio conditions are good, one phase access is more efficient than two-phase access.

A *single-phase access* UL TBF may be summarised as follows:

- **Initiation**. The RLC/MAC layer in the mobile station receives one or more LLC PDUs from the mobile station LLC layer.
- **Access step 1**. The mobile station sends a *packet channel request* on the PRACH.
- **Access step 2**. The GPRS sub-network gives an *uplink assignment* of radio resources on the PCCCH.
- **Identification and LLC PDU transfer**. The mobile station sends the first RLC/MAC block of the segmented LLC PDU and identifies itself by including TLLI.
- **Contention resolution**. The GPRS sub-network returns TLLI to the mobile station. It does this by sending *packet uplink acknowledge* on the DL PACCH. The network must do this as soon as possible once it has received TLLI from the mobile station.
- **Data transfer**. TBF transfer of numbered RLC/MAC blocks which are segments of the LLC PDU.
- **Data transfer completion**. The mobile indicates the end of the TBF with $CV = 0$.
- **Acknowledgment 1**. The GPRS sub-network acknowledges receipt of the numbered RLC/MAC blocks with *final packet uplink acknowledge*. If the RLC/MAC layer is operating in acknowledged mode and any numbered RLC/MAC blocks have *not* been received by the GPRS sub-network, then the DL message will be *packet UL negative acknowledge* and the mobile will retransmit the lost blocks.
- **Acknowledgment 2**. The mobile station acknowledges receipt of the *packet uplink acknowledge* by returning *packet control acknowledgement*.
- **Release**. The assigned radio resources are released by the mobile station and GPRS sub-network.

A *two-phase* UL TBF may be summarised as follows:

- **Initiation**. The RLC/MAC layer in the mobile station receives one or more LLC PDUs from the mobile station LLC layer.
- **Access step 1**. The mobile station sends a *packet channel request* on the PRACH.
- **Access step 2**. The GPRS sub-network gives an *uplink assignment* of radio resources on the PCCCH. This is a single UL RLC/MAC block.
- **Identification and LLC PDU transfer**. The mobile station sends *packet resource request* with its TLLI on the single UL block.
- **Contention resolution**. The GPRS sub-network returns TLLI to the mobile station. It does this by sending *packet uplink acknowledge + packet uplink assignment* on the DL PACCH. The acknowledgment is for the single block TBF, and the assignment gives the radio resources for transfer of the LLC PDU the mobile station is waiting to send.

- **Data transfer**. TBF transfer on the assigned radio resources of numbered RLC/MAC blocks, segments of the LLC PDU.
- **Data transfer completion**. The mobile indicates the end of the TBF with CV = 0.
- **Acknowledgment 1**. The GPRS sub-network acknowledges receipt of the numbered RLC/MAC blocks with *final packet uplink acknowledge*. If the RLC/MAC layer is operating in acknowledged mode and any numbered RLC/MAC blocks have *not* been received by the GPRS sub-network, then the DL message will be *packet UL negative acknowledge* and the mobile will retransmit the lost blocks.
- **Acknowledgment 2**. The mobile station acknowledges receipt of the *packet uplink acknowledge* by returning *packet control acknowledgement*.
- **Release**. The assigned radio resources are released by the mobile station and GPRS sub-network.

Further examples of both uplink and downlink TBFs are given later in this chapter.

4.2 The MAC layer in action

This section is about the MAC layer in GPRS Um operations. It covers the use of logical channels in setting up a packet data transfer between a mobile station and the GPRS sub-network. Expanding on a previous chapter's description of packet data transfer, the mechanism of a TBF is examined in detail. This section covers the MAC in operation and its interaction with the GRR (GPRS radio resource) layer.

4.2.1 Introduction: GPRS attach[2]

As with GSM circuit switched IMSI attach, two-way communication is not possible between a mobile subscriber and the GPRS sub-network unless the mobile station is *GPRS attached*. For a mobile station, the purpose of attach is twofold:

1. To alert the GPRS sub-network to its presence and 'active' condition – that is, that it is switched on and ready to have two-way communication when required.
2. To allow the GPRS sub-network to 'track' the mobile station as it transits location areas for circuit switched operations and routeing areas for GPRS operations – the mobile stations inform the GPRS sub-network when a new area is entered. This allows it to contact the mobile station when an incoming call arrives for that mobile station.

There is, however, an anomaly with GPRS attach that is not evident with CS attach. A GSM mobile station is always ready to receive incoming calls once attached, but a GPRS mobile station may or may not be ready to receive incoming calls after attaching. This difference centres upon IP addresses.

If a GPRS sub-network is using dynamic IP addresses, which are allocated by the GPRS sub-network upon the setting up of an IP call (PDP context activation), then there is no immediately obvious reason for a GPRS mobile station to attach unless it has

---

[2] See Chapter 10 for further details of this GMM function.

been asked to establish an IP call (through the establishment of a PDP context which allocates a dynamic IP address and sets up the QoS profile for Internet communication).

With domestic (home) Internet access, the Internet service provider (ISP) cannot deliver messages to a user unless the user initiates a call to the ISP. This is because communications between the user and the ISP rely on a dynamic IP address given by the ISP *ad hoc* for a single communication initiated by the user. When this communication is terminated, the IP address is removed and the user is of 'no known address'. No communication can be addressed to the user.

With GPRS operations, a dynamic address is allocated when a PDP context (data call set-up) is established, and there is then apparently no point in establishing attach unless this is the result of a request for a PDP context, as without a PDP context the mobile station has 'no known address', but this is considered later.

After attachment and establishment of the PDP context with its associated dynamic address, the mobile station can receive Internet communications. When the communication established through the PDP context is completed, the user keeps the dynamic address, so that the ISP can forward messages to the mobile station user. The sub-network GPRS operators decide for how long the retention of the dynamic addresses are allowed.

A static IP address is the permanent property of a user and if the ISP knows this address, then there is good reason to attach without the establishment of a PDP context.

4.3 'Attach' MAC procedures (ETSI 123.060 section 6)

Figure 4.5 shows a mobile station attaching to the GPRS sub-network. In this example a PBCCH is provided. This GPRS attach is similar to circuit switched IMSI attach, where the mobile station tells the GSM network its identity (TMSI) and its capabilities and that it is switched on, synchronised to a cell within a location area and listening to the CS paging channel.

When a mobile station GPRS attaches, the GPRS sub-network knows that the mobile station will be ready (once a PDP context is established) to use the GPRS sub-network services and (again, after a PDP context is established giving the mobile station an IP address) will alert it within its location or routeing area to an incoming packet via a paging message. This is called GMM (GPRS mobility management) attach.

There are three states for GMM:

1. **Idle** which is not attached. (The mobile station is also in packet idle condition).
2. **Standby** which is attached but with no radio resources allocated for customer data transfer and the ready timer expired. (The mobile station is also in packet idle condition).
3. **Ready** when an MS is attached and has been allocated radio resources for a TBF or has just finished a TBF, released the radio resources and the ready timer is still running. (The mobile station may be in packet idle or packet transfer condition).

The ready timer is started when an LLC PDU is transferred between the LLC layer and the RLC layer. This uplink transfer will start a TBF.

At the end of the TBF the radio resources are released, but the ready timer continues running. The mobile station is in GMM ready condition whilst this timer is running. When the timer expires, the mobile changes to GMM standby condition.

The significance of the ready timer is that, whilst it is running, the GPRS sub-network knows which cell is the mobile station's serving cell. Whilst the ready timer is running,

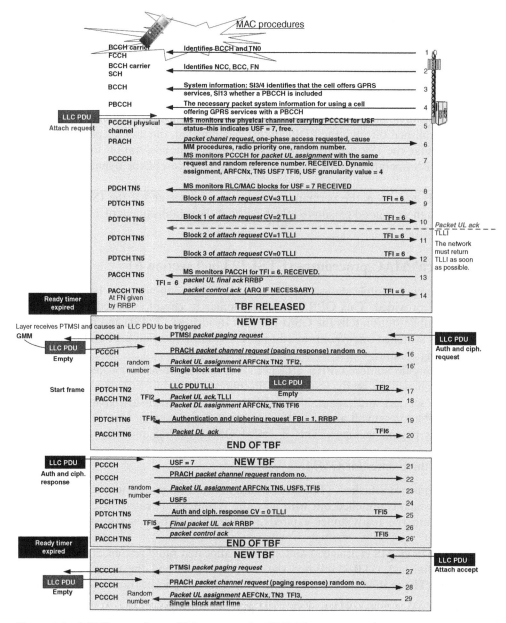

Figure 4.5   MAC procedures, SM requests the GMM layer to attach

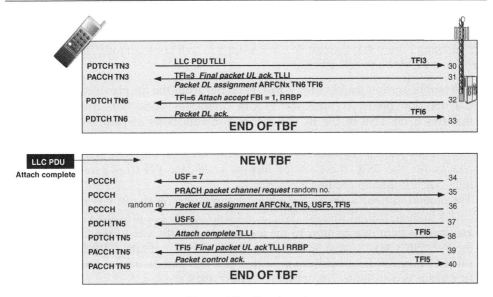

Figure 4.5 Continued

the mobile station must inform the GPRS sub-network when a cell is reselected. With the knowledge of which cell a mobile station is using whilst in the packet idle condition (but the ready timer running), the GPRS sub-network may put a *downlink assignment* message directly onto the paging channel that the mobile station is monitoring. Figure 4.6 shows the effect of the ready timer.

In this context should be mentioned the non-DRX timer. This timer starts immediately an uplink or downlink TBF is completed. Whilst it is running, the mobile station stays fully awake monitoring all downlink packet common control channels (PCCCHs). If this timer were not running, the mobile station would go into discontinuous reception mode (DRX), waking only to read a PCCCH which corresponds to its paging group.

The non-DRX timer will be running in parallel to the ready timer, but will expire before the ready timer. When they are running jointly, the GPRS sub-network knows which cell is the mobile station's serving cell and that the mobile station is listening to all the PCCCHs in that cell. When an incoming packet arrives from the Internet for delivery to this mobile station, the GPRS sub-network operator can put a *downlink assignment* message on any one of the PCCCHs in that cell.

A mobile station must select a PLMN, select a cell using the normal circuit switched procedures and examine the BCCH system information to see whether the cell offers GPRS services.

GPRS services may be offered with or without a packet broadcast control channel (PBCCH). The full procedures for PLMN selection and cell selection and reselection are given in a later chapter.

The procedures shown in Figure 4.5 show a series of TBFs, both uplink and downlink in half duplex mode. Half duplex mode means that, although simultaneous use of uplink

and downlink channels is possible, only one of these is used at a time. If both UL and
DL channels are used simultaneously, then the communication is full duplex.

Examining Figure 4.5, the mobile station starts in 'cell reselect mode', having already
performed PLMN and cell selection.

1. The mobile station searches for a BCCH carrier from the list provided in the initial
   cell selection phase. It finds the FCCH (frequency correction channel), knows
   that this is on TN0 of the BCCH carrier and adjusts its frequency to the BCCH
   carrier.
2. The mobile station decodes the SCH (synchronisation channel) after 4.615 ms. It
   now knows the BSIC (base station identity code), the FN (frame number), and is
   fully synchronised to the BCCH carrier.
3. The mobile station listens to the system information on the BCCH carrier. It discovers from system information type 3 or 4 whether the cell provides GPRS services,
   and from system information 13 the location of the PBCCH (if the GPRS services
   use the PBCCH). System information 3 or 4 may indicate GPRS services but
   PBCCH may not be provided and in that case the mobile station will then use the
   BCCH for system information (SI13) on the GPRS services.
4. The mobile station goes to the PBCCH position indicated by SI13 and discovers
   from the PBCCH the structure of the GPRS control channels.
5. Knowing the GPRS PCCCH structure, the mobile station now monitors the PCCCH
   looking for the uplink status flag (USF) indicating 'uplink free' condition (USF = 7).
   USF = 7 on a physical channel which carries packet common control channels may
   be used to indicate that the next RLC/MAC block in the uplink is free for use as
   a PRACH (packet random access channel). As the RLC/MAC blocks of data are
   sent in a radio burst in each of four frames, then all four bursts must be received
   before the USF can be decoded.
6. The mobile station now sends a *packet channel request* on the free UL PRACH
   channel. One-phase access is used with radio priority one. This request was initiated

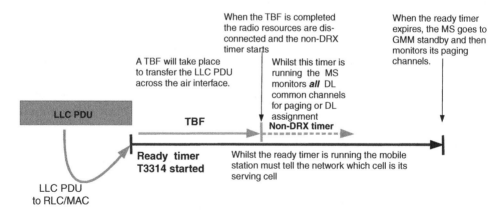

Figure 4.6 This diagram shows the effect of the ready timer and its interaction with a TBF
and the non-DRX timer. As the network knows the mobile station's cell whilst this timer is
running, it can give a downlink assignment directly on the PCCCH

by an LLC PDU received by the RLC/MAC layer. This LLC PDU originates in this case from the GMM layer – *attach request*. This causes the 'single-phase' access request. Full details of the control of mobile station access on the PRACH are given in chapter 7.9.

7. Having sent a *packet channel request* the mobile station monitors all the DL PCCCHs (packet common control channels) looking for a *packet uplink assignment* message which returns the same request and random reference number as was sent on the uplink PRACH. The mobile station acts on the instructions contained in the *packet uplink assignment* which include:

   —The TN and ARFCN (physical channel) to be used (or MA if the assigned channel is frequency hopping).
   —The type of allocation – dynamic, single block or fixed; (dynamic allocations may be used most frequently by GPRS sub-network operators, and this assignment is indeed dynamic). For dynamic allocations the mobile station is told the USF number (range 0–7) that will indicate when the uplink blocks on the allocated physical channel may be used. With this is given a USF granularity number which tells the mobile station the number of uplink RLC/MAC blocks it may transmit when its USF appears on the downlink.
   —Temporary flow indicators (TFIs) for the UL and the DL.
   —A DL TN to monitor for PACCH (packet associated control channel) instructions.
   —Other information is sent by the GPRS sub-network, which includes power control parameters, and the TA (timing advance) the mobile station must apply to full-length information bursts.

8. Switching to the assigned physical channel, the mobile station examines the headers on all downlink RLC/MAC blocks, looking for the USF that has been allocated. One of these blocks carries USF = 7 which has been assigned to this mobile station.

9–12. Recognising its USF, the mobile station sends an *attach request* message on the next four uplink RLC/MAC blocks of the PDTCH (packet data traffic channel). The use of four blocks is determined by the USF assignment, which is given with the USF granularity – one or four blocks, in the *packet uplink assignment* message. The mobile station includes in the UL blocks the TFI and CV (countdown value) in the RLC/MAC block headers and the TLLI (temporary logical link identifier) is included to identify the mobile station to the GPRS sub-network and resolve clashes.

This is the method used for one-phase access. The most important piece of information in this message is the mobile station's identity, PTMSI (packet temporary mobile subscriber identity), which is always sent in conjunction with the RAI (routeing area indication). The RAI includes the MCC (mobile country code), MNC (mobile network code), LAC (location area code) and RAC (routeing area code). IMSI (international mobile subscriber identity) may be sent if no PTMSI is available. For single phase access the GPRS sub-network is obliged to send a *packet uplink acknowledgment* as soon as possible after receiving an uplink RLC/MAC block containing TLLI. A dotted arrow immediately after arrow 10 shows this happening. This acknowledgment is sent solely to return the TLLI.

13. The mobile station monitors the DL physical channel for PACCH and receives its TFI with the message *final packet UL ack*. This also includes an **RRBP**, which the mobile station must use to acknowledge receipt of this control message.

    Note that this procedure is followed even if the RLC link is operating in unacknowledged mode. The RLC/MAC ack serves two purposes: the first is a straightforward indication of the reliability of the radio link; if no 'Ack' is received then the link is suspect; the second, when the RLC link is working in 'connection mode', is to acknowledge numbered RLC/MAC blocks and if any are not received, to retransmit them. If the link is working in 'connectionless' mode, the status of the acknowledged blocks is simply ignored.

14. The mobile station acknowledges receipt of the control signal.

    The TBF is completed; the mobile station monitors its paging channel, or all of the DL PCCCHs for a predetermined period. In this example, the ready timer has expired.

15. The mobile station receives a *packet paging request* on its PCCCH.

16. The mobile station sends a *packet channel request* on the PRACH (after having received USF = 7 on the previous downlink RLC/MAC block). This indicates that the request is in response to a *paging request*. The mobile station monitors the DL PCCCHs and receives a single block *packet UL assignment*.

17. Switching to the assigned physical channel the mobile station transmits on the frame indicated by the assignment.

18. The mobile station now monitors the DL PACCH, recognises its TFI and decodes *final packet UL ack/nack* which includes its TLLI. This also contains a DL assignment.

19. Switching to the assigned DL physical channel the mobile station receives its TFI and the message *authentication and ciphering request*. The final RLC/MAC block carrying this message includes FBI = 1, indicating that this is the final block of the LLC PDU. Accompanying FBI is a parameter RRBP that gives the mobile station an uplink block to respond.

20. The mobile station sends *packet DL acknowledge* on the UL RLC/MAC block indicated by RRBP.

    The TBF is completed.

    After receiving the *authentication and ciphering request*, the mobile station GMM layer will process the information and then generate an LLC PDU in response.

21–22. The mobile station monitors the DL PCCCH and upon receiving USF = 7, sends a *packet channel request*.

23–24. Monitoring all of the DL PCCCHs, the mobile station receives *packet uplink assignment* and switches to the physical channel. On the DL it receives its assigned USF indicating that it should start to send its LLC PDU on the next UL block.

25. The mobile station sends *authentication and ciphering response* on the UL blocks, indicating the completion of the message with CV = 0.

26. The mobile station now stops transmitting and monitors the DL physical channel. It receives its TFI with the message *final packet UL ack*. It responds with *packet control acknowledge* on the assigned RRBP.

    The TBF is completed.

27. The GPRS sub-network now has an LLC PDU for the mobile station and pages it on the PCCCH with PTMSI.
28. The mobile station sends *packet channel request* on the UL PRACH indicating *paging response*.
29. It receives a *packet UL assignment* on the DL PCCCH.
30. Switching to the assigned physical channel, the mobile station waits for the frame in which it will transmit.The mobile station sends an empty LLC PDU containing its TLLI.
31. The mobile station receives a *final packet UL ack*. This contains a *packet DL assignment*.
    The UL TBF is completed and a DL TBF is about to start.
32. The mobile station switches to the assigned DL physical channel, receives its TFI and the DL LLC PDU is transferred to the mobile station on the assigned physical channel. The end of the DL transfer is signalled with FBI = 1 and RRBP.
33. The mobile station acknowledges receipt of the DL TBF.
    The DL TBF is completed.
34–40. Shows the final UL TBF.
    Figure 4.7 gives a physical picture of a part of the above process. The radio bursts are shown from the BTS and mobile station.

4.4   Packet data transfer – PDP context activation[3] (ESTI 123.060 section 9)

PDP context activation means the setting up of a packet data call. This is similar to 'call set-up' in circuit switched operations. The PDP context can remain active (but unused) after completion of the packet data call for which it was established. As the ISP which the subscriber used for the transaction has been given an IP address for the subscriber, it can send Internet packets to the mobile station. This is the concept of 'all-day connection' to the Internet, with charges incurred only when data is passed through the PDP context.

Whether a GPRS sub-network operator will allow all day, or days, or just a part of the day, will depend upon market pressures and the subscription type.

The protocol layer, which initiates PDP context activation in the mobile station is the 'network' layer (the network layer is also called the 'user' and 'application' layer in this book), which asks the SM (session management) layer to establish a PDP context. The SM layer obliges and manages the PDP call set-up. (Protocol layers in the GPRS sub-network are introduced in the next chapter.)

The 'network' layer of the mobile station asks the SM layer to establish a PDP context with the following information:

• The called IP address – oddly this is not always necessary! For example, if the subscriber has just one ISP, then this address can be held in the HLR and when the SGSN receives a *PDP context activation request* which does not include the called IP address, it goes to the HLR and extracts that.

---

[3] See Chapter 11 for details of this SM function.

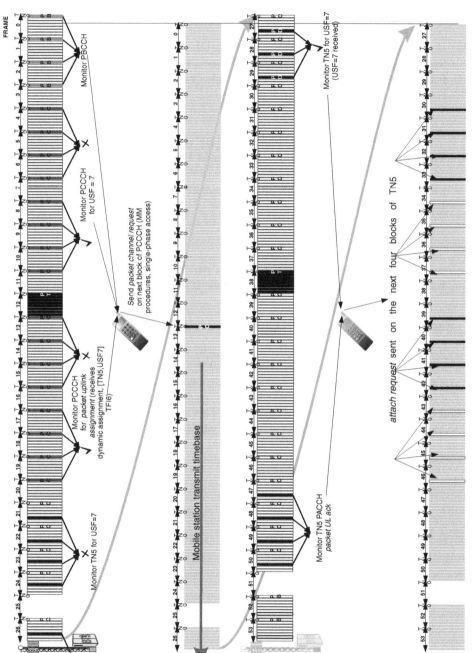

Figure 4.7 A physical appreciation of part of the GPRS attach of Figure 4.5

- Its own IP address – this is only included if the subscriber has a static IP address.
- The quality of service (QoS) profile for the call – again this information may be excluded and retrieved from the HLR by the SGSN.
- The NSAP (network service access point) to be used for the data transfer once the call is established. The 'network' is the 'user' or 'application' layer and the NSAP is the connection point between the 'network' and the GPRS network – 'sub-network'.
- The information may include the access point name (APN), which is the IP address of the GGSN which provides connection to a particular ISP or Intranet.

Figure 4.8 shows the steps leading to *PDP context activation* (packet data call set-up) and the UL customer data flow after the call is set up. The mobile station must be attached before this procedure.

The SM layer has received a command from the application layer to establish a data call. The SM layer has constructed the message *PDP context activation request* and passed this through the GMM to the LLC layer. The LLC layer frames it in unac-knowledged mode, and asks the RLC layer to send it. The RLC layer segments it into RLC blocks and asks the MAC layer to send it over the Um.

1. The mobile station examines the downlink PCCCH RLC/MAC block headers looking for USF = 7, indicating that the next uplink block is free for use as PRACH.
2. It sends a *packet channel request* using the PRACH parameters extracted from the PSI (packet system information) messages. Note that this request is for single-phase access as this TBF will be for SM signalling, which also uses MAC radio priority one. Only customer data packets use radio priorities other than priority one.
3. The mobile station monitors the PCCCH for *packet UL assignment*.
4. If it does not receive the *packet UL assignment* with the same request and random reference number after a reasonable period, it monitors the USF for 'free' again.
5. If it is allowed to do so it retransmits a *packet channel request*.
6. The network responds on the downlink PCCCH with *packet uplink assignment*, or *packet access reject*, or *packet queuing notification*. In this case *packet uplink assignment* gives a dynamic assignment of physical channel TN6 with USF4. The USF granularity has a value of 4 (parameter setting '1') telling the mobile station it can transmit four RLC/MAC blocks after the appearance of USF4. A TFI of 30 will identify uplink and downlink RLC/MAC blocks.
7. The mobile station switches to the assigned physical channel TN6, monitoring the DL RLC/MAC headers for its assigned USF.
8–11. Having received its assigned USF, the mobile station sends the message *PDP context activation request*. In this example it takes four blocks. The *PDP context activation* message includes the QoS parameters requested by the application layer. A countdown value indicator informs the GPRS sub-network when the final block of the LLC PDU is sent (CV = 0).
12. Having completed the UL transfer, the mobile station monitors its allocated PACCH. Recognising TFI = 30, the message *final packet uplink acknowledge + packet downlink assignment* is decoded. The uplink transfer has been received

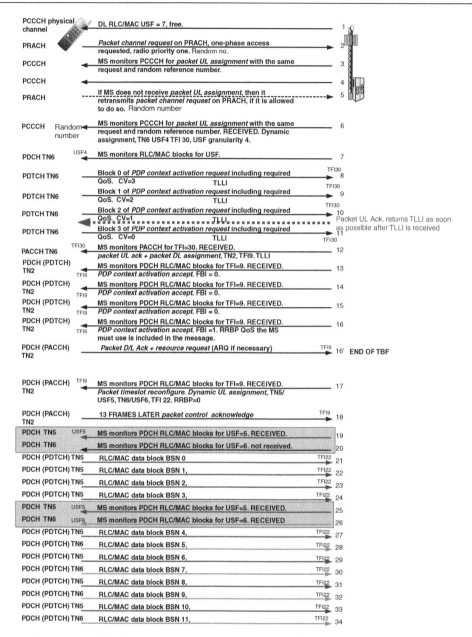

Figure 4.8 MAC procedures for PDP context activation and data transfer

satisfactorily. The GPRS sub-network RLC/MAC layer has received an LLC PDU for downlink transfer. In fact in this case of SM signalling this is highly unlikely at this stage of the signalling!

The downlink assignment nevertheless gives downlink TN2 with TFI9. Note that the GPRS sub-network returns TLLI to the mobile station (which included TLLI on the blocks of the uplink TBF); this is the second and final stage of contention resolution.

13–16. The mobile station switches to the physical channel TN2 and recognising its DL TFI decodes the message *PDP context activation accept* contained in four RLC/ MAC blocks. The last block is indicated by the GPRS sub-network setting FBI = 1. Included with FBI = 1 is an RRBP (relative reserved block period) indicating which uplink block the mobile station must use to send a TBF acknowledgment. The message includes the QoS and radio priority the mobile station must use. The RR function in the GPRS sub-network has also been informed of the QoS and from this will allocate suitable resources for the TBF. The mobile station acknowledges the downlink transfer on the allocated UL PACCH. This completes the SM signalling; the PDP context is activated. It also requests UL resources, which must mean it has an LLC PDU for uplink transfer – at this stage it will have if the SM layer has processed the information received in *PDP context activation accept*.

17. The mobile station monitors its downlink PACCH, recognises its own TFI and decodes the message *packet timeslot reconfigure*. This message allocates uplink resources as requested by the mobile station. More commonly it would be *packet uplink assignment*, and *packet timeslot reconfigure* would normally be used to change already allocated resources, but is included here to illustrate the variety of RLC/MAC messages. The message includes an RRBP for the mobile station to acknowledge this command.

18. The mobile station acknowledges the *packet timeslot reconfigure* with a *packet control acknowledge*.

19–20. The mobile station now switches to the assigned UL allocation and on the DL monitors the USF on TN5 and 6. It receives its USF on TN5 but not on TN6. (Note that the XID negotiation (eXchange IDentities for negotiation of compression parameters) should come before step 21. These are covered in a later chapter.)

21–24. The mobile station sends blocks 0–3 of customer data (SN PDUs).

25–26. Having sent four blocks (the USF granularity allowing the transmission of four blocks with each appearance of USF), the mobile station monitors TN5 and TN6 for its USFs. This time it receives its USF on both physical channels.

27–34. The mobile station sends blocks 4–11 alternately on TN5 and TN6.

Figure 4.9 shows the physical activities for steps 1–34 of Figure 4.8. A shorthand version of the radio bursts and frames is used. Instead of four frames being individually shown for one RLC/MAC block, only one block (or one frame representing four frames) is shown with TNs 0–7 shown only once. This is possible as the use of a TN does not change within the four frames constituting a block.

The top right-hand corner of Figure 4.9 shows the four frames comprising one RLC/MAC block. The blocks shown below this represent these four frames, with the eight lines within each block representing a TN. Smaller rectangles (for example after block 3, FN8–11) are single frames constituting the PTCCH and idle frames which occur on the thirteenth and twenty-sixth frames.

The uplink and downlink blocks are shown and are continued in the third and fourth rows of the diagram. The mobile station is shown reading the downlink PBCCH that in

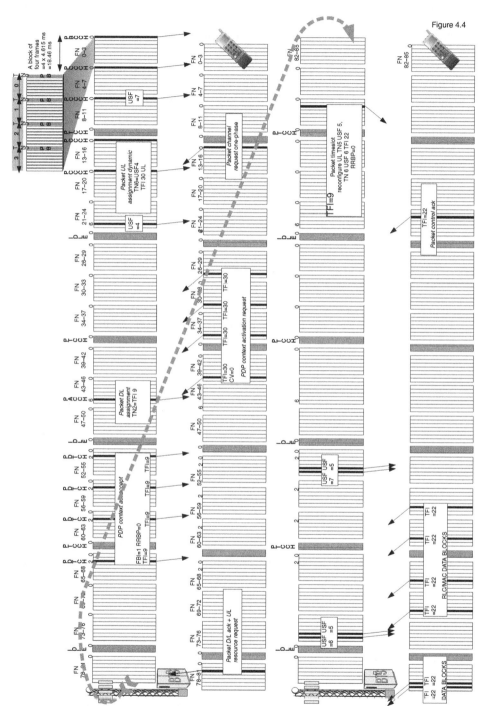

Figure 4.9 A physical view of the PDP context activation and data transfer of Figure 4.8

this example is on TN0 of the radio channel. From this it learns the PCCCH structure and goes to the PCCCH looking for USF = 7, allowing mobile stations to use the uplink PRACH on the next block.

The mobile station receives USF = 7 on block three (FN8–11) and sends a *packet channel request* on the next block (FN13–16). The mobile station then decodes all instances of RLC/MAC blocks on TN0 looking for *packet uplink assignment*, which it receives on block 4 (FN17–20).

The mobile station switches to the assigned channel, TN6, and monitors the downlink blocks for its assigned USF (4). It receives this on block 5 (FN21–24) and transmits four RLC/MAC blocks on the next four uplink blocks 6, 7, 8, 9.

The final block – block 9 – indicates the end of this TBF with CV = 0, and the mobile station monitors its PACCH for acknowledgment of the uplink block transfer. It receives *packet uplink acknowledge + packet downlink assignment*. The mobile station switches to the assigned downlink channel (TN2, TFI9) and receives four RLC/MAC blocks containing the message *PDP context activation accept*. The final downlink block contains *final block indicator* indicating the TBF is complete.

The RRBP (relative reserved block period) is set to zero, indicating that the mobile station should acknowledge after thirteen frames. Thirteen frames after receiving RRBP the mobile station sends *packet downlink acknowledge*. At the same time it asks for uplink resources to send an uplink TBF. The SM signalling to establish a PDP context is completed.

The mobile station now monitors the downlink PACCH looking for *packet uplink assignment*; instead it receives *packet timeslot reconfigure* which contains the uplink dynamic assignment of TN5 and TN6 with their respective USFs. The mobile station is told, with the inclusion of RRBP, to acknowledge this packet control message with a *packet control acknowledge* message.

Thirteen frames later the mobile station acknowledges receipt of the control message. It then switches to the assigned physical channels (TN5 and 6) and monitors the downlink blocks for their respective USFs. It receives the USF for TN5 on block 31 and sends RLC/MAC blocks on blocks 32, 33, 34 and 35. It then monitors the downlink blocks and receives the USFs allowing the uplink use of TN5 and TN6. It then sends eight blocks, four each on TN5 and 6.

4.5   GPRS sub-network originated TBFs

The GPRS sub-network can originate a TBF. If a mobile station has a static IP address and the GPRS sub-network receives an incoming data packet addressed to a static address and there is a PDP context for that mobile station, then it will either:

1. Page the mobile station and upon receiving a response, allocate resources for a downlink TBF. This will happen if the GPRS sub-network does not know which cell the mobile station is in; or
2. Give a downlink assignment directly on the paging channel. This will happen if the GPRS sub-network knows which cell the mobile station is in.

If a mobile station has a dynamic IP address then a PDP context has already been established by the mobile station.

Figures 4.10 and 4.11 illustrate GPRS sub-network originated, sometimes called mobile terminated (MT), downlink packet data transfer. The example shows the procedures for a mobile station with a static address and no previous PDP context established.

The diagrams show the GPRS sub-network paging a mobile station for an incoming IP packet, which requires a PDP context to be established. This paging for IP packet can only happen if the mobile station has a static IP address.

The sequence starts with the GPRS sub-network receiving a TCP/IP data packet; the gateway support node checks with the HLR that the mobile station is reachable, asks for the IP address of the SGSN serving this mobile station and routes the packet to the appropriate serving GPRS support node.

The SGSN sees that there is no PDP context for this mobile station and generates a *paging request* containing the mobile station's PTMSI and IMSI, passing this information to the BSS. The BSS calculates which paging channels the mobile station is listening to and places the paging message on the paging channels of the appropriate groups of cells as indicated by the SGSN.

This procedure is used when the GPRS sub-network does not know which cell the mobile station is camped upon. If it does know the cell, a direct *packet DL assignment* (with TLLI) may be placed on the paging channel.

In Figure 4.10:

1. The paging message carrying the PTMSI of the required mobile station is broadcast. PTMSI is indexed against IMSI in the HLR and SGSN. The mobile station's IP address is also indexed to IMSI.
2. The GMM layer in the mobile station triggers the LLC layer to send an LLC frame in response to the paging request. The LLC layer requests the RLC layer to send it and the RLC layer asks the MAC layer to access the GPRS sub-network. The MAC layer monitors the physical channel carrying PCCCH for USF = 7 (free) in order to send a *packet channel request*.
3. The mobile station sends *packet channel request* indicating a response to paging.
4. The mobile station monitors the PCCCH and receives *packet UL assignment*, allocating a single RLC/MAC block on the UL.
5. The mobile station switches to the new physical channel and on the appropriate frame sends the LLC frame as an RLC/MAC block carrying TLLI.
6. The mobile station then monitors the PACCH for TFI, recognises the TLLI and decodes the *packet timeslot reconfigure* command. This allocates both UL and DL resources and gives the mobile station an RRBP for response.
7. The mobile station acknowledges the allocation with *packet control ack*.
8–11. The mobile station now monitors the allocated DL TN and upon recognising its TFI decodes *PDP context activation command*. FBI indicates the final block of this message = 1 and an RRBP is given for the mobile station's response.
12. The mobile station acknowledges the DL TBF.
13. The mobile station now monitors the DL TN6 for its USF.
14–17. The mobile station now sends *PDP context activation request*. The final block has CV = 0.

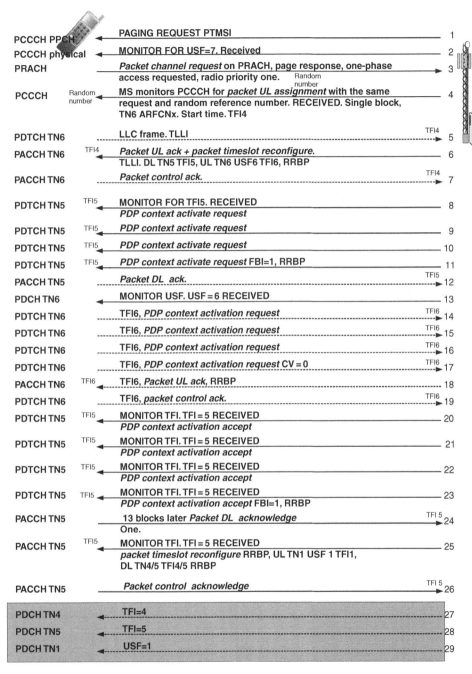

Figure 4.10 MAC procedures, network initiated PDP context activation

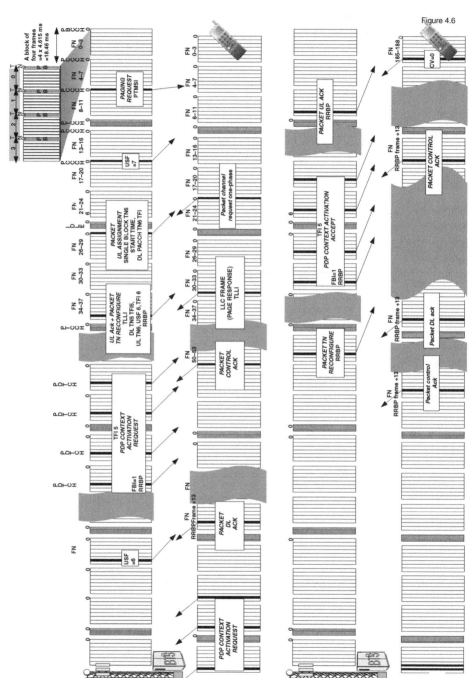

Figure 4.11 A physical view of the network originated PDP context activation of Figure 4.10

18. The mobile station now monitors PACCH, recognises TFI6 and decodes *final packet UL ack*. RRBP is included for the mobile station's response.

19. The mobile station sends *packet control ack* on the block indicated by RRBP.

20–23. The mobile station now monitors the DL PACCH, recognises TFI = 5 and decodes *PDP context activation accept*. This message contains the negotiated QoS parameters and the radio priority to be used in *packet channel request* for SN PDU transfer. FBI = 1 indicates the final DL block and RRBP for the mobile station's reply is included.

24. As RRBP indicated '1', the mobile station sends its response *packet control ack* 13 frames later.

25. The mobile station now reverts to monitoring its PACCH; recognising TFI5, it decodes the message *packet timeslot reconfigure*.

26. The mobile station acknowledges receipt of *packet timeslot reconfigure* with *packet control ack*.

27–29. The mobile station monitors the DL assignment for its TFI, and, at the appropriate stage, for its UL USF.

Figure 4.11 illustrates these procedures from a physical perspective. This figure uses the block structure described for Figure 4.9. The grey separators on Figure 4.11 show a 13-frame separation between the mobile station receiving an RRBP and replying to the RLC/MAC command. This occurs when setting RRBP = 0.

4.6 Alerting the mobile station for a DL TBF

One method of alerting the mobile station by paging on the PCCCH for a DL TBF is illustrated in Figure 4.11. This method is used when the GPRS sub-network GMM layer does not know which cell the mobile station is camped upon. The mobile station's response with an LLC frame (after normal access procedures) serves two purposes, first it positively identifies the mobile station with the TLLI, and secondly, it tells the GPRS sub-network GMM layer in which cell the mobile station is located.

If the GMM layer *does* know the mobile station's cell, then other, more efficient methods can be used. These are illustrated in Figure 4.12.

In Figure 4.12, steps 1–8 show the procedure used if the GPRS sub-network knows which cell the mobile station is in but is not aware of the current timing advance (TA) required. This will only apply in cells of comparatively large radius.

1. The GPRS sub-network calculates which paging channels the mobile station is listening to and sends *packet downlink assignment* on one of the PPCHs. This identifies the mobile station with its TLLI and commands the mobile station to reply at the RRBP with four short bursts.

2–5. The mobile station switches to the assigned physical channel and at the appropriate block transmits four short access bursts.

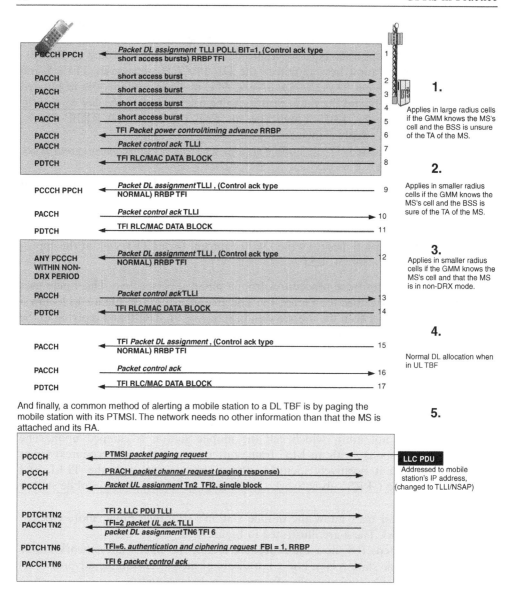

Within the figure region there is this body text:

And finally, a common method of alerting a mobile station to a DL TBF is by paging the mobile station with its PTMSI. The network needs no other information than that the MS is attached and its RA.

Figure 4.12   Ways of alerting a mobile station for a DL TBF

6. The BSS measures the displacement of the short bursts from their correct position. This displacement is a function of the double propagation delay:

—from the BTS to mobile station; and
—from the mobile station to BTS.

  The GPRS sub-network calculates the required TA and sends the TA instructions on the DL PACCH. Transmitted power control parameters that the mobile station will use are also sent with this instruction. The mobile station is given an RRBP for a response.

7. On the downlink the mobile station recognises its TFI and decodes the *packet power control/timing advance* command and acknowledges with *packet control ack* with TLLI on the appropriate UL block.
8. The GPRS sub-network can now send the DL RLC/MAC data blocks containing the *PDP activation request* message, identifying each block with TFI.

   Steps 9–11 show the steps taken in smaller radius cells. The GPRS sub-network knows which cell the mobile station is monitoring and the timing advance may not be important, or the GPRS sub-network is confident of the TA.
9. As in step 1 but a normal acknowledgment is required from the mobile station.
10. Normal single block acknowledgment from the mobile station with TLLI.
11. DL data transfer.

    Steps 12–14 are nearly the same as 9–11 above, but note that the *packet DL assignment* is sent on *any* PCCCH block within the mobile station's non-DRX period. When the non-DRX timer is running, the GPRS sub-network knows that the mobile station will be listening within this period to all PCCCHs in its cell. The mobile station at GPRS attach may negotiate the non-DRX period, otherwise the SI value is used.

    Steps 15–17 may be used to allocate DL resources during an uplink TBF, or to initiate a PDP context for an incoming call whilst another PDP context is active for that mobile station.

4.7   Abbreviations used in this chapter

APN	Access point name
ARFCN	Absolute radio frequency channel number
BCCH	Broadcast control channel
BSN	Block sequence number
BSS	Base station sub-system
BTS	Base transceiver station
CS	Circuit switched
CS	Coding system
CV	Countdown value
DL	Downlink
DRX	Discontinuous reception
FBI	Final block indicator
FCCH	Frequency correction channel
FN	Frame number
GGSN	Gateway GPRS support node
GMM	GPRS mobility management
GPRS	General packet radio service
GRR	GPRS radio resources
GSM	Global system for mobile communication
HLR	Home location register

IMSI	International mobile subscriber identity
IP	Internet protocol
ISP	Internet service provider
LAC	Location area code
LLC	Logical link control
MA	Mobile allocation
MAC	Medium access control
MCC	Mobile country code
MNC	Mobile network code
MS	Mobile station
NSAP	Network service access point
PACCH	Packet associated control channel
PBCCH	Packet broadcast control channel
PCCCH	Packet common control channel
PDCH	Packet data channel
PDP	Packet data protocol
PDTCH	Packet data traffic channel
PDU	Protocol data unit
PLMN	Public land mobile network
PRACH	Packet random access channel
PTCCH	Packet timing control channel
PTMSI	Packet temporary mobile subscriber identity
QoS	Quality of service
RAC	Routeing area code
RAI	Routeing area indication
RLC	Radio link control
RR	Radio resources
RRBP	Relative reserved block period
SACCH	Slow associated control channel
SCH	Synchronisation channel
SGSN	Serving GPRS support node
SM	Session management
SN PDU	Sub-network protocol data unit
TA	Timing advance
TBF	Temporary block flow
TCP/IP	Transmission control protocol/Internet protocol
TFI	Temporary flow identifier
TLLI	Temporary logical link identifier
TN	Timeslot number
UL	Uplink
Um	Air interface
USF	Uplink status flag

5

An Introduction to Protocol Layers Data Flow

(ETSI 123.060)

This chapter introduces the signal flow across the GPRS protocol stack and gives a brief description of the functions of the various layers of the protocol stack. These layers are examined in detail in later chapters as we move up the stack from the physical layer to the SNDCP layer.

5.1 The protocol stack

The GPRS protocol stack (from the mobile station's viewpoint) has two elements:

1. *The customer data flow protocol stack*. The purpose of this stack is to process network protocol data units (N PDUs) (customer data packets) to make them suitable for transfer across the air interface. This part of the stack is called the transmission protocol stack in the specification.
2. *The signalling protocol stack*, which itself has two elements:

   —*Higher layer – layer 3 services signalling*, which is used for session management and GPRS mobility management signalling. This part of the stack is called the signalling protocol stack in the specifications.
   —*RLC/MAC signalling* used for control of the radio resources. (Radio resource (RR) signalling which is included with RLC/MAC signalling is also L3, but this is not services signalling).

A difference between the two signalling types is manifested on the air interface by the higher layer services signalling using the PDTCH and RLC/MAC signalling using the PACCH.

Figures 5.1 and 5.2 illustrate the functions of the layers that make up the protocol stack.

GPRS in Practice: A Companion to the Specifications   Peter McGuiggan
© 2004 John Wiley & Sons, Ltd   ISBN: 0-470-09507-5

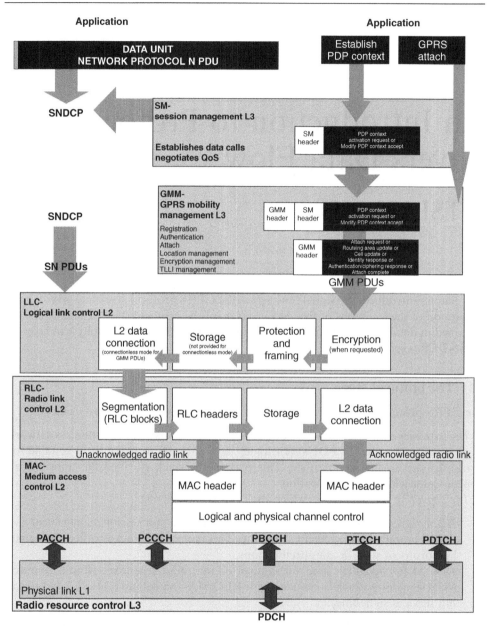

Figure 5.1 The MS GPRS uplink *higher layer service signalling* protocol stack

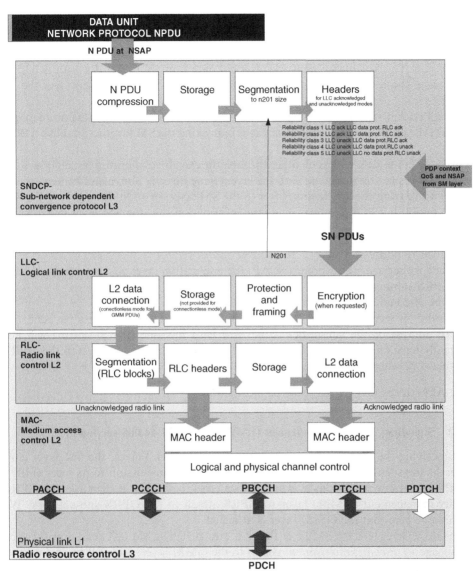

Figure 5.2 The MS GPRS uplink *data transmission* protocol stack

5.1.1   GPRS protocol layers – a brief description

5.1.1.1   The signalling plane

Figure 5.1 illustrates the 'signalling plane' of the GPRS system. At the top of this diagram is a customer data packet from the 'network' or application layer waiting for transmission. This is called a network protocol data unit (N PDU). Its path is barred from transmission, as a PDP context must be established before it is sent over the air interface.

A small packet is shown coming from the application layer to the session management (SM) layer. This is the application layer requesting the SM layer to establish a PDP context.

Layers communicate with each other through the exchange of primitives; all layers use these primitives to communicate with the layers lying directly above and beneath them. This primitive from the application layer to the SM layer, in addition to asking the SM layer to establish a PDP context, sends the information that the SM layer needs to do this, including:

- the called IP address;
- the network service access point (NSAP) the application layer wishes to use to access the GPRS sub-network;
- the type of communication (for example TCP/IP);
- the quality of service required for the communication of customer data packets;
- the IP address of the subscriber if the SIM card has one (the SIM card will not normally have an IP address, but will receive a dynamic IP address from the GPRS sub-network when the PDP context is established).

The APN of the GGSN may also be included.

5.1.1.2   Signalling, session management (SM) layer (GSM 44.008 section 6)

Upon receiving the request from the application ('network') layer, the SM layer constructs the message *PDP context activation request* for transmission to its 'peer entity', the GPRS sub-network SM layer, which is located in the SGSN. It then asks the GPRS mobility management (GMM) layer to GPRS attach to the GPRS sub-network. We shall assume here that the GMM layer is attached.

The GMM layer tells the SM layer that it is attached. The SM layer can now forward the message *PDP context activation request* to the GMM layer, asking it to deal with it. Formally, the SM layer uses the services of the GMM layer to provide a communication link to its GPRS sub-network peer entity.

The GMM layer encapsulates the SM message (which is called an SM PDU) within a GMM frame and passes it down to the LLC layer.

We shall leave further description of other layers until later and stay with the SM layer. The SM layer eventually receives a message from the GPRS sub-network SM layer *PDP context activation request accept* (although it may also be *reject*). This message confirms the quality of service to be used for this communication session, and gives the mobile station a dynamic IP address (if it does not have a static address). It also gives a radio priority which is used by the mobile station when it requests radio resources (on the PRACH) to send a customer data packet (N PDU). The GPRS sub-network uses this priority to decide when, or whether, during the busy hour to give the resources requested. Radio priority is also used by the mobile station in conjunction with a broadcast parameter called persistence value to determine if it is allowed to send further requests (after the first request) for radio resources on the PRACH. This is covered in detail in chapter 7.9.

The mobile station session management layer informs the network layer above (which initiated this process) that a PDP context is now established, giving it the allocated dynamic IP address and the QoS for this PDP context from the GPRS sub-network.

The mobile station network layer may accept the PDP context or reject it; if it rejects it the mobile station SM layer informs the GPRS sub-network SM layer that the PDP context is discontinued; if the mobile station network layer accepts, the SM layer informs the sub-network dependent conversion protocol (SNDCP) layer (on Figure 5.2) that a PDP context is in place. The SNDCP layer now knows the radio priority and QoS levels to be used, and which NSAP to expect the customer PDUs to use.

And that, in outline, is the function of the SM layer. It springs into action once more if asked to modify the PDP context with a better (or worse) QoS and radio priority level or is asked to create a new PDP context (a mobile station can have multiple simultaneous independent data calls, each requiring its own PDP context) or is asked to tear down a data call (remove a PDP context) or detach from the GPRS sub-network.

5.1.1.3 Signalling, mobility management – GMM layer

The GPRS mobility management (GMM) layer's job description includes:

- 'Attaching' to the GPRS sub-network when commanded by the session management or application (network) layer. The attach process alerts the GPRS sub-network to the mobile station's presence and allows it to identify and authenticate it and give the information required for encryption. If the mobile station is attaching in a different routeing area from the one in which it was last attached, then a packet temporary mobile subscriber identity (PTMSI) will be reallocated (although the GPRS sub-network operator may reallocate a PTMSI at any time). From the PTMSI the mobile station and GPRS sub-network generate a local temporary logical link identity (TLLI) that is subsequently used by both the mobile station and GPRS sub-network to identify the mobile station. (A local TLLI is a TLLI generated by the mobile station where the mobile station is in the routeing area (RA) where the PTMSI was allocated; if the mobile station finds itself in a different RA it will use a 'foreign' TLLI until it has performed a routeing area update.) In the mobile station, the GMM layer passes the TLLI to the RLC/MAC layer for inclusion in a TBF, and to the LLC layer to indicate that a valid TLLI is in use. The ciphering key, kc, is passed from the GMM layer to the LLC layer as encryption is performed at the LLC layer.

- Performing a routeing area update when a cell belonging to another RA is reselected or when the RA update timer within the mobile station expires. RA information is passed to the mobile station GMM layer from the RLC/MAC and LLC layers.
- Performing a cell update when the mobile station is in the 'GMM ready' condition and reselects a cell within the same routeing area.
- Detaching from the GPRS sub-network when requested to do so by the session management or application layer. The mobile station can also be ordered by the GPRS sub-network to detach and the mobile station GMM layer will execute the command.
- Providing a signalling conduit for the mobile station SM layer to the GPRS sub-network SM layer.

In our signalling model of Figure 5.1, the GMM layer is requested by the SM layer to attach. The GMM layer constructs the message *attach request* and passes this down to the LLC layer, asking it to deliver the message to the SGSN GMM layer. The LLC does this in the fashion described below. The most important piece of information in this message is the PTMSI of the mobile station. Also included, if available, is the ciphering key sequence number (cksn).

From the PTMSI, the SGSN will be able to identify the IMSI of the mobile station (but only if the PTMSI was allocated by the same GPRS sub-network to which the mobile station is attempting to attach). If the indicated RA is different from the RA of the SGSN to which the mobile station is attempting to attach, then the SGSN will extract the IMSI and subscriber information from the SGSN to which the indicated RA belongs.

From the PTMSI and cksn (and also possibly the PTMSI signature) the GPRS sub-network operator will decide whether to authenticate the mobile station or to go immediately to encryption mode, bypassing authentication.

If the mobile station GMM layer receives the *authentication and encryption* command from the GPRS sub-network, it extracts RAND, passes this to the SIM card which produces the required *signed response* (SRES) and the encryption key kc.

The GMM layer puts SRES into the message *authentication and encryption response*, and passes kc to the LLC layer.

When the mobile station GMM layer receives the message *attach accept* the authentication and encryption process is completed. This message is normally encrypted and may contain a PTMSI reallocation.

The GMM layer informs the SM layer that attach is now completed. The SM layer then sends the message *PDP context activation request* to the GMM layer (asking it to forward the message to the SGSN SM layer).

The GMM layer places this message within a GMM PDU frame and asks the LLC layer to send it to the SGSN GMM layer. The GMM layer is now acting as a conduit for SM messages. All messages emanating from the GMM layer, and these include mobility management signalling and session management signalling, belong to the class higher layer service signalling. You may recall that this higher layer service signalling always uses the logical channel PDTCH on the air interface.

5.1.1.4   Signalling, logical link control (LLC) layer (GSM 44.064)

A detailed explanation of the LLC protocol is given in Chapter 9.
The mobile station LLC has the task of:

- Establishing an L2 data link with its corresponding GPRS sub-network layer (peer entity) – LLC layer, using the TLLI as the mobile station address. This data link can operate in connection or connectionless transmission mode; only customer data blocks use the connection transmission mode, and only then if they have a negotiated QoS reliability class of 1 or 2.

  In connection mode the LLC layers provide a highly reliable data link between the mobile station and GPRS sub-network for delivery of SN PDUs. SN PDUs are customer data packets processed by the SNDCP layer and are delivered in connection mode (acknowledged – I – data) or connectionless mode (unacknowledged – UI – unitdata) dependent upon the GPRS sub-network allocated customer QoS. SN PDUs are placed in appropriate LLC DCL (data communications link) frames.

- Handling multiple network N PDU sessions through the four customer data packet access points. These access points are QoS 1, 2, 3 and 4.

- Framing signalling messages delivered by the GMM layer for transmission in unacknowledged mode. SN PDUs of reliability class 3, 4 and 5 are also delivered in unacknowledged mode.

- Framing SMS messages in unacknowledged transmission mode.

- Ciphering and deciphering of SN PDUs and GMM and SMS messages.

Continuing the story of our *attach request* message from the GMM layer; when the mobile station LLC layer receives this message from the GMM layer, it is placed into an LLC unacknowledged information (UI) frame for unacknowledged (connectionless) transmission and the radio resource (RR) layer is requested to deliver this to the GPRS sub-network LLC layer. When the GPRS sub-network LLC layer receives it, it is checked for errors, stripped out of the LLC frame and passed to the GPRS sub-network GMM layer.

When the *authentication and ciphering request* is sent from the GPRS sub-network to the mobile station LLC layer, it is examined for errors, stripped of its LLC wrapping and passed to the mobile station GMM layer.

The *authentication and ciphering response* message is dealt with in the same manner as *attach request*. The mobile station LLC layer receives the ciphering key, kc, from the mobile station GMM layer. It will then, when requested, encipher all messages from higher layers and decipher messages sent by the GPRS sub-network LLC layer.

The next message *attach accept* received from the GPRS sub-network LLC layer is checked for errors, deciphered (the GPRS sub-network LLC layer having activated cipher mode) and delivered to the mobile station GMM layer. The mobile station GMM layer constructs TLLI from the PTMSI contained in this message and delivers the TLLI to the mobile station LLC and RLC layers.

The message *attach complete* is ciphered, framed and passed to the RR layer for delivery to the GPRS sub-network.

The mobile station LLC layer now has a valid TLLI and is allowed to send and receive SN PDUs when requested. In due course this will be requested by the SNDCP layer, perhaps for full LLC protected mode, in which case the LLC layer will establish asynchronous balanced mode (ABM) operation with its GPRS sub-network counterpart.

5.1.1.5   Signalling, radio link control/medium access control (RLC/MAC) layer

The RLC/MAC layer is a sub-layer of the radio resource layer. The RLC layer has the responsibility:

- To receive LLC protocol data units from the LLC layer for transmission over the air interface and to send to the LLC layer LLC PDUs that have been assembled from RLC/MAC blocks received from the air interface.
- To segment LLC PDUs received from the LLC layer into RLC data blocks, and reassemble RLC data blocks received from the air interface into LLC PDUs for transfer to the LLC layer.
- To segment and reassemble RLC/MAC control messages into and out of RLC/MAC control blocks.
- To use, where appropriate, acknowledged mode transmission (ARQ) for retransmission of RLC/MAC data blocks received with errors.

The purpose of the medium access control (MAC) component of the RLC/MAC layer is:

- From the GPRS sub-network viewpoint, to allow shared use of the radio medium (physical channels) between multiple mobile stations, and from the point of view of a single mobile station, allow sharing of the radio resources between multiple PDP contexts (data calls).
- To control the transmission of temporary block flow of data packets carrying signalling or customer data between mobile stations and the GPRS sub-network.
- To allow contention resolution and arbitration for mobile station access to GPRS sub-network resources.
- To allow queuing and scheduling for mobile station terminating calls. To receive PBCCHs (packet broadcast control channels) and PCCCHs (packet common control channels), allowing mobile stations to autonomously select and reselect cells.
- To place and extract the correct information on the correct logical channel on the correct physical channel at the correct time.

Continuing our story, the LLC encapsulates the GMM message *attach request* and asks the RR layer to deliver it to the GPRS sub-network LLC layer.

The RR layer springs into action sending a *packet channel request* at the appropriate moment on the packet random access channel (PRACH) logical channel. It then keeps the receiver open on all of the downlink PCCCHs listening for a *packet uplink assignment* message. It decodes this message, passing its content to the MAC layer. The mobile station RR layer now knows the type of assignment and instructs the MAC layer to act upon this information which will include:

- the physical channels to be used;
- uplink status flags (USF) if dynamic uplink allocation is used;

- USF granularity if dynamic uplink allocation is used;
- the start frame for transmission and a bit map of block allocations if the assignment is fixed.

If the uplink assignment is fixed, it will transmit the *attach request* message as RLC/ MAC blocks at the scheduled time. If the uplink assignment is dynamic, it will transmit when the USF indicates permission.

The RLC layer frames the message into RLC/MAC blocks, which, FEC encoded and interleaved, are transmitted on the assigned uplink packet data traffic channel (PDTCH) when the MAC detects USF (for dynamic allocation).

The mobile station RLC/MAC layer then examines the downlink RLC/MAC block headers (after deinterleaving and decoding) on the assigned PDCH looking for its assigned temporary flow indicator. When it detects this it examines the SAP and passes the reassembled blocks to the LLC layer. The RLC layer may be asked by the LLC layer to communicate in acknowledged or unacknowledged mode.

When GMM procedures are completed, the mobile station may have been given a new packet temporary mobile subscriber identity (PTMSI).

When *PDP context activation* is completed, the mobile station SNDCP layer may start to send sub-network (SN) PDUs to the LLC layer.

The mobile station LLC layer will pass the resulting LLC PDU to the RR layer and the MAC will put the RLC blocks on to a PDTCH on the correct physical channel at the assigned time.

Radio resource signalling across the air interface uses the packet associated control channel (PACCH). A GPRS physical channel (PDCH) used as a PDTCH or PACCH is signalled by the RLC/MAC header. Radio resource signalling on a PACCH can interrupt a PDTCH at any time. The PCCCHs also use RR signalling to control access by mobile stations to the GPRS services.

5.1.1.6   Transmission, sub-network dependent conversion protocol (SNDCP) layer (GSM 44.065)

We have now completed the introduction to the 'signalling' protocol stack of Figure 5.1.

A PDP context is established and the network layer (application layer) is informed of this by the SM layer and accepts (or rejects) the QoS negotiated. The SNDCP layer is informed by the SM layer of the QoS and is now ready to receive customer data blocks from the network (application) layer. All at this moment focuses on the SNDCP layer.

The SNDCP is responsible for:

- Converting and segmenting the external network formats (or N PDUs – network protocol data units) into the sub-network (GPRS) formats (SN PDUs – sub-network protocol data units).
- Compression of the N PDUs to make for more efficient data transmission within the GPRS sub-network.
- Negotiation of the compression parameters to be used in communication between a mobile station and the sub-network; this is done using the XID SNDCP – exchange identity primitives.

- Managing multiple PDP context PDU transfers, ensuring that the N PDUs from each PDP context are transmitted to the LLC layer in sufficient time to maintain the QoS for each PDP context.
- Storing of a segmented N PDU when the LLC layer is operating in acknowledged mode until the LLC layer informs the SNDCP layer that the 'peer entity' SNDCP layer has received the segments. In the receive direction the SNDCP stores all segments and reassembles them as N PDUs before passing them to the network layer.
- Numbering of segments of N PDUs when the LLC layer is operating in unacknowledged mode, and, in the receive direction, 'filling in' missing numbered blocks (with a string of zeroes) before passing them to the network layer.

The mobile station SNDCP layer is now informed by the SM layer that a PDP context is activated and the user (network) layer is also informed. The user (network) layer proceeds to send N PDUs to the SNDCP layer.

One of these might be carried by an XID primitive, which is sent to the GPRS sub-network SNDCP layer to negotiate compression parameters, and when compression is agreed, data primitives, carrying customer data (N PDUs) are transferred. The SNDCP segments the customer data messages (probably carried in TCP/IP format), frames them into SN PDUs and sends them to the LLC layer with a request for acknowledged mode transmission at LLC level. This depends upon the QoS in use. PDP contexts allocated a QoS with reliability classes 1 and 2 will request LLC acknowledged mode transmission.

The processed N PDUs coming out of the SNDCP are called SN PDUs (sub-network PDUs.) These PDUs, with a maximum length of 1520 octets are individually passed to the LLC layer with a primitive LL-DATA-REQ (for acknowledged transfer) or LL-UNITDATA-REQ (for connectionless mode). The LLC layer encrypts the SN PDU (if requested) and places it in a layer 2 data link frame which can (if required) guarantee delivery to the far end LLC layer.

The framed, encrypted SN PDU is called an LLC PDU and has a maximum length of 1560 octets; 1520 octets for the SN PDU and 40 octets of LLC over head.

Each of these LLC PDUs is individually passed down to the RLC/MAC layer with primitives GRR-DATA-REQ or GRR-UNITDATA-REQ, which respectively request the RLC layer to send the LLC PDU in acknowledged mode or unacknowledged mode (across the air interface between RLC peer entities).

The maximum length of an RLC/MAC block is 23 octets, which are used as follows:

- MAC header: 1 octet;
- RLC header and control octets: 2 or 3 or 4 or 6 octets;
- LLC PDU data: 20 or 19 or 18 or 16 octets (for CS1 operation).

The RLC layer segments the LLC PDUs into blocks of this size, up to 20 octets plus three header octets, giving a total of 23 octets for coding system 1. These blocks are called RLC/MAC blocks. Each RLC/MAC block is sent to the physical layer where it is encoded using one of the coding schemes CS1 to CS4 (more octets can be carried in the RLC/MAC block before encoding if the encoding system used is CS2–4.) After encoding, the encoded bits are interleaved and placed within a radio burst in four frames of a physical channel on the air interface – this, after modulation, is an RLC/MAC block on the air interface.

Figure 5.3 gives a view of the PDU flow through the 'transmission plane' of the protocol stack. The mobile station 'user' layer (more correctly the 'network' layer) sends to the SNDCP layer an N PDU. The SNDCP layer processes the N PDU, if necessary segmenting it into a number of sub-network (SN) PDU blocks, each of which the LLC layer accepts as single blocks.

The size of N PDU that the LLC accepts is determined by the parameter n201 within the LLC layer. This has a maximum value of 1520 octets, but it is a variable, as the LLC layer in the mobile station negotiates with the GPRS sub-network LLC layer to agree upon a value for n201. (1520 octets is the maximum size for TCP/IP blocks interfacing to an Ethernet LAN.)

The SNDCP may have to segment the N PDUs, dependent upon n201. In Figure 5.3, the SNDCP layer does segment, and nine segmented blocks are shown leaving the SNDCP layer. Each of these blocks contains an SNDCP header and each block is delivered separately on the primitive LL-DATA-REQ or LL-UNITDATA-REQ. These blocks are called SN PDUs, sub-network protocol data units.

The LLC layer processes each SN PDU and sends each one as an LLC PDU to the RLC/MAC layer.

The maximum possible size for an LLC PDU is 1560 octets. Each LLC PDU is conveyed to the RLC/MAC layer upon a primitive GRR-DATA-REQ or GRR-UNITDATA-REQ. The RLC/MAC layer processes an LLC PDU for transmission over the air interface.

The LLC PDU is segmented into a number of RLC/MAC blocks. The maximum number of octets of LLC PDU that an RLC/MAC block can carry depends upon the channel coding system that is in use in the physical layer. For CS1, the maximum is 20 octets. This determines the number of RLC/MAC blocks required to transfer the LLC PDU across the air interface. Each RLC/MAC block requires four contiguous frames on the air interface and it takes about 18 ms to send an RLC/MAC block.

Figure 5.4 shows the overall protocol stacks for GPRS sub-network services. In the figure, the protocol stack is displaced to show both the 'transmission plane' and the 'signalling plane'.

The path labelled '1' is the path that network PDUs take, bypassing the SM and GMM layers. The path labelled '2' is that used to establish a session which allows the transfer of the network PDUs of '1' (network PDUs cannot be sent over the network until a session is established – the arrow labelled '5' is the signalling between SNDCP and the SM layer indicating the establishment of a session).

If the SM is asked to establish a session for N PDU transfer via path '2' and the mobile station is not 'attached', then the GMM layer will firstly attach.

The path labelled '3' is that used for attaching the mobile station to the GPRS network independently of SM.

The path labelled '4' is a limited communication between RLC and GMM, allowing the GMM layer to transfer the temporary logical link identity (TLLI) to the RLC layer.

A similar, but not identical structure applies to the SGSN structure.

Figure 5.5 gives the mobile station protocol stack in more detail, and we will be using a version of this protocol stack in the descriptions which follow.

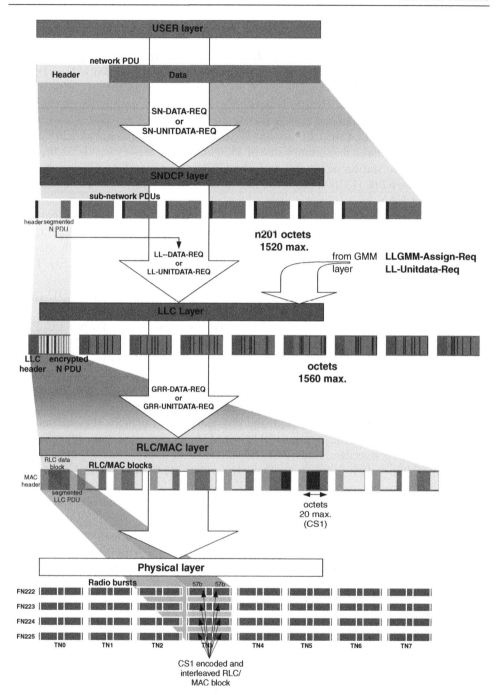

Figure 5.3 Customer data (N PDUs) through the protocol stack

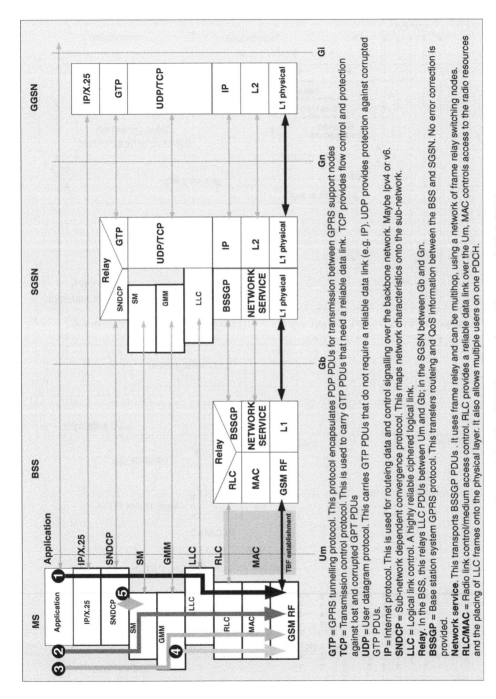

Figure 5.4 The GPRS network architecture, MS-GGSN

GTP = GPRS tunnelling protocol. This protocol encapsulates PDP PDUs for transmission between GPRS support nodes
TCP = Transmission control protocol. This is used to carry GTP PDUs that need a reliable data link. TCP provides flow control and protection against lost and corrupted GPT PDUs
UDP = User datagram protocol. This carries GTP PDUs that do not require a reliable data link (e.g. IP). UDP provides protection against corrupted GTP PDUs.
IP = Internet protocol. This is used for routeing data and control signalling over the backbone network. Maybe Ipv4 or v6.
SNDCP = Sub-network dependent convergence protocol. This maps network characteristics onto the sub-network.
LLC = Logical link control. A highly reliable ciphered logical link.
Relay. In the BSS, this relays LLC PDUs between Um and Gb; in the SGSN between Gb and Gn.
BSSGP = Base station system GPRS protocol. This transfers routeing and QoS information between the BSS and SGSN. No error correction is provided.
Network service. This transports BSSGP PDUs . It uses frame relay and can be multihop, using a network of frame relay switching nodes.
RLC/MAC = Radio link control/medium access control. RLC provides a reliable data link over the Um, MAC controls access to the radio resources and the placing of LLC frames onto the physical layer. It also allows multiple users on one PDCH.

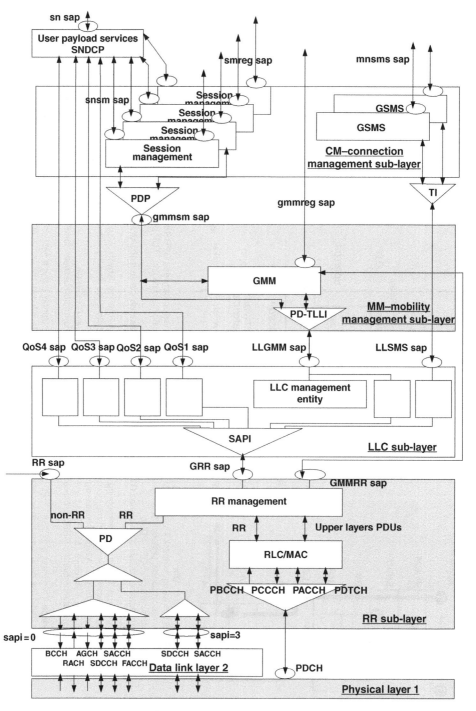

Figure 5.5   The GPRS protocol architecture MS class C

5.2 GPRS signal flow – arrow diagrams (GSM 44.060 sections 5, 7, 8)

Note that the following diagrams are included to illustrate the interactions between the protocol stack layers. Many of the functions described here are covered in more detail in later chapters, where the functions of each layer are examined more closely.

The arrow diagrams of Figure 5.6 onwards show layer 3 PDUs transported across the air interface. A layer 3 service signalling or customer data PDU is always encapsulated within a layer 2 LLC PDU and is transported across the air interface using a TBF. To show the full TBF each time an LLC PDU is transported across the air interface would make the arrow diagrams rather unwieldy. The full TBF mechanism is shown in Figure 4.5 and has been demonstrated in chapter 4.

5.2.1 GPRS attach[1]

A mobile station is GPRS attached when the mobile station has registered with GMM in the GPRS sub-network SGSN. The SGSN then knows that the mobile station is within GPRS coverage and the routeing area it is currently using.

If the mobile station establishes a PDP context with the SGSN, the SGSN can now send *packet paging* messages on the PCCCH-PPCH (this applies if a PBCCH exists in the cell. If there is no PBCCH then the mobile station uses the GSM 51-frame multiframe CCCH structure of the cell to access GPRS services – the common control channel paging channel (CCCH PCH) that is used for CS paging is also used for GPRS paging).

After attaching, the mobile station releases the radio resources and stays in GMM ready condition until the ready timer expires, when it changes to GMM standby. The mobile station RLC/MAC layer in this mode might receive an LLC PDU for an uplink data transfer temporary block flow. It then moves to the GMM ready condition when the ready timer is started.

The GMM layer is responsible for mobility management functions, which include 'attach' to the GPRS sub-network, so the attach requests from the mobile station always terminate on the GPRS sub-network GMM layer.

In Figure 5.6 the primitive GMM-REG-ATTACH initiates the procedure. The GMM layer prepares an *attach request* message, which contains the parameters required by the SGSN. These are shown in the GMM PDU of Figure 5.6. The most important parameter is the PTMSI (or, if this is not available, the IMSI), which will identify the mobile station to the GPRS sub-network. Also included with this primitive are the:

- TLLI;
- old RAI (the RA in which this old TLLI was calculated – if the mobile station is still in that same RA then 'old' means 'current');
- SAPI (the SAPI for GMM = 1);
- cause;
- radio priority (radio priority = 1 for GMM messages);

---

[1] See Section 10.2 for further details.

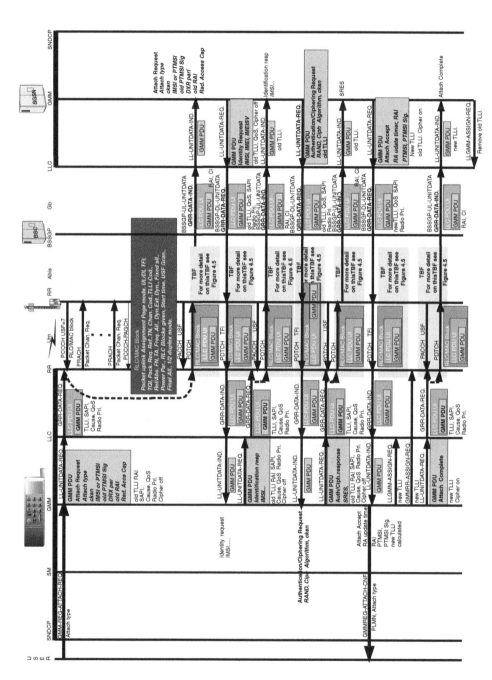

Figure 5.6 Protocol stack flow diagrams – GPRS Attach

- ciphering on or off (always off for this procedure);
- radio network capabilities of the mobile station (these include the MS power class, single-GSM radio band or dual-GSM radio band or triple-GSM radio band capability, multislot class, split paging capability).

The *attach request* message is forwarded to the LLC layer with the primitive LL-UNITDATA-REQ. This primitive requests the LLC layer to forward the GMM PDU to the GPRS sub-network LLC layer in asynchronous disconnect mode (ADM), or unacknowledged mode.

The LLC layer encapsulates the GMM PDU within an LLC PDU and requests the RR layer with the primitive GRR-DATA-REQ to forward this PDU to the GPRS sub-network LLC layer. This primitive is asking the RR layer to transfer the PDU in acknowledged mode. Whether the RR layer does this depends upon the instructions it has received from the cell's packet system information; if this has indicated that unacknowledged RLC/MAC transfers must be used, then that will be used.

On receiving this primitive, the RR layer must establish communication with the cell as indicated.

The *packet uplink assignment* includes options for fixed allocation, dynamic allocation, extended dynamic allocation and single block allocation:

- **Fixed allocation** gives the mobile station a fixed number of RLC/MAC blocks for an uplink TBF with the start frame. Uplink and downlink TFIs are allocated. In fixed allocations a DL_CONTROL_TN, a downlink physical channel is also allocated for use as PACCH. The mobile station monitors this PACCH channel for downlink control commands during an uplink TBF.
- **Dynamic allocation** tells the mobile station that it may only use the uplink physical channel (or channels for multislot transmission) for RLC/MAC block transfer when an RLC/MAC block on the same downlink physical channel carries in its header the USF indicator which is allocated to the mobile station. The mobile station may then transmit on the uplink physical channel in the block immediately following the downlink block carrying its USF. The quantity of blocks it may then transmit may be one block or four blocks, as indicated by the USF granularity. Uplink and downlink TFIs are allocated.
- **Extended dynamic allocation** tells the mobile station that if it is allocated more than one physical channel for an uplink TBF, then, if the allocated USF appears on an RLC/MAC block of the downlink lowest timeslot number of the allocation, it may transmit RLC/MAC blocks on all of the allocated uplink physical channels, This differs from normal dynamic allocation, where a USF indicator is allocated to each of the physical channels assigned to the mobile station, and the mobile station must read the USF indicator on each downlink assigned physical channel to see if it is allowed to transmit on the corresponding uplink physical channel.
- **Single block allocation** is given if the mobile station has indicated in its *packet channel request* that it requires two-phase access. The single block allocation will tell the mobile station which frequency, TN and frame it must use to send TLLI and *packet resource request*, which will tell the GPRS sub-network what radio resources the mobile station requires. Single block allocations are also used to send measurement reports, and in response to a *packet channel request* which indicates that the mobile station has

been paged, or requires two-phase access. In addition, the GPRS sub-network may allocate only a single block despite the mobile station requesting single-phase access. A single uplink block allocated in that case changes the single-phase access to two-phase access.

The mobile station MAC is responsible for the LLC PDU segments being placed on the correct frequency, TN and block.

The GPRS sub-network RR layer receives the RLC/MAC blocks carrying the LLC PDU segments with the message *attach request*. As these were encoded (using one of the encoding systems CS1 to CS4) and interleaved by the mobile station physical layer, then each block must be deinterleaved and decoded and the LLC PDU reassembled.

The service access point indicator (SAPI) of the LLC PDU can then be read and the reassembled LLC PDU is passed to the LLC layer using the primitive GRR-DATA-IND.

The SGSN LLC layer extracts the GMM PDU and passes it to the GMM layer. The GMM layer now receives the *attach request* message and associated parameters. In this example, the PTMSI apparently means nothing to the SGSN, so it generates an *identity request* asking the mobile station to send its IMSI. The PTMSI received by the GPRS sub-network when a mobile station tries to attach will mean nothing if the mobile station is now roaming, having been previously attached to another network. In that case the GPRS sub-network will ask the mobile station to send its IMSI to allow identification. The identity request is transferred using LL-UNITDATA and GRR-DATA requests.

The *identity request* LLC PDU is transferred downlink across the air interface with the TFI in the RLC/MAC block header indicating the mobile station that is addressed for this TBF.

The mobile station GMM layer, having received *identity request*, responds with an *identity response* containing its IMSI. LL-UNITDATA and GRR-DATA request primitives carry this message. The LLC PDU carrying *identity response* is transferred uplink across the air interface when a downlink RLC/MAC block indicates the correct USF.

The SGSN, on receiving the IMSI, can now identify whether the mobile station is allowed to use the GPRS services on this GPRS sub-network. It receives a set of triplets, ciphering key, kc, a random number RAND and another number called signed response (SRES). These are generated by the authentication centre (AUC) at the request of the SGSN (via the HLR). They are the same algorithms as are used in CS operations. If the mobile station is roaming, the triplets are generated by the AUC in the home network (HPLMN).

The SGSN now generates an *authentication and ciphering request* message. This contains RAND and the ciphering algorithm to be used by the mobile station and GPRS sub-network LLC layers, and the ciphering key sequence number (cksn). The cksn is sent by the mobile station with the *attach request*. If the cksn sent by the mobile station corresponds to the cksn held by the SGSN for that mobile station, and the PTMSI is correct, then this gives the GPRS sub-network the option of skipping the authentication procedure and going straight into ciphered mode.

The parameters kc and ciphering algorithm are also transferred to the GPRS sub-network LLC layer where the ciphering is performed.

The *authentication and ciphering request* is received by the mobile station GMM layer and RAND is passed to the SIM which calculates SRES, the signed response. This is derived from parameters RAND and ki, the customer identification key.

The mobile station SIM card and the AUC hold ki. It is used with RAND to derive SRES. Upon the GPRS sub-network GMM layer receiving the correct SRES response from the mobile station, authentication is completed.

The GPRS sub-network GMM layer generates *attach accept* and sends this with an RA update timer value (for periodic routeing area updates), the RAI and a new value of PTMSI, with the option of a PTMSI signature. The PTMSI signature is a number associated with PTMSI; if the mobile station is allocated a PTMSI signature it must include it with the *attach request*. The GPRS sub-network, on receiving a valid PTMSI signature has a double verification of PTMSI and may skip authentication.

The GPRS sub-network LLC layer ciphers the information content of the GMM PDU *attach accept*. This ciphered message is transferred to the LLC layer of the mobile station in a TBF. It is deciphered there and passed to the mobile station GMM layer which stores the PTMSI and its signature, notes the RA update timer and uses the RAI and PTMSI to calculate a local TLLI.

The mobile station network layer is informed that the mobile station is now attached and the new TLLI is assigned to the LLC layer and the RR layer (the TLLI is included in some RLC/MAC blocks).

The mobile station GMM layer then sends *attach complete* and asks the LLC layer to deliver it in ciphered mode. When the GPRS sub-network GMM layer receives this, the attachment is completed.

With this verification, the SGSN GMM layer can disable the old TLLI that was in use and inform the RR and LLC layers to use the new TLLI. The mobile station is now in a position to do the following:

- Monitor its calculated packet paging channels for downlink paging messages. (However, no paging messages will be received until a PDP context is established);
- Request uplink resources for uplink TBF;
- Perform cell reselection using C1, C31 and C32 criteria. (When a mobile station is GPRS attached to a cell which has a PBCCH, then GPRS cell reselection is used and not CS cell reselection which uses the C1 and C2 criteria).

5.2.2   *Mobile originated PDP context activation and TBF [2] (ETSI 123.060 section 9)*

PDP context activation involves the SM layer setting up the protocols and parameters to allow the transfer of network PDUs which carry customer packet data between the mobile station and network.

The mobile station user (network) layer indicates that it wishes to send packet data by sending the primitive SMREG-PDP-ACTIVATE-REQ to the SM layer. This contains a description of the customer PDP type, the destination IP address, the required QoS parameters and the network service access point (NSAP) it intends to use to transfer

---

[2] See Section 11.1 for further details.

customer data. This primitive is received by the mobile station session management layer, which must first establish that the mobile station is GPRS attached.

The GMMSM-ESTABLISH (req. and cnf.) primitives ascertain that the mobile station is GPRS attached. If the GMM layer is not GPRS attached when it receives this request it will go into the attach procedures covered earlier (and covered in detail in a later chapter).

The mobile station SM layer now sets up the PDP context. It starts by generating the SM PDU *PDP context activation request* for delivery to its SGSN peer. The contents of this message, which is sent to the mobile station GMM layer with a GMMSM-UNI-TDATA-REQ primitive, are shown in Figure 5.7.

The mobile station GMM layer encapsulates the message in a GMM PDU and sends this with the TLLI and a 'ciphering on' indication to the LLC layer in an LL-UNI-TDATA-REQ primitive. This message reaches the GPRS sub-network GMM layer on a TBF across the air interface. The SGSN SM layer extracts, if necessary, the subscriber parameters from the home location register (HLR), and asks the appropriate GGSN to set up a connection to the indicated ISP address.

The GGSN allocates a dynamic IP address to the mobile station and communicates through the ISP to establish a PDP context. If this is successful, it signals back to the SGSN that communication is established to the indicated address and tells the SGSN the mobile's dynamic IP address.

The SGSN prepares its reply *PDP context activation accept*, including the QoS parameters that will be used for this PDP context. The 'negotiated' QoS may be different to that requested by the mobile station; these QoS parameters will depend upon the radio and other resources which the GPRS sub-network is capable of providing at the time of the request, and possibly subscription class.

This message is received by the mobile station SM layer. The mobile station SM layer indicates to the mobile station SNDCP and network layers that a PDP context is activated. If the mobile station network layer agrees with the QoS profile, the PDP context is accepted.

The mobile station SNDCP layer is now in a position to negotiate compression parameters for PDP transfer with the SGSN. It receives the mobile station network layer primitive SN-XID-REQ, where XID means exchange identity.

Whether the SNDCP uses the primitive LL-XID-REQ, or LL-ESTABLISH-REQ, depends upon the QoS reliability. A reliability level better than 3 uses LL-ESTABLISH-REQ, and 3 or worse uses LL-XID-REQ. The difference in result of the two primitives is that for LL-ESTABLISH-REQ, full asynchronous balanced mode (ABM) (acknowledged LLC mode) is established by the LLC layer, and for LL-XID-REQ, asynchronous disconnect mode (ADM) (unacknowledged mode) is used. For this example LL-ESTABLISH-REQ is used.

The LLC layer generates an LLC set asynchronous balanced mode (SABM) frame as a result of receiving this primitive.

The LLC layer may also negotiate its own LLC XID parameters at this point, either using the SABM or, if an LL-XID-REQ is received, by generating an LLC XID frame, which will carry both the SNDCP and LLC XID parameters for negotiation.

The LLC PDU is received by the SGSN LLC layer which deciphers and passes on the SNDCP PDU.

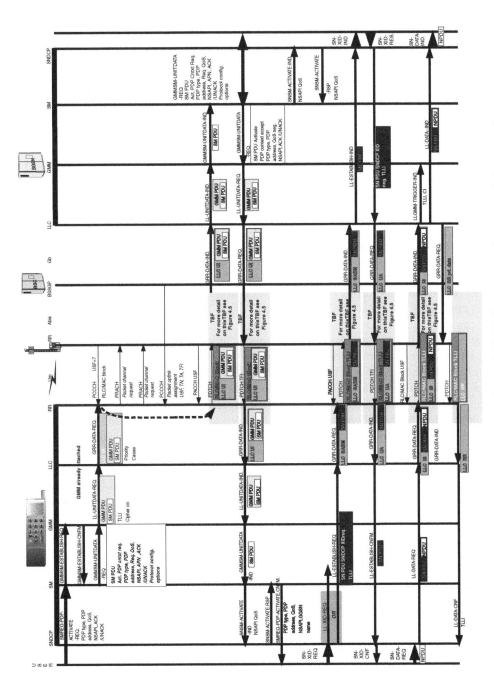

Figure 5.7 Protocol stack signal flow – MO PDP context activation and data transfer

The LLC layer waits until it has received an LL-ESTABLISH-RESP to its LL-ESTABLISH-IND primitive to the SNDCP before replying to LLC SABM with LLC un-numbered acknowledge (UA).

On receiving the LL-ESTABLISH-IND primitive from the LLC layer, the SGSN SNDCP layer indicates the required SNDCP compression parameters to the network layer with primitive SN-XID-IND. The GPRS sub-network responds with the compression parameters that will be used with an SN-XID-RESP primitive. This causes the SNDCP layer in the SGSN to generate the primitive LL-ESTABLISH-RESP, which carries the negotiated parameters. This is ciphered by the LLC layer and placed in an LLC UA frame.

The exchange of SABM and UA establish an ABM acknowledged mode data link between the two LLC end points. (The UA may also carry the negotiated LLC XID parameters.)

The SN PDU carrying the negotiated SNDCP parameters is delivered to the mobile station SNDCP layer by LL-ESTABLISH-CNFM, which also indicates acknowledged ABM transfer at the LLC layer is established.

The mobile station SNDCP layer delivers the negotiated compression parameters to the user (network) layer through the SN-XID-CNFM primitive. If this is acceptable, the user (network) layer responds to the original SM layer SNSM-ACTIVATE-IND with SNSM-ACTIVATE-RESP, indicating to the SM layer that it is now fully prepared to deliver SN PDUs containing user (network, or customer) data. The mobile station user (network) layer can now send N PDUs to the SNDCP layer with the primitive SN-DATA-REQ.

The SNDCP layer compresses and segments (if necessary) the N PDUs into SN PDUs and delivers them to the LLC layer with the primitive LL-DATA-REQ. This primitive asks the LLC layer to deliver the SN PDU in acknowledged mode. This is transferred as a TBF on a PDTCH over the air interface.

The SN PDU is deciphered by the SGSN LLC layer and delivered to the SGSN SNDCP layer in an LL-DATA-IND primitive. The LLC layer also informs the GMM layer that an SN PDU is delivered by sending the LLGMM-TRIGGER-IND. This includes the TLLI.

The trigger primitive starts the ready timer in the SGSN. A similar timer is started in the mobile station when the mobile station delivers an LLC PDU to the RLC/MAC layer.

At some point the SGSN LLC layer acknowledges receipt of the LLC information and supervisory (IS) frame by sending an appropriate response frame – in this case a receiver ready (RR) frame.

The mobile station LLC layer will indicate to the SNDCP layer that the SN PDU has been delivered with an LL-DATA-CNFM primitive.

During the PDTCH TBF, RLC/MAC blocks carry the TLLI if the access is one-phase. The first GPRS sub-network acknowledgment returns the TLLI. This resolves any contention where two mobile stations may have tried to access the same logical channel at the same time.

5.2.3 *Paging and MT PDP transfer (GSM 44.060 section 6)*

A mobile station will have a dynamic IP address if a PDP context exists. A mobile station can be paged if a PDP context exists. Paging can happen without the prior existence of a PDP context for those mobile station subscribers with a static IP address.

If the GPRS sub-network operator offers SMS over GPRS the mobile station may be paged whilst having no IP address.

In the example shown in Figure 5.8, the mobile station is in GPRS *MM standby* and the GPRS sub-network does not know which cell it is in. It does, however, know the mobile station's routeing area (RA) and must page the mobile station to discover which cell within this RA the mobile station is camped upon.

Because the GPRS sub-network GMM layer has lost track of the mobile station's cell, it sends an LLGMM-SUSPEND-REQ primitive to the LLC layer, disabling transfer of SN PDUs to that mobile station. GMM PDU transfer over the air interface is still allowed. The suspension of SN PDUs will persist until the GMM layer sends the primitive GMM-RESUME-REQ when it knows which cell the mobile station is using.

The GPRS sub-network receives an N PDU for that mobile station and requests the SNDCP to send it with the SN-DATA-REQ primitive. The SNDCP tries to send the segmented, compressed N PDU as an SN PDU by sending the primitive LL-DATA-REQ to the LLC layer.

The SN PDU is accepted by the LLC layer and acknowledged with the LL-DATASENT-IND primitive. However, the LLC layer cannot forward the SN PDU because of the SUSPEND imposed by the GMM layer. It requests the GMM layer to page the mobile station with the primitive LLGMM-PAGE-IND. The GPRS sub-network GMM layer now requests the RR layer to page the mobile station, giving the PTMSI and IMSI to enable the RR layer to calculate which packet paging channels (PPCHs) to use. The GMM layer also indicates the group of cells or the RA which must be paged.

The RR layer puts the paging request on to the appropriate paging group channel for each cell. The mobile station 'wakes up' and receives the paging message. The content of the received paging message is transferred to the GMM layer of the mobile station.

The mobile station GMM layer sends the primitive LLGMM-TRIGGER-REQ with the reason 'paging response'. This causes the LLC layer to send an empty LLC frame to the RLC/MAC layer. This layer creates an RLC/MAC block from the LLC PDU and adds the TLLI. The GPRS sub-network GMM layer receives this and tells its LLC layer to resume normal operations with the primitive LLGMM-RESUME-REQ.

At this point an ABM link does not exist and one must be established. Normal downlink PDP transfer then takes place with appropriate acknowledgments.

5.3 Temporary block flow acknowledged

Figure 5.9(a) shows the TBF procedure when the mobile station has no allocated resources for TBF transfer, and an LLC PDU (which will be carrying layer 3 PDUs from either the SNDCP layer or the GMM layer) arrives at the RLC /MAC layer. In this case the LLC PDU is carrying a GMM PDU and the total length of the LLC PDU is 80 octets. As each RLC/MAC block on the air interface can carry only 20 octets when using CS1, the RLC layer must calculate how many RLC/MAC blocks will be needed by this TBF to transfer the LLC PDU. It calculates that five RLC/MAC blocks are needed in this case (these calculations are covered in more detail in Chapter 8).

The type of *packet channel request* sent to the GPRS sub-network, whether short access or one-phase access or two-phase access is determined by the LLC PDU type.

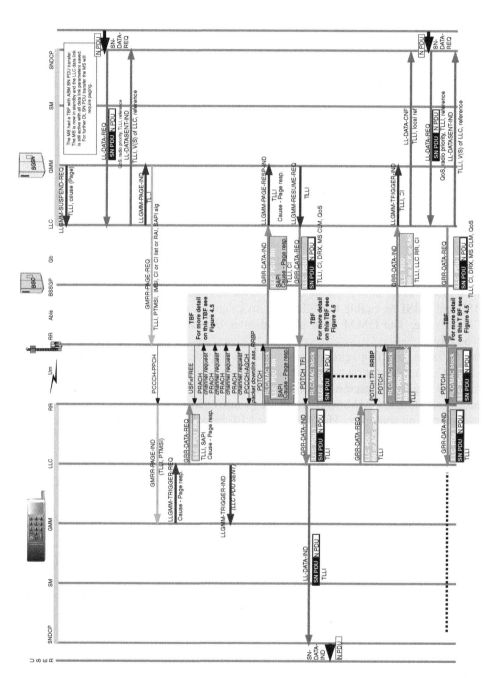

Figure 5.8 Protocol stack signal flow – paging and MT data flow

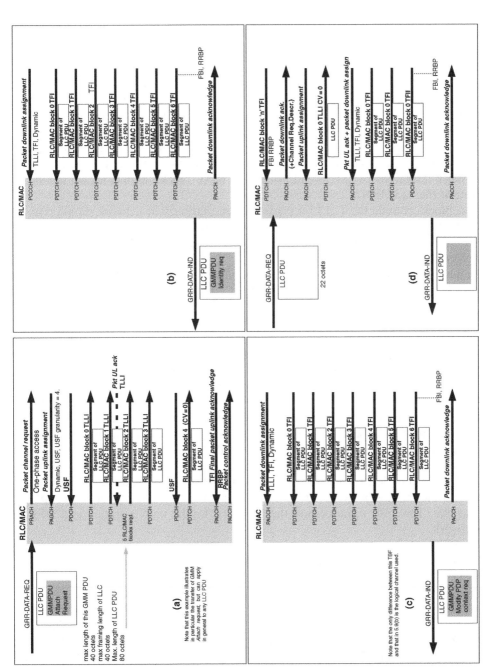

Figure 5.9 Protocol stack signal flow – various TBFs

RLC passes the blocks to the MAC layer, which gets the TBF resources to transfer the LLC PDU across the air interface by sending, at the appropriate time, a *packet channel request*. This request may be for one- or two-phase access but in this case it asks for one-phase access. The mobile station MAC layer then monitors all downlink PCCCHs looking for a *packet uplink assignment* message on a packet access grant channel (PAGCH).

The PAGCH tells the mobile station that the assignment is dynamic and gives a USF number. The USF granularity indicates that four RLC/MAC blocks may be sent consecutively after the appearance of USF. It is also told which frequency and timeslot number (TN) to use and is given a TFI and a downlink physical channel to monitor for PACCH messages.

The mobile station MAC layer switches to and monitors the assigned downlink physical channel (PDCH), examining each RLC/MAC block header for the assigned USF. When it receives its USF it sends four RLC/MAC blocks, each block containing a segment of the LLC PDU.

These blocks contain the TLLI for contention resolution. The network should return the TLLI for contention resolution. The mobile station then stops transmitting and waits for the next appearance on the downlink physical channel of its assigned USF and the procedure is repeated.

The countdown value (CV) is decrementing, and when it reaches '0', the GPRS sub-network knows that the final block of the LLC PDU transfer has been sent.

If all the blocks have been received correctly, the GPRS sub-network sends an RLC/MAC control block containing *final packet uplink acknowledge*. This acknowledges the sent blocks. If any of the blocks are not acknowledged, the mobile station will retransmit them, until positive acknowledgment is received.

If the GPRS sub-network has not received a block sent by the mobile station, it will know because each RLC/MAC block is numbered with a block sequence number (BSN). If any have not been received, then the acknowledgment sent by the GPRS sub-network will be *packet uplink negative acknowledge*, which indicates the missing blocks that must be resent by the mobile station. If necessary, additional uplink resources will be given to the mobile station at this point, allowing it to resend the missing blocks.

The TBF is completed when the mobile station RLC layer sends an RLC control block with the message *packet control acknowledge* in response to a *final packet uplink acknowledge*. The TBF resources are then released by the mobile station and GPRS sub-network.

The GPRS sub-network may, instead of sending *final packet uplink acknowledge*, send *packet uplink acknowledge + packet downlink assignment* to allow a GPRS sub-network LLC PDU to be transferred on the downlink to the mobile station.

During a TBF transferring an uplink LLC PDU across the air interface, the mobile station RLC/MAC layer may receive another LLC PDU. It can include in the *packet control acknowledge* message a request for further uplink resources to transfer this new LLC PDU. This request for additional uplink radio resources can also be sent at any time during an uplink TBF by interrupting the data transfer on the PDTCH and substituting a resource request on the uplink PACCH.

Figure 5.9(b) shows a downlink TBF. The mobile station is in the MM ready condition (attached, with a valid PTMSI and TLLI and the ready timer running) and is monitoring a packet common control channel (PCCCH). This will be its allocated packet-paging channel (or all the downlink PCCCHs if the mobile station is in non-DRX mode at this time).

The downlink message on the PCCCH identifies the mobile station with its TLLI and gives the mobile station an assignment, which includes a physical channel and TFI. The mobile station switches to and monitors the assigned physical channel, decoding all the RLC/MAC blocks on that PDCH looking for its TFI.

The RLC reassembles the downlink RLC/MAC blocks into an LLC PDU from the GPRS sub-network. Upon receiving a block with the final block indicator (FBI) bit set to '1', the mobile station knows that this is the final block in the LLC PDU transfer from the GPRS sub-network. If all blocks are received correctly, the mobile station sends an RLC control block message *packet downlink acknowledge* to the GPRS sub-network on the frame indicated by the downlink RRBP.

Figure 5.9(c) is similar, except that the logical channel on which the mobile station receives the *packet downlink assignment* message is the PACCH. This means that the mobile station is in an uplink TBF, and the GPRS sub-network wants to transfer a downlink LLC PDU to the mobile station.

If the mobile station has completed its uplink TBF, the downlink transfer takes place as described above; if it has not completed the uplink TBF, it may suspend the uplink TBF until the downlink TBF is completed (for half duplex operation). For full duplex operation, the uplink and downlink TBFs will take place (quasi) simultaneously.

Figure 5.9(d) shows half duplex operation with the final stage of a downlink TBF with an LLC PDU waiting at the mobile station to be sent as an uplink TBF (the FBI bit on the downlink is set to '1').

The mobile station sends the RLC control block message *packet downlink acknowledge* on the frame indicated by RRBP, and also indicates that it requires uplink TBF resources. These are granted with *packet uplink assignment* on the PACCH.

The mobile station sends its LLC PDU in an uplink TBF. When this is completed, the GPRS sub-network, which has LLC PDUs waiting to be transferred, instead of sending *packet uplink acknowledge* sends *packet uplink acknowledge + packet downlink assignment*.

5.4 Abbreviations used in this chapter

ABM	Asynchronous balanced mode
ADM	Asynchronous disconnect mode
APN	Access point name
ARQ	Automatic request for retransmission
AUC	Authentication centre
BSN	Block sequence number
CCCH	Common control channel
CS	Circuit switched
CV	Countdown Value
DCL	Data communications link
DL	Downlink
FBI	Final block indicator

FEC	Forward error correction
GGSN	Gateway GPRS support node
GMM	GPRS mobility management
GPRS	General packet radio service
HLR	Home location register
IMSI	International mobile subscriber identity
IP	Internet protocol
IS	Information/supervisory
ISP	Internet service provider
LLC	Logical link control
MAC	Medium access control
N PDU	Network protocol data unit
NSAP	Network service access point
PACCH	Packet associated control channel
PAGCH	Packet access grant channel
PBCCH	Packet broadcast control channel
PCH	Paging channel
PCCCH	Packet common control channel
PDCH	Packet data channel
PDP	Packet data protocol
PDTCH	Packet data traffic channel
PDU	Protocol data unit
PPCH	Packet paging channel
PRACH	Packet random access channel
PSI	Packet system information
PTMSI	Packet temporary mobile subscriber identity
QoS	Quality of service
RA	Routeing area
RAI	Routeing area indication
RAND	Authentication random number
RLC	Radio link control
RR	Radio resources
RR	Receiver ready
RRBP	Relative reserved block period
SABM	Set asynchronous balanced mode
SAP	Service access point
SAPI	Service access point identity
SGSN	Serving GPRS support node
SIM	Subscriber identity module
SM	Session management
SMS	Short message service
SNDCP	Sub-network dependent convergence protocol
SN PDU	Sub-network protocol data unit
SRES	Signed response for authentication
TBF	Temporary block flow
TCP/IP	Transmission control protocol/Internet protocol

TFI	Temporary flow identifier
TLLI	Temporary logical link identifier
TN	Timeslot number
UA	Unnumbered acknowledge
UI	Unacknowledged information
UL	Uplink
USF	Uplink status flag
XID	Exchange identities

6

GPRS Mobile Station Characteristics
(GSM 45.002)

6.1 Mobile station types

Figure 6.1 illustrates the GPRS MS classes A, B and C.

- *MS class A* allows simultaneous attach, simultaneous activation, simultaneous monitor, simultaneous invocation and simultaneous traffic for GSM GPRS and CS (circuit switched) services.
- *MS class B* allows simultaneous attach, simultaneous activation and simultaneous monitor. GPRS circuit is put on hold when a CS circuit is invoked. Simultaneous traffic for GSM GPRS and CS services is not allowed. The subscriber can only send or receive calls on a CS and GPRS circuit sequentially. Selection of one of the services is performed automatically with a GPRS service going to hold when a CS service is activated, or the subscriber may be offered the choice.
- *MS class C* allows only non-simultaneous attach, non-simultaneous activation, non-simultaneous invocation and non-simultaneous traffic for GSM GPRS and CS services. The GPRS or CS service is activated as an alternative to the other. The status of the non-activated service (GPRS or CS) is detached – non-reachable when one of the services is activated.

6.2 GPRS mobile equipment (ME) and subscriber profiles (ETSI 123.060 section 15)

6.2.1 Subscriber profile

Currently (August 2004) subscriber QoS profiles are not used. The GPRS sub-network simply gives a mobile station the best it can offer when an Internet call is initiated.

GPRS in Practice: A Companion to the Specifications   Peter McGuiggan
© 2004 John Wiley & Sons, Ltd   ISBN: 0-470-09507-5

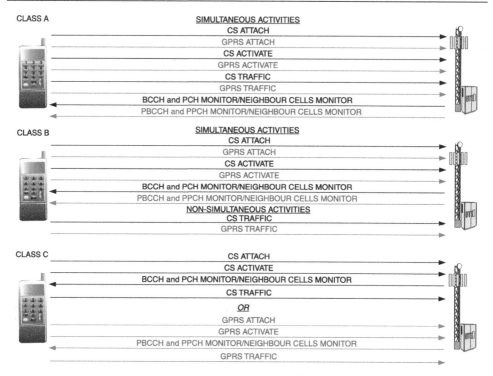

Figure 6.1 GPRS Mobile station types

As QoS profiles are not used, there is no price discrimination for subscribers using the GPRS services.

Figures 6.2 and 6.3 show the classes of subscriber QoS profile. The subscriber may negotiate with the GPRS service provider to purchase a profile from this set of parameters (suitably packaged in a way that is easily understood by the subscriber!)

A QoS profile cannot be guaranteed and the MS user layer and GPRS sub-network may renegotiate the QoS profiles when the subscriber initiates an Internet call.

The *service reliability* parameter is translated into five classifications. Reliability class 1 and 2 give acknowledged mode of operation on the LLC link and RLC link. Reliability class 4 and 5 give unacknowledged LLC and RLC operation, 3 gives LLC unack, RLC ack.

These parameters are 'negotiated' with the GPRS sub-network in that the service requirements are sent by the MS to the GPRS sub-network. The GPRS sub-network will grant the required profile if it has the capability, but in periods of congestion during busy hours, may offer an inferior set of parameters. If the MS user layer does not find the offered QoS parameters acceptable, it may discontinue the Internet session. The offered QoS profile also gives a *radio priority*. This is sent by the MS to the RLC/MAC layer where it determines the access procedures, and it may also be sent to the GPRS sub-network in the *packet channel request*, in which case the network may ignore requests of low radio priority during busy hours.

In the future each subscriber will have a 'profile' comprising a selection from the parameters of Figures 6.2 and 6.3 for each subscribed service. The subscriber profile will

SERVICE PRECEDENCE High, Normal or Low
SERVICE RELIABILITY

Reliability class	Lost SDU probability	Duplicate SDU probability	Out of sequence SDU probability	Corrupt SDU probability	Example of application
1	10^{-9}	10^{-9}	10^{-9}	10^{-9}	Limited error tolerance, no error correction capability
2	10^{-4}	10^{-5}	10^{-5}	10^{-5}	Good error tolerance, limited error correction capability
3	10^{-2}	10^{-5}	10^{-5}	10^{-2}	Very good error tolerance, error correction capability

SERVICE DELAY This includes all the GSM network delays, but not external networks

Delay class	Delay maximum value			
	SDU size 128 octets		SDU size 1024 octets	
	Mean transfer delays	95 percentile delays	Mean transfer delays	95 percentile delays
1 Predictive	<0.5	<1.5	<2	<7
2 Predictive	<5	<25	<15	<75
3 Predictive	<50	<250	<75	<375
4 Best effort	unspecified			

Figure 6.2 GPRS subscription profile. Each subscriber is allocated a service profile upon taking out a subscription. This may be renegotiated between the network and MS when a service is requested. This profile includes the type of services PTP-CLNS and/or PTP-CONS, and the QoS parameters as decided by the network operator which may include service priority, service reliability, service delays and service data throughput

then be stored in the SIM card and in the HPLMN HLR. The QoS profile stored on the SIM will be used upon the mobile station requesting a PDP context for a subscribed service.

6.3 Mobile equipment multislot capabilities (GSM 45.002 Annexe B)

Figure 6.4 shows the specification for ME multislot capability. Figure 6.5 illustrates the application of the specification for a mobile station multislot class 10 transferring uplink and downlink LLC PDUs.
From Figure 6.4:

- *Multislot classes 1–12 type 1* have a multislot capability in the UL and DL directions and may use this capability (quasi) simultaneously. This group of multislot classes may use half duplex or full duplex communication.
    The GSM system has a standard three-burst separation between the mobile station receiving and transmitting on the same timeslot, and we shall shortly examine how this separation, in conjunction with the multislot characteristics, affects a GPRS mobile station's operations.
- *Multislot classes 19–29 type 1* are less sophisticated than the previous group, and in the current GPRS phase, these will use only half duplex operation. There is a very good reason why this group of mobile stations are limited to half duplex operation.

PEAK THROUGHPUT CLASS
Measured at the Gi and R reference points

Peak throughput class	Peak throughput/octets per second
1	Up to 1000 (8 kb/s)
2	Up to 2000 (16 kb/s)
3	Up to 4000 (32 kb/s)
4	Up to 8000 (64 kb/s)
5	Up to 16 000 (128 kb/s)
6	Up to 32 000 (256 kb/s)
7	Up to 64 000 (512 kb/s)
8	Up to 128 000 (1024 kb/s)
9	Up to 256 000 (2048 kb/s)

MEAN THROUGHPUT CLASS
Measured at the Gi and R reference points

Mean throughput class	Mean throughput/octets per hour
1	Best effort
2	100 (~0.22 bits/s)
3	200 (~0.44 bits/s)
4	500 (~1.11 bits/s)
5	1000 (~2.2 bits/s)
6	2000 (~4.4 bits/s)
7	5000 (~11.1 bits/s)
8	10 000 (~22 bits/s)
9	20 000 (~44 bits/s)
10	50 000 (~111 bits/s)
11	100 000 (~0.22 k bits/s)
12	200 000 (~0.44 k bits/s)
13	500 000 (~1.11 k bits/s)
14	1000 000 (~2.2 k bits/s)
15	2000 000 (~4.4 k bits/s)
16	5000 000 (~11.1 k bits/s)
17	10 000 000 (~22 k bits/s)
18	20 000 000 (~44 k bits/s)
19	50 000 000 (~111 k bits/s)

Figure 6.3 Subscription profile – service data throughput

If we examine multislot classification 26, for example, we see that it is capable of transmitting on four uplink physical channels and receiving on eight downlink physical channels. Simultaneous transmission and reception of such magnitude is possible only if the mobile is capable of transmitting and receiving at the same time. This particular group does not have this capability, and the specification limits their operation to half duplex.

• *Multislot classes 13–18 type 2* are the most sophisticated group of mobile stations. These have the capability to simultaneously transmit and receive, requiring duplexers, splitters and filters to separate the transmit and receive paths. They have the capability for full duplex operation.

Figures 6.5 and 6.6 illustrate the application of the specification parameters illustrated in Figure 6.4. These are applied to uplink and downlink allocations for a type 1 multislot class 10 MS. Figure 6.5 illustrates their application to a valid uplink and downlink allocation when measurements are required, and Figure 6.6 their application to an invalid allocation when measurements are required.

Let us now examine Figure 6.5 in more detail.

The MS type 1 multislot class 10 has been allocated PDTCHs on TN0, 1, 2, 3 on the downlink DL, and TN2 on the uplink UL (one up, four down).

Multislot capability continued.........

Tta
The time needed for an MS to measure neighbour cells and then to change frequency ready to transmit.
For TYPE 1 MSs:
a) the minimum number of TSs necessary between the end of a transmitted TS and the start of the next transmitted TS when measurements are required, *OR*
b) the minimum number of TSs necessary between the end of a received TS and the start of the next transmitted TS.
This does not take account of TA.
Not applicable to type 2 MSs.

Ttb
The time needed for an MS to get ready to transmit. Only applicable when neighbour cell measurements are not required.
For TYPE 1 MSs:
a) the minimum number of TSs necessary between the end of a received TS and the beginning of the next transmit TS when the frequency is changed between these two TSs. *OR*
b) the minimum number of TSs between the end of a transmitted TS and the next transmit TS when the frequency is changed between these two TSs.
This does not take account of TA.
For TYPE 2 MSs, it is the time between the end of the last transmit burst in a frame and the first transmit burst in the next frame.

Tra
The time needed for an MS to measure neighbour cells and then to change frequency ready to receive. For TYPE 1 MSs:
a) the minimum number of TSs necessary between the end of a transmit TS and the start of the next receive TS where neighbour cell measurements are performed between these two TSs *OR*
b) the minimum number of TSs necessary between the end of a receive TS and the start of the next receive TS where neighbour cell measurements are performed between these two TSs.
For TYPE 2 MSs, it is the time between the end of the last receive burst in a frame and the first receive burst in the next frame.

Trb
The time needed for an MS to get ready to receive. Only used when neighbour cell measurements are not required. For TYPE 1 MSs:
a) the minimum number of TSs necessary between the end of a transmit TS and the start of the next receive TS where the frequency is changed between these two TSs. *OR*
b) the minimum number of TSs necessary between the end of a receive TS and the start of the next receive TS where the frequency is changed between these two TSs.
For TYPE 2 MSs, it is the time between the end of the last receive burst in a frame and the first receive burst in the next frame.

Note that this script is a paraphrasing
of the specification

MS MULTISLOT CAPABILITIES

Multislot class	Maximum number of slots			Minimum number of slots				Type	
	RX	TX	Sum	Tta	Ttb	Tra	Trb		
1	1	1	2	3	2	4	2	1	Half duplex / Full duplex
2	2	1	3	3	2	3	1	1	
3	2	2	3	3	2	3	1	1	
4	3	1	4	3	1	3	1	1	
5	2	2	4	3	1	3	1	1	
6	3	2	4	3	1	3	1	1	
7	3	3	4	3	1	3	1	1	
8	4	1	5	3	1	2	1	1	
9	3	2	5	3	1	2	1	1	
10	4	2	5	3	1	2	1	1	
11	4	3	5	3	1	2	1	1	
12	4	4	5	2	1	2	1	1	
13	3	3	N/A	N/A	a	3	a	2	Full duplex / Half duplex Transmit and receive at the same time
14	4	4	N/A	N/A	a	3	a	2	
15	5	5	N/A	N/A	a	3	a	2	
16	6	6	N/A	N/A	a	2	a	2	
17	7	7	N/A	N/A	a	1	0	2	
18	8	8	N/A	N/A	0	0	0	2	
19	6	2	N/A	3	b	2	c	1	Half duplex
20	6	3	N/A	3	b	2	c	1	
21	6	4	N/A	3	b	2	c	1	
22	6	4	N/A	2	b	2	c	1	
23	6	6	N/A	2	b	2	c	1	
24	8	2	N/A	3	b	2	c	1	
25	8	3	N/A	3	b	2	c	1	
26	8	4	N/A	3	b	2	c	1	
27	8	4	N/A	2	b	2	c	1	
28	8	6	N/A	2	b	2	c	1	
29	8	8	N/A	2	b	2	c	1	

a = 1 with frequency hopping, 0 without.   b=1 with frequency hopping or change from RX to TX
c = 1 with FH or change from TX to RX, 0 without FH or no change from RX to RX.   b=0 without FH or no change from RX to TX

Type 1 MSs are not required to transmit and receive at the same time.
Type 2 MSs are required to transmit and receive at the same time.
For HSCSD only multislot classes 1–18 are recognised. An MS with higher capability must use class 1–18.

RX. The *maximum* number of RX TS the MS can use per TDMA frame. The RX TS need not be contiguous.
Type 1 MSs are allocated TSs within the window of size TS0–TS RX and will not transmit between receive TSs within a frame.
TX. The *maximum* number of TX TS the MS can use per TDMA frame. The TX TSs need not be contiguous.
Type 1 MSs are allocated TSs within the window of size TS0–TS TX and will not receive between transmit TSs within a frame.
SUM. The total number of UL and DL TSs that can be used by the MS. The MS must be able to use all combinations from 1TS to SUM TSs.

Figure 6.4 ME capabilities – multislot classes

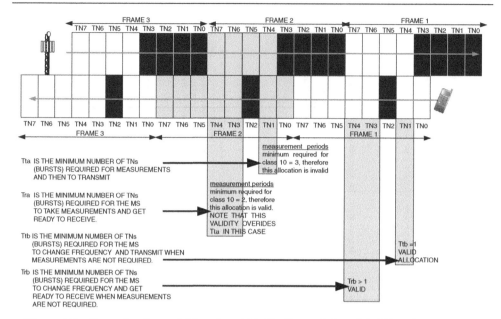

Figure 6.5 Example of a multislot class 10 MS allocated four DL TS and one UL TS

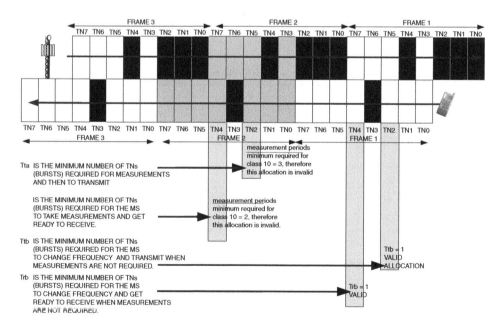

Figure 6.6 Example of a multislot class 10 MS allocated four DL TS and one UL TS

The parameter *Tta* in this case gives the mobile station one burst to finish receiving from the GPRS sub-network on TN3, change frequency to a neighbour cell BCCH radio channel, measure the receive level, and change frequency again ready to transmit to the GPRS sub-network. Unsurprisingly, *Tta* in this case is invalid. This is seen by examining the required *Tta* value for multislot class 10 in the table of Figure 6.4. There should be a gap of three radio bursts between the mobile station receiving from and transmitting to the BTS for *Tta* to be valid. The one-burst gap of Figure 6.5 is therefore not valid.

The next parameter to consider is *Tra*, the time available to the MS to perform measurements between transmitting to the BTS and then being ready to receive from the BTS. This gap is used to change frequency and perform measurements on a neighbour cell BCCH carrier. The MS in Figure 6.5 is shown transmitting on TN2, and then there is a two-burst gap before it has to change frequency ready to receive from the BTS. The table entry in Figure 6.4 shows that a gap of two radio bursts is sufficient for the mobile station to complete this task. In this example, the period *Tra* is therefore valid for neighbour cell measurements.

This two-burst *Tra* period occurs during every 4.615 ms radio frame, allowing the mobile station to get a good measure of the received level of the neighbour cell.

This particular allocation, four down on TN0 to 3 and one up on TN2 can be validly allocated to this multislot class 10 mobile station by the GPRS sub-network.

The two remaining parameters of the specification, *Trb* and *Ttb*, are not concerned with measurement periods. They indicate how quickly the synthesiser must change frequency from receive to transmit (*Ttb*), and from transmit to receive (*Trb*).

Looking at Figure 6.5 and comparing these two periods with those specified for multislot class 10 in Figure 6.4, this particular allocation of uplink and downlink timeslots is valid.

Figure 6.6 illustrates an invalid allocation for a multislot class 10 mobile station. The allocation is four down, TN0, 1, 2, and 4, and one up, TN3. *Tta* and *Tra* are each one-burst periods, and both are invalid for neighbour cell measurements. The GPRS sub-network should not give this allocation to a multislot class 10 mobile station.

Figure 6.7 illustrates some practical examples, taken from the specification.

In the examples shown, the convention used is that the radio bursts are travelling from the BTS toward the mobile station, therefore the oldest time is on the right-hand side of the paper and time is travelling from right to left, the opposite of the normal convention. This applies to both the BTS and mobile station as the MS is synchronised to the BTS and the three-burst offset between the MS receiving and transmitting is indicated.

The first example shows a multislot class 4 mobile station in a DL TBF transfer on TN0 to 3. In the first DL block transfer, the mobile station is 'polled' on TN1 to send a TBF DL packet acknowledgment. The response frame is indicated by an RRBP (relative reserved block period) of value 0. The possible values range from 0 to 3, with parameter value 0 indicating the reserved block is N (the frame number in which the command is issued) $+ 13$ frames. In this example the mobile station responds on frame 13 as the RRBP is in frame 0. Other values are $1 = (N + 17 \text{ or } 18)$, $2 = (N + 21 \text{ or } 22)$ and $3 = (N + 26)$.

Parameters *Ttb* and *Tra* are shown and should be compared with Figure 6.4.

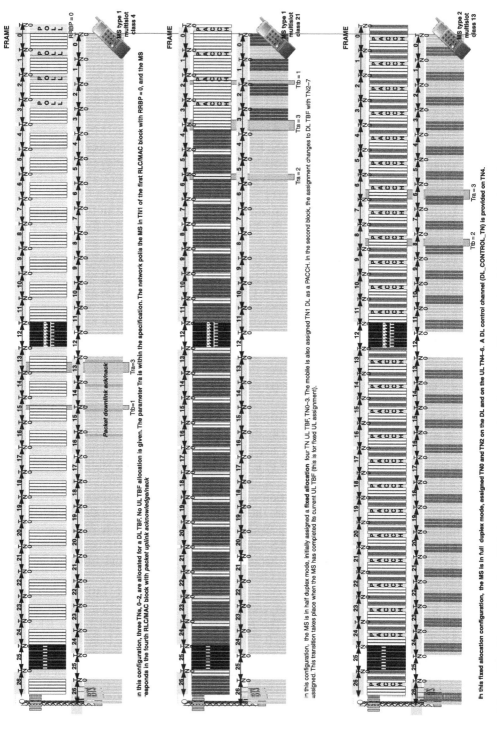

In this configuration, three TNs, 0–2, are allocated for a DL TBF. No UL TBF allocation is given. The network polls the MS in TN1 of the first RLC/MAC block with RRBP = 0, and the MS responds in the fourth RLC/MAC block with packet uplink acknowledge/nack

In this configuration, the MS is in half duplex mode, initially assigned a **fixed allocation** four TN UL TBF, TN0–3. The mobile is also assigned TN1 DL as a PACCH. In the second block, the assignment changes to DL TBF with TN2–7 assigned. This transition takes place when the MS has completed its current UL TBF (this is for fixed UL assignment).

In this fixed allocation configuration, the MS is in full duplex mode, assigned TN0 and TN2 on the DL and on the UL TN4–6. A DL control channel (DL_CONTROL_TN) is provided on TN4.

Figure 6.7 Examples of multislot parameter constraints in action (Tra etc should be compared to the parameters in Figure 6.4)

The second example shows a multislot class 21 mobile station in half duplex mode with initially an uplink allocation of TN0 to 3. Given with this fixed allocation is the DL_CONTROL_TN, which is monitored by the mobile station for PACCH control messages during the fixed UL TBF.

In this case, on the first RLC/MAC block, the mobile station receives a *packet downlink assignment* on the PACCH; the mobile station therefore completes its fixed UL TBF and then switches to the allocated DL TNs for the DL TBF. The allocated DL TNs are TN2 to 7.

Ttb, Tta and *Tra* are shown and should be compared with the parameters in Figure 6.4.

The final example shows a multislot class 13 mobile station in full duplex mode. TN4 to 6 are allocated on the UL and TN0 to 2 on the DL. In addition, TN4 on the DL is designated as PACCH. (This could also be used as a DL PDTCH).

Tra and *Ttb* are shown and should be compared with the parameters in Figure 6.4.

Finally we return to our multislot class 10 mobile station with a valid allocation of four down on TN0–3 and one up on TN2. Figure 6.8 shows this mobile station in action with the valid allocation. The two-burst measurement period in every radio frame is clearly shown.

In addition to measuring the receive level of the neighbour cell BCCH frequency, the mobile station must identify the neighbour to make the measurements valid. Because of GSM frequency reuse patterns, the BCCH frequency that the mobile station is trying to measure could be another neighbour cell using the same frequency. Indeed, the level measurements made could be false if the mobile station is receiving two cells, both transmitting on the same frequency.

The mobile station must decode the synchronisation channel (SCH) of the neighbour BCCH carrier to validate the measurements taken. The SCH carries the base station identity code (BSIC) and the frame number (FN). Additionally, the SCH has an extended training sequence code (TSC), giving it a C/I advantage over other logical channels.

The PDTCH the mobile station is using has a 52-frame multiframe structure with idle frames occurring with every 26 frames. During the idle frame the mobile station does not transmit or receive, so has an extra eight burst periods in which it might 'capture' the neighbour cell SCH. This eight-burst period is sometimes called the *BSIC search period*.

The neighbour BCCH radio carrier TN0 that carries the SCH has a 51-frame multiframe period, which means that for every 52-frame period, there is a slippage of one frame between the SCH and the BSIC search period.

It can be shown that the maximum distance between the BSIC search period and the SCH is four frames, and it will take $4 \times$ (52-frame multiframe period) for the SCH to 'creep' into the BSIC search period. As a 52-frame period is very nearly 240 ms, then the mobile station is guaranteed to capture the SCH within 4×240 ms, about 0.96 s. This is the reason for the 51-frame multiframe, 52-frame multiframe and the idle frames.

Note that this search period is a 'one-off'. Once the mobile station has found the neighbour cell BCCH carrier SCH, then it knows precisely the relationship of the neighbour cell TN0 to its own cell's TN0 and the mobile station can go to the neighbour cell SCH at any suitable time.

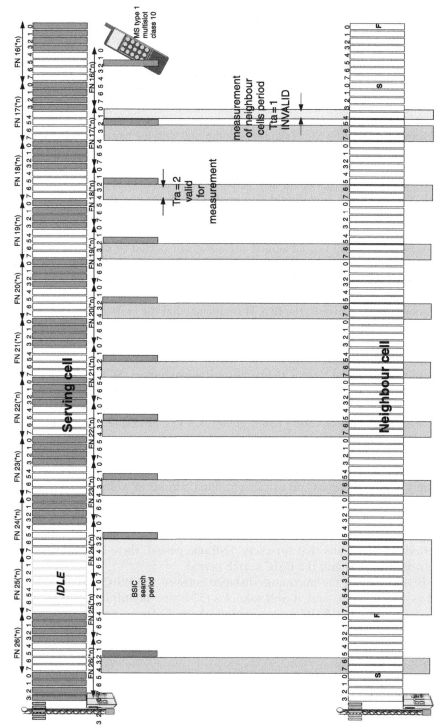

Figure 6.8   Illustration of MS multislot class 10 operational constraints

6.4   Abbreviations used in this chapter

BCCH	Broadcast control channel
BSIC	Base station identity code
BTS	Base transceiver station
C/I	Carrier to interference ratio
CS	Circuit switched
DL	Downlink
GPRS	General packet radio service
GSM	Global system for mobile communication
HLR	Home location register
HPLMN	Home public land mobile network
LLC	Logical link control
MAC	Medium access control
ME	Mobile equipment
MS	Mobile station
PACCH	Packet associated control channel
PDP	Packet data protocol
PDTCH	Packet data traffic channel
PDU	Protocol data unit
QoS	Quality of service
RLC	Radio link control
RRBP	Relative reserved block period
SCH	Synchronisation channel
SIM	Subscriber identity module
TBF	Temporary block flow
TN	Timeslot number
TSC	Training sequence code
UL	Uplink

7

Operations in the Physical Layers

This chapter is about how the GPRS mobile station interacts with the radio resources of the GPRS sub-network. Some of these interactions are in common with the GSM circuit switched operations and others are GPRS specific. The topics covered include:

- **PLMN selection** – this is in common with GSM circuit switched operations and describes how the mobile station selects a GSM network when switched on.
- **Cell selection and cell reselection**. A mobile station must select the best available cell of a selected PLMN. This is called cell selection. After this it must periodically measure its neighbour cells and select the best. This is cell reselection. This may be the same as for GSM circuit switched operations or specifically for GPRS operations. The circumstances in which they differ and how this results in differing modes of operation are covered in this chapter.
- **Neighbour cell measurements**. These are always performed by the GPRS mobile station (with a possible exception when the mobile station is registered as static and always uses the same cell). Sometimes the measurements are reported to the GPRS sub-network, sometimes not. This chapter explains these varying circumstances and looks in a general way at the problem of neighbour cell measurement.
- **Extended measurements**. These are neighbour cell measurements specifically requested by the GPRS sub-network and measured and reported by the GPRS mobile station. They include level measurements and interference measurements.
- **Timing advance**. A fast moving GPRS mobile station may be in a downlink TBF without transmitting to the GPRS sub-network for a period of time that is significant in terms of timing advance. The GPRS system must provide a method of timing advance control, which compensates for extended periods of not receiving any transmission from the mobile station.
- **Power control**. A similar problem exists as for TA control – there may be a significant period in which the GPRS sub-network receives no transmission from the mobile station.

GPRS in Practice: A Companion to the Specifications   Peter McGuiggan
© 2004 John Wiley & Sons, Ltd   ISBN: 0-470-09507-5

- **Channel encoding**. There are four systems of channel encoding in GPRS, CS1 through CS4. This chapter covers these encoding systems.
- **Paging**. The physical layer in the GPRS sub-network and the mobile station must calculate the paging channels belonging to a GPRS mobile station, and the mobile station must 'wake up' at the correct time to listen to these paging channels.
- **DRX management**. DRX means 'discontinuous reception'. GPRS always uses DRX when the mobile station is in the GMM standby state. There is an additional meaning for DRX in GPRS, the time a mobile station must 'stay awake' after the completion of data transfer across the air interface, the non-DRX period.
- **Access control**. Just as the GSM circuit switched network controls how mobile stations access the network services, so does the GPRS sub-network, but in a more sophisticated fashion.
- **Contention resolution**. In any communication system which allows random access attempts on common channels to the communication services, there is a probability of collisions which must be resolved, and these collisions may result in two or more mobile stations attempting to access the same dedicated channel which also must be resolved.
- **Frequency hopping**. A GPRS physical channel (PDCH) is defined by a timeslot and a frequency, or a timeslot and a range of frequencies if the physical channel is frequency hopping. This section looks at frequency hopping (identical to circuit switched frequency hopping) and the parameters that the physical layer must interpret to follow the correct sequence.

7.1 Physical layers

In GPRS, the physical layer is a sub-layer of the radio resource (RR) layer. The physical layer is divided into two parts, the physical link layer and the physical RF layer. The functions of the two physical layers are outlined in Figure 7.1. This chapter gives a detailed description of the functions of the physical link layer. The physical RF layer functions are not covered (GMSK modulation is introduced in Appendix 1).

7.2 PLMN selection (GSM 43.002 section 3)

Whether a mobile station is circuit switched only or has GPRS capabilities, the first hurdle it must jump after being switched on is PLMN selection. This is the process of selecting the best PLMN available in the area where it is switched on. If it is in the home PLMN service area, then it is programmed to select its 'home' PLMN irrespective of whether other PLMNs are the better choice. This does vary if the mobile station was in a foreign network when last switched off. In that case, the mobile station searches for the last PLMN upon which it was registered.

PLMN selection and the immediately following procedure, cell selection, are both circuit switched (CS) mode functions and they are included here to give a full picture leading to the GPRS cell reselection process.

The section on GPRS cell reselection uses the GPRS packet system information sent on the PBCCH or PACCH. If the cell does not have a PBCCH channel then the mobile station uses the GSM cell reselection procedures whilst in GPRS mode. The GPRS cell

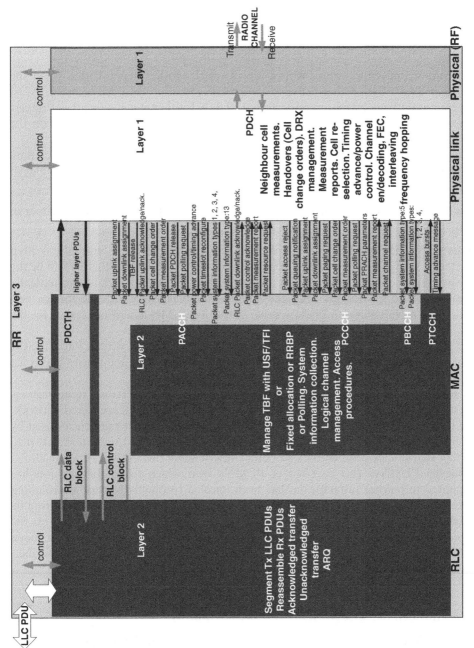

Figure 7.1   The MS physical layer in context

reselection parameters are included in Figure A1.1 of Appendix 1 which shows the contents lists of the system information messages.

Figure 7.2 shows PLMN selection in operation in the automatic and manual modes. It will select:

- If it can find its '*last registered*' or, secondly, its '*home*' network, one of these is selected.
- If it cannot find either of the above it will select another network that is on the '*preferred PLMN*' list if that network has a receive level of −85 dBm or more.
- If it cannot find a network on the preferred PLMN list then it will select the strongest network that it can find with the condition that the network must not be on the '*forbidden PLMN*' list.
- In *manual mode* the mobile station displays all the networks that it can find, including those on the forbidden list. The subscriber then manually selects one of the displayed networks.

The mobile station has a number of lists stored in the SIM card, which assists it in finding a PLMN.

- **Preferred PLMN list**. This is a list of preferred roaming networks normally written into the SIM card by the HPLMN network operator. At the top of this list will be the network on which the mobile station was last registered (this network is automatically placed at the top of the list when the mobile station registers with the foreign network). The mobile station will always search for this PLMN as a priority, then for the HPLMN, then for other networks on the preferred list.
- **BCCH list**. This is a list of BCCH carrier frequencies, normally written into the SIM from the system information of the last registered network. The frequencies in this list will be scanned whilst looking for the last registered network.
- **BA range**. BA stands for BCCH allocation, and this list of just two frequencies is again normally extracted from the system information of the last registered network. The two frequencies are the limits of the range of BCCH carrier frequencies used by the last registered network. If the BCCH list fails to find a last registered network, then all frequencies between the end points described by this list will be scanned.
- **Forbidden PLMNs**. This is a list of PLMNs with which the mobile station has attempted to register (attach), but has been rejected (probably because no roaming agreement is in place between the HPLMN and that foreign network). If, during its search for a PLMN, the mobile station finds a PLMN which is registered on the forbidden list, it will reject it from the PLMN selection process. In manual mode PLMN selection, the mobile station presents all discovered PLMNs to the subscriber, including forbidden PLMNs.

7.3 Initial cell selection (GSM 43.022 section 4)

Initial cell selection is the process of the mobile station finding the best possible cell on a newly selected PLMN. The mobile station has selected a PLMN and finds itself monitoring a particular cell of the selected PLMN. It reads the broadcast control channel (BCCH) system information of this cell. One important part of this system

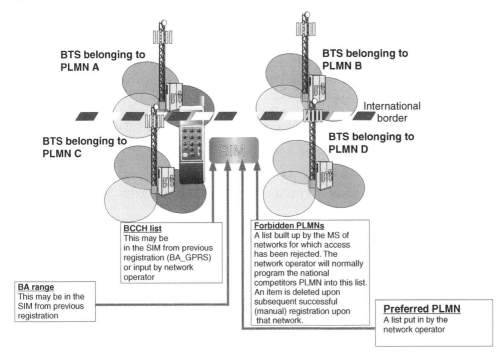

Figure 7.2 PLMN selection. The MS, upon switch-on will attempt to find firstly its last registered PLMN and then its home PLMN. If it has the BCCH list in the SIM it will scan those frequencies looking for the last registered PLMN, decoding SI on the BCCH carriers to find which PLMN they belong to. If there is no list but there is a BCCH_RANGE in the SIM, it will scan that frequency range looking for the last registered PLMN. If neither of these lists exist, or if the outcome of scanning them is unsuccessful in finding the last registered PLMN or HPLMN, the MS will scan all 124 GSM-P frequencies, or all 374 DCS frequencies looking for the last registered PLMN. If the last registered PLMN is not found, and the MS is in automatic selection mode, it will select the HPLMN, and if neither the last registered nor the HPLMN is found it will select the PLMN with the strongest carrier that belongs to the preferred PLMN list. If neither the last registered nor the HPLMN nor a preferred list PLMN is found it will select the strongest PLMN found, and after performing *cell selection*, will camp on the best cell, register, and if the registration is successful, display that PLMN to the subscriber. If the registration is unsuccessful it will attempt the next strongest PLMN; if all are unsuccessful it will select the strongest cell of any PLMN and use it for emergency calls if necessary. If the MS is in manual mode and is unsuccessful in finding the last registered or HPLMN, then all PLMNs found will be displayed to the subscriber. The subscriber will select one of those displayed and the MS will then perform *cell selection* and attempt to register on that PLMN. If the registration is unsuccessful, the MS will register that PLMN upon the 'Forbidden PLMN' list and the subscriber must then manually select another PLMN. In manual mode all PLMNs are displayed including forbidden PLMNs and the subscriber may attempt to register on a displayed forbidden PLMN; if this latter registration is successful, the PLMN is deleted from the forbidden PLMN list

information is the BA (BCCH allocation) which is a list of neighbouring cells to the one it is monitoring. This list gives the BCCH frequency for each of the neighbouring cells. The mobile station now measures the received level of each neighbour cell on this list and obtains from each cell the cell selection parameters which must be used when considering that cell for selection. The mobile station calculates, from the received level and the cell selection parameters for each cell, which of the cells is the best for selection.

The best cell is found by calculating the *C1 cell selection criterion* for each cell. This uses the received level and cell selection parameters of each cell. The best cell has the largest value of C1.

$$C1 = RLA - RXLEV\_ACCESS\_MIN$$
$$- \max(MS\_TXPWR\_ACC\_MAX\_CCH - MSPC), 0$$

where

RLA = the average received level from the cell.

RXLEV_ACCESS_MIN = the minimum receive level that a mobile station must receivefrom a cell before it is allowed to select that cell. This parameter is received from the system information of the cell that is measured.

MS_TXPWR_ACC_MAX_CCH = the maximum power that a mobile station is allowed to transmit when requesting access to the network services (*channel request* on the RACH). This parameter is received from the system information of the cell that is measured.

MSPC = mobile station power class, the maximum possible transmitted power of the mobile station, typically + 33 dBm.

The mobile station selects and 'camps on' the cell with the largest positive value of C1 as long as that cell has no other prohibitions in force (for example, a cell may bar all mobile stations or selected mobile stations from using the cell for access attempts; if that is the case, the cell is excluded from the selection process). From the system information of the newly selected cell, the mobile station calculates which circuit switched paging channel it must monitor for paging messages. It will then execute *IMSI attach* to inform the network of its presence; alternatively LAU (location area update), or periodic LA update will achieve the same end. Cell selection is illustrated in Figures 7.3 and 7.4.

Cell selection is, in reality, the process of initially selecting the cell with the least uplink path loss for the mobile station, taking into account whether that cell permits access and the priority (high or low) for access to that cell (a high access priority will make the mobile station select that cell in preference to one of low priority).

Cell selection is initiated when a mobile station is switched on in GSM coverage or when temporary loss of radio coverage from the network occurs, or when moving from one PLMN to another.

Figure 7.3 shows the C1 criterion. We shall consider this in more detail when looking at GPRS cell reselection.

CHOSEN PLMN

3. The MS compares the calculated C1 criteria for each of the measured cells and 'camps on' the cell with the largest C1 value. This value must be greater than or equal to zero. 'Camping on' means the MS selects a control channel, listens to its paging group and registers with the network.

2. The MS reads the neighbour cells BCCHs SI on the cells of the BA list, measures the RLA for each cell, the RXLEV_ACCESS_MIN and MS_TXWR_MAX_CCH for each cell. It checks if each cell is barred for access and the access priority, high or low. It calculates the C1 criterion for each cell.

1. The MS reads the BCCH SI on the cell of the chosen PLMN, measures the RLA for this cell, the RXLEV_ACCESS_MIN and MS_TXWR_MAX_CCH for this cell, and the neighbour cell list BA_BCCH. It checks if this cell is barred for access and the access priority, high or low.

C1 = RLA – RXLEV_ACCESS_MIN – max{(MS_TXPWR_ACC_MAX_CCCH – MSPC), 0}

RLA = average received level
RXLEV_ACCESS_MIN = the minimum RLA the MS must receive to access the cell
MS_TXPWR_ACC_MAX_CCCH = the max power the MS is allowed to transmit on
                                            RACH
MSPC = the MS power class

Figure 7.3   Cell selection

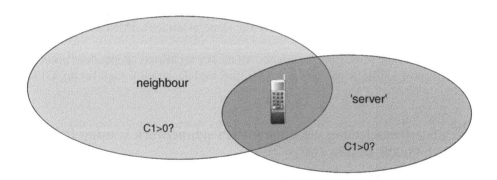

C1 = RLA – RXLEV_ACCESS_MIN – max(MS_TXPWR_MAX_CCH – MSPC, 0)
RLA = Receive level average dBm
RXLEV_ACCESS_MIN = the lowest receive level an MS must be receiving to select a cell dBm
MS_TXPWR_MAX_CCH = RACH transmit level dBm
MSPC = mobile station power class dBm
max(A, 0); if A >0 then max(A, 0) = A, otherwise 0.

C1 neighbour >C1 'server', then select neighbour, otherwise stay on 'server'

'Camp' on selected cell
Select CCCH
Select PCH
Register with the network on selected cell:
IMSI attach
LA update
Peiodiodic LA update
Go into cell reselect mode.

Figure 7.4   Cell selection

The initial cell selection applies to both the GSM and GPRS modes of operation and this could mean that the mobile station selects a cell that does not provide GPRS services. In that case the mobile station will not be able to GPRS attach and will not receive GPRS paging messages as the SGSN will be unaware of its existence.

If the selected cell provides only GSM services, the mobile station IMSI attaches through:

1. **Location area update**. If it has been previously attached to the network the SIM will store the last LA identifier before detaching from the network. On selecting a cell after switching on again, if this stored LAI is different from the LAI of the selected cell, it will perform a location area update (IMSI attach).
2. **IMSI attach**. If, during the previous attach, the network indicated IMSI attach/detach was in operation, the mobile station will IMSI attach.
3. **Periodic location area update**. If, during the previous attach, the network indicated periodic LA updating was in operation and the mobile station's timer has expired, it will perform LA update (IMSI attach).

If the selected cell provides GSM and GPRS services, the mobile station IMSI attaches through:

1. **IMSI attach**. The mobile station will always IMSI (GSM) attach (or alternatively perform LA update, which is effectively the same as attach). IMSI attach may be combined with
2. **GPRS attach** at the same time (combined attach) or separately, dependent upon the mobile station GPRS class and whether the mobile station requires to GPRS attach.

7.4   GPRS cell reselection when the GPRS sub-network is using PBCCH (GSM 45.008 section 10)

Cell reselection is the process of a mobile station ensuring that it is always on the best cell. After the initial cell selection, it repeats the measurements of neighbour cells about every five seconds, and then calculates the best cell for reselection using cell reselection criteria. So, having initially 'camped' upon the best cell using the cell selection criteria, the mobile station then transfers to cell reselection mode. Whether it does this using the GPRS cell reselection criteria (C1, C31 and C32) or the CS cell reselection criteria (C1 and C2), depends upon whether the mobile station is in the GPRS attached state and the cell having a PBCCH.

Unlike circuit switched cell reselection which is only performed in the idle condition (that is when there is no dedicated channel allocated to the mobile station), GPRS cell reselection is performed in the active condition (when the mobile station has been given radio resources for a packet transfer) and in the packet idle condition (when the mobile station has not been given radio resources for a packet transfer). Figure 7.5 shows comparisons between CS and GPRS cell reselection.

The GPRS cell reselection conditions depend upon the status of the NCO – network control order (from the system information or packet system information). NCO has the states NCO0, NCO1, NCO2:

- **NCO0** – the mobile station controls cell reselection in active and packet idle conditions.
- **NCO1** – the mobile station controls cell reselection in active and packet idle conditions with the measurement results sent in reports to the GPRS sub-network when in the active condition.
- **NCO2** – the GPRS mobile station controls cell reselection in packet idle condition, but for the active, packet transfer condition the GPRS sub-network acts upon the mobile station's measurement reports to control which cell the mobile station will use. This is done by the GPRS sub-network sending *cell change order* (a form of GPRS handover). NCO2 is similar to the circuit switched method of cell reselection.

A mobile station can only be in packet idle or active mode if it is GPRS attached; the mobile station will then be in the GMM standby (packet idle) or GMM ready condition (in GMM ready condition, the mobile station will be in a TBF or have just finished a TBF and have reverted to packet idle condition. GMM ready condition is determined by a timer which starts when a TBF is started and expires some time after a TBF has been completed).

There are three GMM states for the mobile station:

1. **GMM idle** when the mobile station is not GPRS attached. (The MS is also in packet idle mode).
2. **GMM standby** when the mobile station is GPRS attached but has no radio resources for packet transfer and the ready timer has expired. (The MS is also in packet idle mode).
3. **GMM ready** when the ready timer is running. In this state the mobile has the radio resources for a packet transfer, or it may have just finished a packet transfer. (The MS may be in packet idle or packet transfer mode).

GPRS attach can be initiated by the session management layer or directly by the user (network) layer in the mobile station.

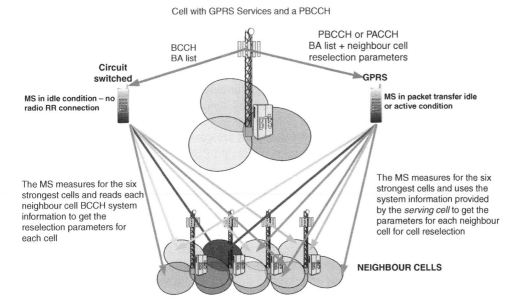

Figure 7.5   A comparison between circuit switched and GPRS cell reselection measurements

A mobile station may attach without the requirement for a subsequent *PDP context activation*. If this is the case, then the mobile station performs extra work (such as monitoring its packet-paging channel and periodically reinspecting the packet system information) all to no avail, as it will not receive paging messages unless a PDP context is established.

The justifications for GPRS attaching without an immediate PDP context are:

- It speeds up the establishment of a PDP context when this is required (a PDP context will be required when the mobile station user wishes to connect to the Internet).
- If the GPRS sub-network operator is providing SMS over GPRS, then the mobile station must be GPRS attached but a PDP context is not necessary for the transfer of SMS messages over the GPRS sub-network.

The GPRS sub-network has the capability to force mobile stations to the *GPRS detach* condition, and this may be the result of congestion control considerations, or a mobile station's volume of data transfer whilst attached. The volume of data transfer determines the revenue that the GPRS sub-network operator earns from the subscriber. A mobile station that is GPRS attached but does not bring in the money is, after a predetermined period, likely to be forcibly detached.

GPRS attach may be initiated by the mobile station SM layer when the application layer (more correctly the network layer) indicates to the SM layer a requirement to establish a PDP context. GPRS attach can also be initiated by the mobile station network layer directly ordering the GMM layer to attach.

If the cell has a PBCCH, when a mobile station GPRS attaches *then the GPRS cell reselection criteria become operational.*

7.4.1   Conditions for GPRS cell reselection

There are other conditions, in addition to the mobile station being GPRS attached, before GPRS cell reselection may be used. These are summarised in Figure 7.6.

If one of the conditions is absent, the mobile station will use the circuit switched cell reselection parameters, C1 and C2. (The C2 cell reselection criterion is illustrated in Figure 7.7).

For the case where a class A mobile station in circuit switched dedicated mode is used, the network will control which cell is in use by sending, if necessary, handover (cell change order) commands.

The details of circuit switched cell reselection are covered later as a GPRS attached mobile station will use this mode when a cell has no PBCCH.

7.4.2   GPRS cell reselection parameters (GSM 45.008 section 10)

Figure 7.8 illustrates a GPRS mobile station that has selected a cell and is in the process of GPRS cell reselection. The parameters it uses for reselection are shown in the diagram. Most of these parameters are obtained from the packet system information (PSI). Packet system information parameters are included in Appendix 2.

Packet system information message types 3 or 4 tell the mobile station if the cell offers GPRS services and packet system information type 13 tells the mobile station if the cell has a PBCCH and the location of the physical channel carrying the PBCCH.

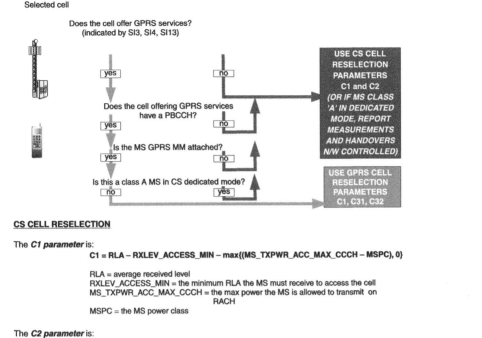

CS CELL RESELECTION

The **C1 parameter** is:

$$C1 = RLA - RXLEV\_ACCESS\_MIN - max\{(MS\_TXPWR\_ACC\_MAX\_CCCH - MSPC), 0\}$$

RLA = average received level
RXLEV_ACCESS_MIN = the minimum RLA the MS must receive to access the cell
MS_TXPWR_ACC_MAX_CCCH = the max power the MS is allowed to transmit on
RACH

MSPC = the MS power class

The **C2 parameter** is:

$$C2 = C1 + CELL\_RESELECT\_OFFSET - (TEMP\_OFFSET \textit{ for } PENALTY\_TIME)$$

CELL_RESELECT_OFFSET = a dB weighting applied to a cell which may be positive or negative.
TEMP_OFFSET = a positive dB weighting applied to a cell for the time PENALTY_TIME
PENALTY_TIME = a timer set in the MS by the neighbour cell; on its expiry, the TEMP_OFFSET is removed.
TEMP_OFFSET is only applied to neighbour cells.

Figure 7.6 GPRS or CS cell reselection? When an MS is in the IMSI attached condition, it performs CS cell reselection so that it is always on the best cell (from the network operator's point of view) for the provision of circuit switched services. To do this it measures the neighbour cells broadcast in the BA_BCCH list in SI messages. For each of the ARFCN/BSIC pair, it measures the average receive level RLA and decodes the neighbour cell BCCH for SI on cell reselect parameters. Using the measurements and the reselection parameters from the serving cell and neighbour cells, it calculates C1 and C2 and, other cell reselection parameters permitting, reselects the cell with the largest positive value of C2. *If the MS and cell are in a position to use the GPRS cell reselection parameters, then these are used instead of the GSM CS cell reselection parameters.* Whether the MS uses GPRS cell reselection criteria depends upon a number of conditions

The serving cell PBCCH provides a full set of reselection parameters for the serving cell and each neighbour cell, so to reselect a cell, the mobile station must simply measure the received level (and BSIC) from the server and each neighbour cell. The parameters for the server and each neighbour cell include:

- *Cell_Bar_Access.* This tells the mobile station whether or not a cell is barred. The cell may be barred to all mobile stations or to some mobile stations of certain classes. If a cell is barred, then the affected mobile station may not attempt access (*packet channel request*) to that cell, and the cell is not considered for reselection.

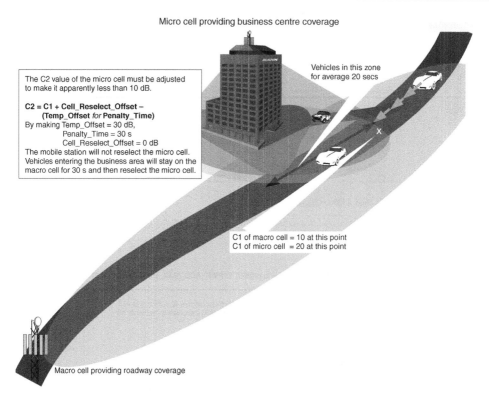

Micro cell providing business centre coverage

The C2 value of the micro cell must be adjusted to make it apparently less than 10 dB.

C2 = C1 + Cell_Reselect_Offset –
(Temp_Offset *for* Penalty_Time)
By making Temp_Offset = 30 dB,
        Penalty_Time = 30 s
        Cell_Reselect_Offset = 0 dB
The mobile station will not reselect the micro cell.
Vehicles entering the business area will stay on the
macro cell for 30 s and then reselect the micro cell.

Vehicles in this zone
for average 20 secs

C1 of macro cell = 10 at this point
C1 of micro cell = 20 at this point

Macro cell providing roadway coverage

Figure 7.7 The use of the C2 criterion to 'steer' mobile stations into reselecting the desired cell. This diagram illustrates a use of the GSM C2 cell reselection criterion (a similar set of parameters is applied to the GPRS C31 and C32 criteria). The purpose is to prevent fast-moving mobile stations from reselecting small radius cells, as if they so do, a call set-up may be lost as the mobile station rapidly leaves the small cell coverage area.

- *GPRS_RxLevel_Access_Min.* This is the minimum receive power in dBm that a mobile station must receive from a cell in order to reselect that cell.
- *GPRS_MS_TxPOWER_MAX_CCH.* The maximum power a mobile station may transmit when sending *packet channel request* messages on the PRACH to the GPRS sub-network.
- *GPRS_PRIORITY_CLASS.* A priority classification for a cell; a cell which has a higher priority will be preferred for reselection over one of a lower priority. The priority range is 0–7, with 7 as the highest priority. This parameter acts upon the list of best neighbour cells, eliminating all cells that have lower than the highest priority if the highest priority cell in the list passes the C1 and C31 tests.
- *GPRS_HCS_THR.* The hierarchical cell structure threshold. This is a receive level threshold for a cell. It may be used flexibly by a GPRS sub-network operator to favour or disfavour certain cells. The way the mobile station acts upon this parameter is illustrated in the sections that follow.
- *GPRS_CELL_RESELECT_HYST.* An offset which may be applied to neighbour cells when a mobile station is in packet transfer mode or GMM ready condition. This

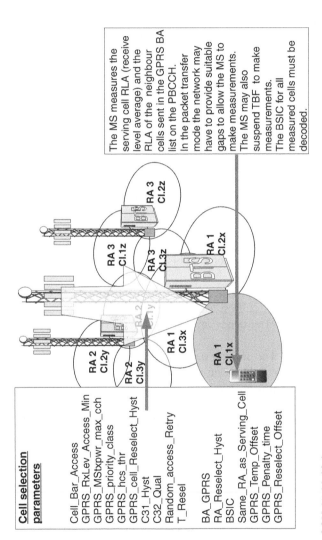

Cell selection parameters

Cell_Bar_Access
GPRS_RxLev_Access_Min
GPRS_MStxpwr_max_cch
GPRS_priority_class
GPRS_hcs_thr
GPRS_cell_Reselect_Hyst
C31_Hyst
C32_Qual
Random_access_Retry
T_Resel

BA_GPRS
RA_Reselect_Hyst
BSIC
Same_RA_as_Serving_Cell
GPRS_Temp_Offset
GPRS_Penalty_time
GPRS_Reselect_Offset

The MS measures the serving cell RLA (receive level average) and the RLA of the neighbour cells sent in the GPRS BA list on the PBCCH.
In the packet transfer mode the network may have to provide suitable gaps to allow the MS to make measurements.
The MS may also suspend TBF to make measurements.
The BSIC for all measured cells must be decoded.

GSM 05.08 10.1.
in packet idle mode
The MS monitors all BCCH carriers contained in the BA_GPRS list, taking one measurement for each BCCH carrier at least every four seconds. RLA is the received level averaged over 5 s or five consecutive paging blocks.
A list of the six strongest BCCH carriers is updated at least once per RLA average period.
The MS attempts to decode the full PBCCH data of the serving cell at least every 30 seconds. The MS attempts to decode the BSIC of the six strongest neighbour cells every 10 seconds. or 14 consecutive paging blocks.

in packet transfer mode
The MS monitors all BCCH carriers contained in the BA_GPRS list, and the BCCH carrier of the serving cell, measuring these consecutively in every TDMA frame.

Figure 7.8 GPRS cell reselection

is the margin in dB by which a neighbour cell must be better than the serving cell in order for the neighbour cell to be reselected. The aim of this parameter is to avoid cell reselection if the neighbour cell is only marginally better than the serving cell. Without this parameter, the mobile station might 'bounce' between cells that have only a small difference, being forced to execute *cell update* each time it bounces.

- *C31_HYST*. C31 is one of the GPRS reselection qualifiers. This indicator, if set, tells the mobile station to apply GPRS_RESELECT_HYSTERESIS, the hysteresis is the margin by which a neighbour must be better than the server.
- *C32_QUALIFIER*. This indicator tells the mobile station when to apply GPRS_ RESELECT_OFFSET, which is a parameter the GPRS sub-network operator uses to favour or disfavour cells which have passed the C1 and C31 tests and share the highest priority with another cell which has also passed the C1 and C31 tests.
- *RANDOM_ACCESS_RETRY*. This indicates whether a mobile station is allowed to try to access another cell should access to the best cell fail for any reason.
- *T_RESEL*. This timer applies if the mobile station has previously had an abnormal release from the cell. The mobile station is not allowed to reselect the cell within this period.
- *BA_GPRS*. The BCCH allocation, that is a list of the BCCH frequencies of neighbour cells.
- *RA_RESELECT_HYST*. The margin in dB by which a neighbour cell in a different routeing area to the serving cell must be better than the serving cell in order for the mobile station to reselect that neighbour cell. For example, if such a neighbour is calculated to be 5 dB better than the serving cell, but the RA_RESELECT_HYST is 10 dB, then the neighbour cell will not be selected. This applies in GMM ready and GMM standby conditions.
- *BSIC*. The base station identity code, broadcast on the synchronisation channel (SCH) of the BCCH carrier. The mobile station uses this to identify that the frequency it is measuring is in fact the required frequency and not an interfering carrier of the same frequency.
- *SAME_RA_AS_SERVING_CELL*. The serving cell tells the mobile station whether the neighbour cell to be measured is in the same routeing area as the serving cell or not. Hence, the mobile station will know whether RA_RESELECT_HYST should be applied.
- *GPRS_TEMP_OFFSET*. This is a margin, in dB, which is applied to a neighbour cell for a time period (GPRS_PENALTY_TIME), making the neighbour cell less attractive than the serving cell for that time period.
- *GPRS_PENALTY_TIME*. The period for which GPRS_TEMP_OFFSET is applied to a neighbour cell.
- *GPRS_RESELECT_OFFSET*. An offset applied by the GPRS sub-network operator to cells to make them less or more attractive for reselection.
- *RLA (receive level average)*. The mobile station measured and averaged level of reception for the serving and each of the neighbour cells.

7.4.3 GPRS cell reselection criteria (GSM 45.008 section 10)

There are three criteria for GPRS cell reselection, C1, C31 and C32.

7.4.3.1   C1 criterion

The C1 criterion is similar to the circuit switched C1 criterion.

$$C1 = RLA - GPRS\_RXLEV\_ACCESS\_MIN - \max$$
$$(GPRS\_MS\_TXPWR\_MAX\_CCH - MSPC, 0)$$

where

RLA = receive level average (dBm).

GPRS_RXLEV_ACCESS_MIN = the lowest receive level a mobile
station must be receiving to select a cell (dBm).

GPRS_MS_TXPWR_MAX_CCH = PRACH transmit level (dBm).

MSPC = mobile station power class (dBm).

Note: a cell may only be selected if C1 > 0

The mobile station calculates the C1 value for the serving cell and each of the neighbour cells. This is illustrated in Figures 7.9 and 7.10, where some calculations are shown for illustration.

7.4.3.2   C31 criterion

This criterion uses the hierarchical cell structure (HCS) parameters for the serving (s) and neighbour (n) cells.

$$C31(s) = RLA(s) - HCS\_THR(s)$$
$$C31(n) = RLA(n) - HCS\_THR(n) - \{GPRS\_TEMP\_OFFSET(n)$$
$$\text{for } GPRS\_PENALTY\_TIME(n)\} \times L(n)$$

where

RLA = the receive level average for the cell.

HCS_THR = the receive level for a cell of a given priority.

GPRS_TEMP_OFFSET = an offset in dB for GPRS_Penalty_TIME.

GPRS_Penalty_TIME = the time for which GPRS_TEMP_OFFSET applies.
This timer is started when the mobile station measures a neighbour cell
which uses this parameter.

$L = 0$ if PRIORITY_CLASS(n) = PRIORITY_CLASS(s).

$L = 1$ if PRIORITY_CLASS(n) is not equal to PRIORITY_CLASS(s).

(n) = neighbour cell.

(s) = serving cell.

Note: a cell may only be selected if C31 > 0

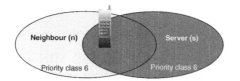

1. Apply C1

C1(s) = RLA – G_RXLEV_ACCESS_MIN – max(G_MS_TXPWR_MAX – MSPC, 0)
    (–85)   (–110)   (40)   (33) = 18

C1(n) = RLA – G_RXLEV_ACCESS_MIN – max(G_MS_TXPWR_MAX – MSPC, 0)
    (–70)   (–110)   (30)   (33) =40

IS C ≥ 0 for both cells?

2. Apply C31

           (range –110..–48 dB)

C31(s) = RLA – HCS_THR
      (–85)   (–90)   = 5

                  (range 0..60 dB, inf)   (range 10–320 s)   (L = 0 if priority class(n)=priority class(s), else =1)

C31(n) = RLA – HCS_THR – **(G_TEMP_OFFSET** for **PENALTY-TIME)**L(n)
      (–70)   (–80)   [(40)   20s]   0 =10

If C31 ≥0 for cells, apply C32 to highest priority cells (priority = 0 (lowest) to 7)

3. Apply C32

C32(s) = C1(s)
  (18)   CONDITIONAL ON QUALIFIER

         (range –52..+48 dB: only applied if C32_QUAL=1, then only to (n) with highest RLA)   (range –52..+48 dB, used if RA(n)≠ RA(s))

C32(n) = C1(n) + **[G_RESELECT_OFFSET]** – (G_TEMP_OFFSET for PENALTY-TIME){1 – L)(n)} - [RA_RESELECT_HYST]
    (40)   (+10)   [(40)   (20s)]   (1)   (10)   = 0 for 20s then 40

Figure 7.9 An illustration of GPRS cell reselection by a mobile station in MM standby condition. This diagram demonstrates the calculation of the C31 and C32 criteria for reselection between two cells of the same priority class. Illustrative values of the parameters and receive levels are shown

The C31 criterion is similar in form to the GSM circuit switched C2 criterion. Figure 7.7 illustrates the action of the GSM C2 criterion.

The mobile station calculates the C31 values for the serving cell and for neighbour cells with 0 < C1.

Some examples of the application of C31 are shown in Figures 7.9 and 7.10.

7.4.3.3   C32 criterion

The C32 criterion discriminates between cells of the same highest priority class which have passed the C31 criterion.

$$C32(s) = C1(s)$$
$$C32(n) = C1(n) + GPRS\_RESELECT\_OFFSET(n) - \{GPRS\_TEMP\_OFFSET(n)$$
$$\text{for } GPRS\_PENALTY\_TIME(n)\}(1 - L\ (n))$$

where GPRS_RESELECT_OFFSET = dB, positive or negative.

Figures 7.9 and 7.10 illustrate the application of C1, C31 and C32, with calculations, applied to one serving cell and one neighbour cell. C31 and C32 qualifiers are also illustrated in action in these figures.

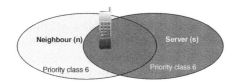

1. Apply C1

C1(s) = RLA – G_RXLEV_ACCESS_MIN – max(G_MS_TXPWR_MAX – MSPC, 0)
     (–85)   (–110)   (40)   (33)   = 18

C1(n) = RLA – G_RXLEV_ACCESS_MIN – max(G_MS_TXPWR_MAX – MSPC, 0)
     (–70)   (–110)   (30)   (33)   = 40

IS C1 ≥ 0 for both cells?

2. Apply C31

(range –110..–48dB)

C31(s) = RLA – HCS_THR   QUALIFIER
    (–85)   (–90) = 5   (L = 0if priority class(n)=
                        priority class(s), else =1)   **0–14 dB if C31_HST is set and RA(n)=RA(s)**

                (range 0.60 dB, inf)   (range 10–320 s)

C31(n) = RLA – HCS_THR – **(G_TEMP_OFFSET** for **PENALTY-TIME)**L(n)   -   **G_RESELECT_HYST**
    (–70)   (–80)   [(40)   20s]   0   -   10   = 0

If C31 ≥ 0 for cells, apply C32 to highest priority cells (priority = 0 (lowest) to 7)

3. Apply C32

0-14 dB if C31_HST is set and RA(n)=RA(s)

C32(s) = C1(s)   **G_RESELECT_HYST**
  (18)   CONDITIONAL ON QUALIFIER   **OR**
    (range –52..+48 dB: only applied if C32_QUAL=1, then only to (n) with highest RLA)   (range –52..+48 dB, used if RA(n)≠ RA(s))

C32(n) = C1(n) + **[G_RESELECT_OFFSET]** – (G_TEMP_OFFSET for PENALTY-TIME){1 – L(n)} - **[RA_RESELECT_HYST]**
  (40)   (+10)   [(40)   (20s)]   (1)   (10)   = 0 for 20s then 40

Figure 7.10 An illustration of GPRS cell reselection by a mobile station in MM Ready condition. This diagram demonstrates the calculation of the C31 & C32 criteria for reselection between two cells of the same priority class. Illustrative values of the parameters and receive levels are shown.

The effect of C1, C31 and C32 is summarised below.

7.4.3.4 Application of C1, C31 and C32

1. Is C1 ≥ 0 for the neighbour cells and the serving cell?
2. Eliminate from the reselection process all those cells with C1 < 0.
3. Calculate C31 for those cells with C1 ≥ 0. Apply qualifiers if appropriate. Is C31 ≥ 0?
4. Eliminate all cells with C31 < 0.
5. If all the cells have C31 < 0, then take all cells to step 10.
6. Rank cells by priority (0–7, 7 = highest priority) with C31 ≥ 0.
7. Eliminate all those cells except for the cells of the highest priority in the ranked list.
8. If there is only one cell with the highest priority, then that cell is selected.
9. If there is more than one cell of the highest priority, calculate C32 for these cells, applying qualifiers if appropriate.
10. Select the cell with the highest C32 value.
11. If two or more cells have the same C32 value and one of these cells is the serving cell then stay in the serving cell.
12. If two or more cells have the same C32 value and none of these cells is the serving cell then select any cell.

How does all this work out in practice? Figure 7.11(a–d) shows a hypothetical network of macro cells, micro cells and pico cells. Now *micro cells* and *pico cells* will have the TEMP_OFFSET for PENALTY_TIME parameters in operation for cell reselection. They will also tend to be in the higher cell priority classes compared to the macro cells. In general, the *macro* cells will not have TEMP_OFFSET for PENALTY_TIME applied and will tend to be in the lower cell priority classes.

If the serving cell is a *micro* cell, assuming that *all* the cells pass the C1 > 0 criterion, then they *all* go forward to the C31 criterion. After this has been applied we may be left with the lower priority cells passing the C31 test and micro cells of the same cell priority as the serving cell passing the C31 test (in addition to the serving cell). But micro cells and pico cells of a different cell priority to the serving cell will most probably have failed the C31 test. This arises because the TEMP_OFFSET for PENALTY_TIME is not applied to micro or pico cells of the *same priority* as the serving cell, therefore they will probably pass the C31 test, but TEMP_OFFSET for PENALTY_TIME is applied to cells of *different priority* classes, and this will probably drag down their C31 values to less than zero.

Going forward to the C32 test will be cells of the same or lower priority compared to the serving cell. Cells of lower priority are automatically disqualified at the C32 stage, leaving just the serving cell and neighbours of the same priority (and possibly higher priority).

The TEMP_OFFSET for PENALTY_TIME is now applied to the neighbours.

The mobile station will stay on the serving cell until one of the PENALTY_TIME parameters in a neighbour cell expires. If nothing else has changed after this period of time, then the mobile station will select this neighbour if its C32 value is greater than the serving cell's (plus any cell or routeing area select hysteresis).

Figure 7.11(a–d) shows the situation for the mobile station having sequentially various cells as its serving cell. This works out quite well, but there may be a situation where none of the cells (including the serving cell) passes the C31 test. In that case, *all* cells go forward to the C32 test. At this stage, cells of higher priority are apparently favoured! This arises because, during the C31 test, cells of higher priority than the serving cell have had TEMP_OFFSET for PENALTY_TIME applied, whereas cells of lower or the same priority have not. When we then move forward to C32, cells of higher priority do not have TEMP_OFFSET for PENALTY_TIME applied, but cells of the same priority as the serving cell do have it applied. Cells of lower priority are probably disqualified; therefore the mobile station will select a cell of higher priority.

This is not a favourable situation as the mobile station may be fast moving, and select a low-radius, higher priority cell, and rapidly move through it, perhaps losing a TBF set-up in the new cell. UMTS avoids this pitfall by making the mobile station decide whether it is a fast moving mobile station. If the mobile station has executed a certain number of cell reselections within a certain period of time, then it regards itself as a fast moving mobile station and will attempt to reselect only neighbour cells of lower or the same priority.

7.4.4 *GPRS cell reselection when the GPRS sub-network is not using PBCCH (GSM circuit switched cell reselection) (GSM 45.002 section 6)*

If the network offers GPRS without using PBCCH, then the GPRS mobile station must use the circuit switched C2 reselection criterion. This is given by:

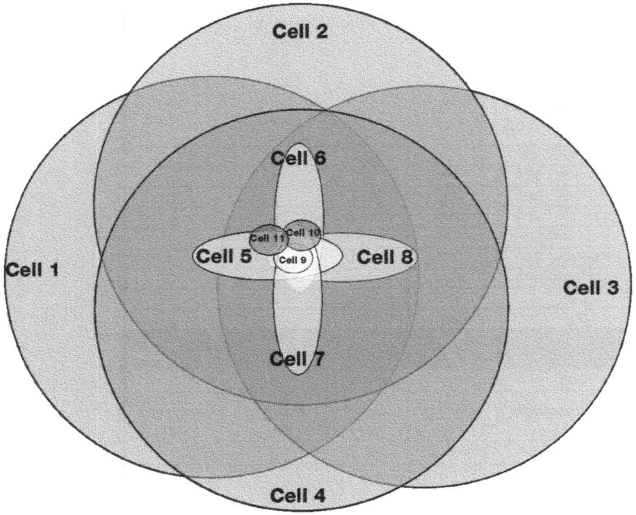

Figures 7.9 and 7.10 show the arithmetic of cell reselection, but not the concept. This part attempts to show the concept.
To the left are shown a number of cells, cell 1 to cell 11.

Cells 1–4 are macro cells: these *will not* normally have TEMP_OFFSET applied. They will normally have the **LOWER PRIORITY CLASS.**

Cells 5–8 are micro cells: these normally *will* have TEMP_OFFSET applied. They will normally be in the **HIGHER PRIORITY CLASS.**

Cells 9–11 are pico cells: these normally *will* have TEMP_OFFSET applied. They will normally be in the **HIGHEST PRIORITY CLASS.**

We shall examine the cell reselection process with the mobile station using various cells as the **serving cell.**

Cell number	Cell priority	Cell type	C1 > 0?	Temp offset applied in C31?	C31 > 0	To C32?	Temp offset applied in C32?	Which cell is selected?
1	1	macro	yes	no	yes			
2	1	macro	yes	no	yes			
3	2	macro	yes	no	yes			
4	3	macro	yes	no	yes			
5	6	micro	yes	no (see note 1)	yes	yes	yes	
6	6	micro	yes	no (see note 1)	yes	yes	yes	
7	6	micro	yes	no (see note 1)	yes	yes	yes	
8	6	micro	yes	n/a	yes	yes	n/a	Serving cell
9	7	pico	yes	yes (see Note 2)	no			
10	7	pico	yes	yes (see Note 2)	no			
11	7	pico	yes	yes (see Note 2)	no			

Note 1
As cells 5–7 are the same priority class as the serving cell TEMP_OFFSET is not applied

Note 2
As cells 9–11 are the same priority class as the serving cell TEMP_OFFSET is applied

The MS remains on server until the TEMP_OFFSET penalty timer expires in a better neighbour cell

Figure 7.11(a) An example of the application of C31 and C32 with a micro cell as the serving cell

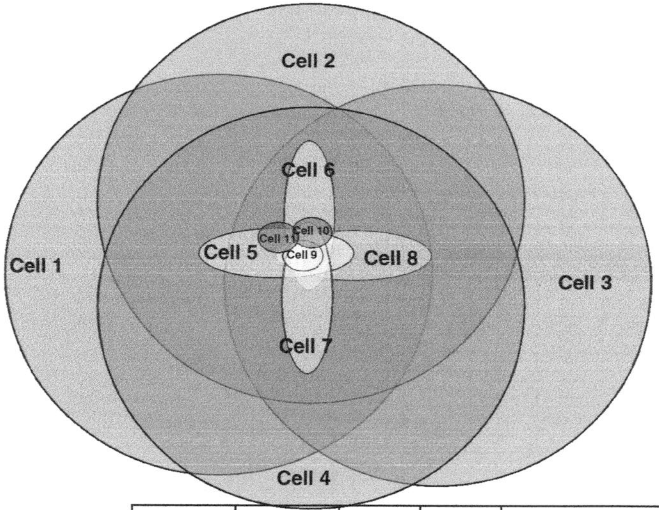

Figures 7.9 anc 7.10 show the arithmetic of cell reselection, but not the concept. This part attempts to show the concept.
To the left are shown a number of cells, cell1 to cell 11.

Cells 1–4 are macro cells: these *will not* normally have **TEMP_OFFSET** applied. They will normally have the **LOWER PRIORITY CLASS.**

Cells 5–8 are micro cells: these normally *will* have **TEMP_OFFSET** applied. They will normally be in the **HIGHER PRIORITY CLASS.**

Cells 9–11 are pico cells: these normally *will* have **TEMP_OFFSET** applied. They will normally be in the **HIGHEST PRIORITY CLASS.**

Cell number	Cell priority	Cell type	C1>0?	Temp offset applied in C31?	C31>0	To C32?	Temp offset applied in C32?	Which cell is selected?
1	1	macro	yes	no	yes	no		
2	**1**	**macro**	**yes**	**no**	**yes**	**no**		*(Serving cell)*
3	2	macro	yes	no	yes	yes		
4	3	macro	yes	no	yes	yes		
5	6	micro	yes	yes (see note 2)	no	no		
6	6	micro	yes	yes (see note 2)	no	no		
7	6	micro	yes	yes (see note 2)	no	no		
8	6	micro	yes	yes (see note 2)	no	no		
9	7	pico	yes	yes (see note 2)	no	no		
10	7	pico	yes	yes (see note 2)	no	no		
11	7	pico	yes	yes (see note 2)	no	no		

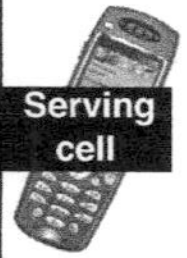

Note 2
as cells 5–11 are not the same priority class as the serving cell TEMP_OFFSET is applied

The MS selects cell 4

Figure 7.11(b)   An example of the application of C31 and C32 with a macro cell as the serving cell

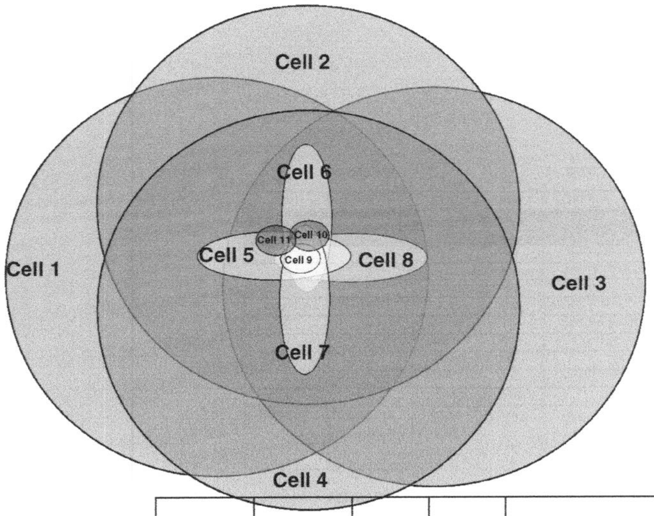

Figures 7.9 and 7.10 show the arithmetic of cell reselection, but not the concept. This part attempts to show the concept. To the left are shown a number of cells, cell 1 to cell 11.

Cells 1–4 are macro cells: these *will not* normally have **TEMP_OFFSET** applied. They will normally have the **LOWER PRIORITY CLASS**.

Cells 5–8 are micro cells: these normally *will* have **TEMP_OFFSET** applied. They will normally be in the **HIGHER PRIORITY CLASS**.

Cells 9–11 are pico cells: these normally *will* have **TEMP_OFFSET** applied. They will normally be in the **HIGHEST PRIORITY CLASS**.

Cell number	Cell priority	Cell type	C1>0?	Temp offset applied in C31?	C31>0	To C32?	Temp offset applied in C32?	Which cell is selected?	Serving cell
1	1	macro	yes	no	yes	no	no		
2	**1**	**macro**	**yes**	**no**	**yes**	**no**	**no**		Serving cell
3	2	macro	yes	no	yes	no	no		
4	3	macro	yes	no	yes	no	no		
5	6	micro	yes	yes (see note 2)	no	no	no		
6	6	micro	yes	yes (see note 2)	no	no	no		
7	6	micro	yes	yes (see note 2)	no	no	no		
8	6	micro	yes	yes (see note 2)	no	no	no		
9	7	pico	yes	yes (see note 2)	no	no	no		
10	7	pico	yes	yes (see note 2)	no	no	no		
11	7	pico	yes	yes (see note 2)	no	no	no		

Note 2
as cells 5–11 are the same priority class as the serving cell TEMP_OFFSET is applied

The MS selects cells 4

Figure 7.11(c) An example of the application of C31 and C32 with a macro cell as the serving cell

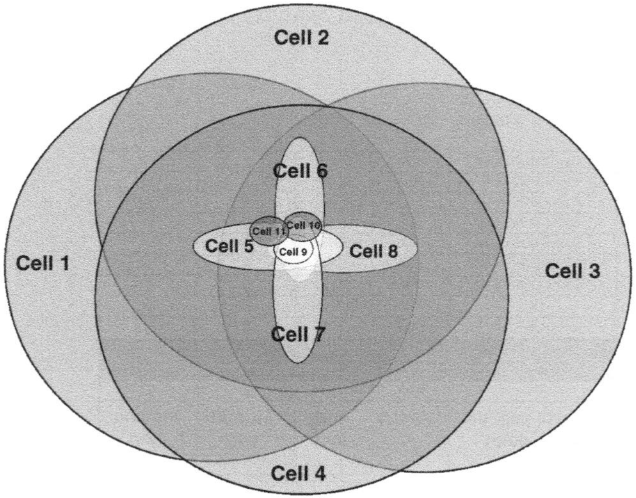

Figures 7.9 and 7.10 show the arithmetic of cell reselection, but not the concept. This part attempts to show the concept.
To the left are shown a number of cells, cell 1 to cell 11.

Cells 1-4 are macro cells: these *will not* normally have **TEMP_OFFSET** applied. They will normally have the **LOWER PRIORITY CLASS**.

Cells 5-8 are macro cells: these normally *will* have **TEMP_OFFSET** applied. They will normally be in the **HIGHER PRIORITY CLASS**.

Cells 9-11 are pico cells: these normally *will* have **TEMP_OFFSET** applied. They will normally be in the **HIGHEST PRIORITY CLASS**.

Cell number	Cell priority	Cell type	C1>0?	Temp offset applied in C31?	C31>0	To C32?	Temp offset applied in C32?	Which cell is selected?
1	1	macro	yes	no	yes	no		
2	1	macro	yes	no	yes	no		
3	2	macro	yes	no	yes	no		
4	3	macro	yes	no	no	no		
5	6	micro	yes	yes (see note 2)	no	no		
6	6	micro	yes	yes (see note 2)	no	no		
7	6	micro	yes	yes (see note 2)	no	no		
8	6	micro	yes	yes (see note 2)	no	no		
9	7	pico	yes	no (see note 1)	yes	yes	yes	
10	7	pico	no	n/a	no	no	n/a	Serving cell
11	7	pico	yes	no (see note 1)	yes	yes	yes	

Note 1
as cells 9 and 11 are the same priority class as the serving cell TEMP_OFFSET is not applied
Note 2
as cells 5–8 are not the same priority class as the serving cell TEMP_OFFSET is applied

The MS selects either cell 9 or 11 (whichever is best) after expiry of their PENALTY_TIMERS. Note that **no cell is selected** until cell 9 or 11 becomes valid.

Figure 7.11(d) An example of the application of C31 and C32 with a pico cell as the serving cell

$$C2 = C1 + CELL\_RESELECT\_OFFSET - (TEMP - OFFSET \text{ for } PENALTY\_TIME)$$

where

C1, see Section 7.4.3.1.

CELL_RESELECT_OFFSET is a positive or negative factor in
dB broadcast by the neighbour cell.

TEMP_OFFSET is a positive dB factor broadcast by the neighbour cell.

PENALTY_TIME is a factor, units seconds, broadcast by the neighbour cell.

Figure 7.7 illustrates an application of the C2 criterion.

There is potentially a problem for mobile stations in packet transfer using the C2 criterion for cell reselection. As cell reselection is not determined by service considerations, then the physical layer of the mobile station may decide that the best neighbour cell is one that does not offer GPRS service. When the GMM layer is prompted to perform a cell update, it reads the BCCH information of the new cell, discovers that the cell does not offer GPRS services and so cannot perform a cell update. The physical layer insists that this neighbour is the one to be reselected and the GMM layer cannot act on this.

Meanwhile, the TBF which was in progress on the old serving cell is lost, and the LLC PDU that was being transferred cannot now be transferred!

The mobile station must inform its user that the GPRS service cannot now be provided, and the CS cell reselection takes over.

Similarly, in the GMM standby condition, a cell may be reselected which does not offer GPRS services and the customer will not be able to use packet services, even though cells that offer GPRS services and have a C1 value greater than zero may surround the cell!

7.5 Discontinuous reception (DRX) and paging in a cell with PBCCH

This section examines the procedures for packet paging – the procedures initiated by the GPRS sub-network that are used in some circumstances to establish a downlink TBF. The steps followed in this section are:

1. The mobile station, having selected a GPRS cell which has a packet broadcast control channel (PBCCH) and is attached to the GPRS sub-network, selects a *PCCCH group*. A PCCCH group is only of interest if more than one physical channel is carrying PCCCHs. In that case the mobile station must decide which physical channel to use – there are two (or more) groups of PCCCHs, group 0 on the lowest timeslot number physical channel, and group 1 on the next highest TN physical channel (and so on).
2. How the mobile station selects a *paging group* from the PCCCH group. A paging group is a paging channel; it is called a group because a group of mobile station IMSIs will use the same paging channel.
3. How the mobile station selects paging groups with *split paging* considerations. Split paging is the process whereby a mobile station decides to use more than one paging channel.
4. The paging procedures.

7.5.1   Determining the PCCCH Group (deciding which physical channel to use)

If a mobile station reads from packet system information that there is more than one GPRS physical channel PDCH carrying PCCCHs, then it must decide which of these physical channels to use for listening for paging messages on the downlink and sending access bursts (*packet channel request*) on the uplink. The GPRS sub-network takes the same decision to send a paging message on a particular physical TN carrying PCCCHs. This is done mutually by using the mobile station's IMSI.

Figure 7.12 illustrates the principle of PCCCH TN selection; in this example, the PBCCH is on TN0 of a radio carrier (it can be on any TN of any carrier within a cell, except of course TN0 of the broadcast control channel (BCCH) radio carrier).

There are always PCCCHs on the same TN as the PBCCH (TN0 in the illustration), and the GPRS sub-network operator in this case has also decided to place more PCCCHs on TN1. There are two PCCCH *groups*, the group on TN0 and the group on TN1. The PSI on the PBCCH will tell the mobile station the location of the PCCCH groups. The mobile station must decide which of these two groups to use.

Figure 7.12 shows how this is done, in this case for a mobile station with an IMSI of 67891234. Having selected a PCCCH group, the mobile station expects to receive downlink paging messages on a paging channel within this group (the GPRS sub-network does exactly the same calculation in deciding upon which paging channel the paging message must be placed) and when it does, responds by sending an uplink PRACH burst containing *packet channel request* on the TN of the selected PCCCH group.

When the GPRS sub-network RR layer receives a request to page a mobile station, the request includes the mobile station's IMSI; from this the RR layer must calculate the PCCCH group for each cell broadcasting the paging message.

In circuit switched operations, a network operator using a BCCH carrier in the non-combined mode has the option of putting CCCHs on timeslots two, four and six in addition to timeslot zero of the BCCH carrier. Additional timeslots are rarely used in practice for common control channels and it is unlikely that PCCCHs will be found on timeslots other than that carrying the PBCCH.

7.5.2   Determining the paging group (GSM 45.002 section 6)

After determining the PCCCH group, the mobile station (and the GPRS sub-network when a paging message is waiting for transmission) must calculate the *paging group* to which the subscriber's IMSI belongs. A paging group is a paging channel, which a group of IMSIs will use. If a cell has, say, twelve paging channels, then all the GSM IMSIs in the world will (theoretically) belong to one of these twelve paging channels; all the world's IMSIs are divided into twelve groups, the paging groups.

Figure 7.13 illustrates the principle of paging group selection for a cell that has a PBCCH, and this section illustrates how the paging groups are calculated.

One of the major differences between GPRS paging groups and circuit switched paging groups is that the paging group for GPRS includes the parameter *SPLIT_ PAG_CYCLE* that determines the number of paging channels a mobile station uses. In the case of Figure 7.13 with a SPLIT_PAG_CYCLE parameter value of 2, the mobile

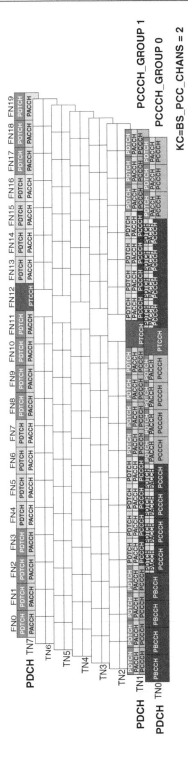

KC=BS_PCC_CHANS = 2

PCCCH_GROUP 0

PCCCH_GROUP 1

The MS knows from PSI1 or PSI2 the number of TN (physical channels – PDCHs) carrying PCCCHs. The number of physical channels carrying PCCCHs is the parameter KC. Using this and IMSI, the MS calculates which TN carrying PCCCH it should use. This is the PCCCH that will carry paging messages for that IMSI, and is also the PCCCH that the MS will use for *packet channel requests*.
The basis of the calculation is shown below:

PCCCH_GROUP = [(IMSI mod 1000)mod(KC x N)]divN

KC = number of physical slots carrying PCCCH
N = 1 for PCCCH,   (9 – BS_AG_BLKS_RES)BS_PA_MFRMS for CCCH not combined
                  (3 – BS_AG_BLKS_RES)BS_PA_MFRMS for CCCH combined with SDCCH

Applying this to an MS with IMSI 67891234, for the configuration shown in the diagram where KC = 2. IMSI mod 1000 is simply the three least significant digits of the IMSI = 234; [234mod (KC x N)]divN = [234mod2]div1 = [0]div1=0 (The div function takes the integer result of division, ignoring remainder or decimal fractions.)

PCCCH_GROUP = (234mod2)div1
PCCCH_GROUP = 0 (The mod function performs integer division, taking the remainder as the result)

In this case the MS will select PCCCH_GROUP 0, which is always on the lowest numbered TN carrying PCCCH

Figure 7.12 PCCCH group selection

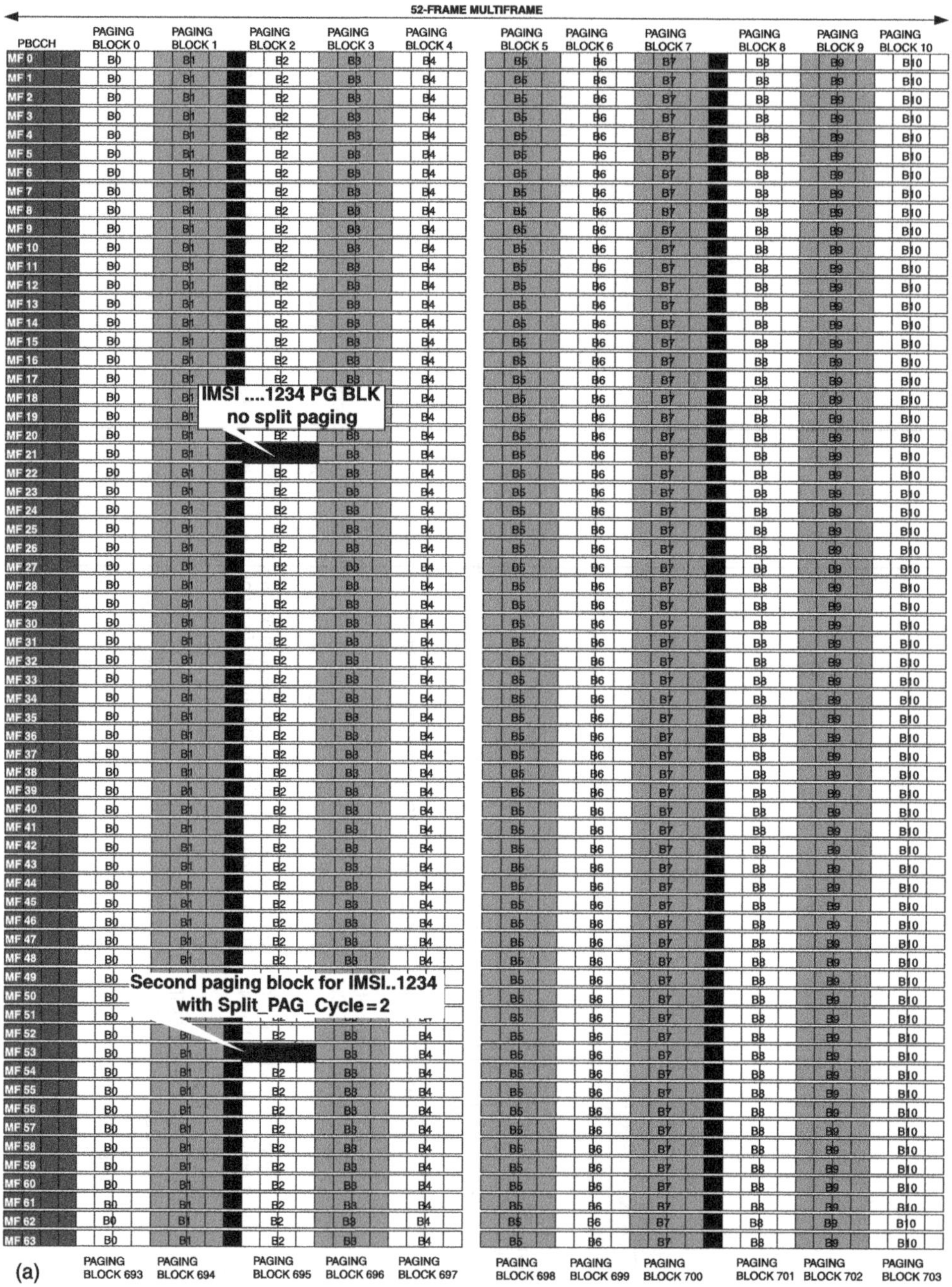

52-FRAME MULTIFRAME

IMSI ....1234 PG BLK no split paging

Second paging block for IMSI..1234 with Split_PAG_Cycle = 2

(a) PAGING BLOCK 693   PAGING BLOCK 694   PAGING BLOCK 695   PAGING BLOCK 696   PAGING BLOCK 697   PAGING BLOCK 698   PAGING BLOCK 699   PAGING BLOCK 700   PAGING BLOCK 701   PAGING BLOCK 702   PAGING BLOCK 703

Figure 7.13 PCCH paging block structure (a) This diagram shows sixty-four 52-frame multiframes, each multiframe carrying one PBCCH block and 11 PCCCH blocks, that is BS_PAG_BLKS_RES = 0. Two paging channels are selected for this mobile station whose IMSI ends in 1234 and which is using a split paging cycle value of 2. The calculations are shown in Figure 7.14

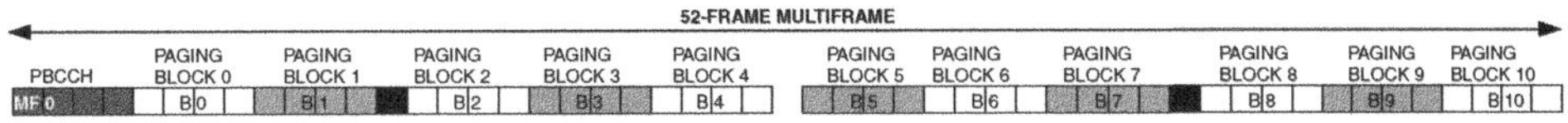

(b)

Figure 7.13 (b) This diagram shows a 52-frame multiframe with, in the first position, a PBCCH block. Eleven PCCCH blocks follow this, and because none of these is reserved for other uses, all eleven are available as GPRS paging channels, numbered from PCH0 to PCH10 as shown. The cycle before these eleven paging channels are repeated is 64 52-frame multiframes. As each of these sixty-four multiframes have eleven paging channels, then the total number of paging channels is 704, as illustrated in (a) which shows the complete sixty-four multiframes. The mobile station must calculate, based upon its IMSI and the split paging cycle code, which of these paging channels to listen to for paging messages. The GPRS sub-network must perform the same calculations to put the paging messages on the channels that the mobile station will be listening to

station listens to two paging channels; the maximum possible number of paging channels a GPRS mobile station may have to monitor is 352.

Figure 7.13 shows the basic PCCCH paging block structure. This is based upon sixty-four 52-frame multiframes. In the example shown, the first block of each 52-frame multiframe is carrying a packet broadcast control channel (PBCCH). As there are 12 RLC/MAC blocks in a 52-frame multiframe, the maximum number of paging blocks in this multiframe is eleven. This is the case in Figure 7.13 as none of the blocks are reserved for purposes other than paging (i.e. packet access grant (PAGCH) blocks). In this example, the PSI parameter BS_PAG_BLKS_RES = 0, which somewhat oddly means that no base station paging blocks are reserved for anything other than paging!

BS_PAG_BLKS_RES gives the number of blocks in a 52-frame multiframe (on a TN carrying PCCCHs) which are reserved for uses *other* than paging channels, that is they are reserved for exclusive use as PAGCHs – packet access grant channels. If none are reserved as PAGCHs then all are available for use as paging channels.

As the total number of paging blocks is based upon a cycle of 64 52-frame multiframes, then in this case there is a total of $11 \times 64 = 704$ paging blocks (or channels).

A mobile station attaching to the GPRS sub-network negotiates in the *attach request* message a parameter called the DRX parameter, which includes *SPLIT_PAG_CYCLE* and *NON_DRX_TIMER*. The ranges of values of these parameters are:

SPLIT_PAG_CYCLE values: 0 = No DRX, 1–64, 71, 72, 74, 75, 77, 79, 80, 83,

86, 88, 90, 92, 96, 101, 103, 107, ..., 352

(see GSM 04.08 10.5.5.6)

NON_DRX_TIMER values: No DRX after transfer state, 1 s, 2 s, 4 s, ···, 64 s

7.5.3 Selection of paging group from PCCCHs

A paging group is a group of IMSIs which will all listen to the same paging block (or channel). The paging channel for a group of IMSIs is given by:

$$
\begin{aligned}
\text{PAGING\_GROUP} = [[(\textit{IMSI mod}1000) \textit{ div } (KC \times N)] \times N \\
+ (\text{IMSI mod}1000) \, \text{mod}N + \max\{(m \times M) \\
\text{div}\text{SPLIT\_PAG\_CYCLE}, \text{m}\}] \, \text{mod}M
\end{aligned}
$$

where

$m = 0, \ldots,$ min (M, SPLIT_PAG_CYCLE) $- 1$

$KC =$ number of physical channels carrying PCCCH

$M =$ number of paging blocks available on the one PCCCH

$\quad = (12 - \text{BS\_PAG\_BLKS\_RES} - \text{BS\_PBCCH\_BLKS})$ 64 for PCCCH

$\quad = (9 - \text{BS\_AG\_BLKS\_RES})$ 64 for non-combined CCCH

$\quad = (3 - \text{BS\_AG\_BLKS\_RES})$ 64 or combined CCCH

$N = 1$ for PCCCH

$N = (9 - \text{BS\_AG\_BLKS\_RES})$ BS_PG_MFRMS for non-combined CCCH

$N = (3 - \text{BS\_AG\_BLKS\_RES})$ BS_PG_MFRMS for combined CCCH

Applying these rules in the case for:

> IMSI $= 1234,$
>
> split paging cycle $= 2,$
>
> one PBCCH per 52-frame multiframe,
>
> all the remaining 11 blocks in a 52-frame multiframe being used
>
> for paging channels, giving $M = 704,$

The limiting case is for $m =$ min (704, 2) $- 1 = 1$, therefore $m = 1$ is the limit.

The working illustrating that there are two paging channels in this case, paging channels 234 and 586, is shown in Figure 7.14 and these two paging channels are shown in Figure 7.13.

7.5.4 Selection of paging group from CCCHs when there is no PBCCH in a cell

Figure 7.14 also shows the equations used by a GPRS mobile station when there is no PBCCH in a cell that offers GPRS services. The multiframe repetition cycle for the GPRS paging channels is still sixty-four multiframes, but in this case 51-frame multiframes, and not 52-frame multiframes as in the previous example.

The paging cycle for the GSM circuit switched paging channel is variable and is determined by the GPRS sub-network operator with the broadcast parameter BS_PG_MFRS, which may be interpreted as base station paging multiframe repetitions.

A paging group is a group of IMSIs which will all listen to the same paging block. The paging block is given by:

$$\text{PAGING\_GROUP} = [[(\text{IMSI mod } 1000)\text{div}(KC \times N)] \times N + (\text{IMSI mod } 1000)\text{mod}N + \max\{(m \times M)\text{divSPLIT\_PAG\_CYCLE}, m\}]\text{mod}M$$

for $m = 0,....\min(M, \text{SPLIT\_PAG\_CYCLE}) - 1$

KC = number of physical channels carrying PCCCH
M = number of paging blocks available on the one PCCCH = (**12** − BS_PAG_BLKS_RES − **BS_PBCCH_BLKS**)64 for PCCCH
= (9 − BS_AG_BLKS_RES)64 for non-combined CCCH
= (3 − BS_AG_BLKS_RES)64 for combined CCCH

N = 1 for PCCCH
N = (9 − BS_AG_BLKS_RES)BS_PG_MFRMS for non-combined CCCH
N = (3 − BS_AG_BLKS_RES)BS_PG_MFRMS for combined CCCH

Applying these rules in the case for **IMSI.......1234, split paging cycle = 2**, one PBCCH per 52-frame multiframe, all the remaining 11 blocks in a 52-frame multiframe being used for paging channels, giving M =704
The limiting case is for $m = \min(704, 2) - 1 = 1$, therefore $m = 1$ is the limit

For the range of m, 0 to the limit of 1

$\text{PAGING\_GROUP} = [[(\text{IMSI mod } 1000)\text{div}(KC \times N)] \times N +$ .... (i)
$(\text{IMSI mod } 1000)\text{mod}N +$ .... (ii)
$\max\{(m \times M)\text{divSPLIT\_PAG\_CYCLE}, m\}$ .... (iii) $]\text{mod}M$

For m = 0
$\text{PAGING\_GROUP} = [(234)\text{div}(1) \times 1$ .... (i) = $\underline{234}$
$+ (234)\text{mod}1$ .... (ii) = 0
$+ \max\{(0 \times 704)\text{div2}, 0\}$ .... (iii) = 0
$]\text{mod}704$
= 234mod704 = $\underline{234}$

For m = 1
$\text{PAGING\_GROUP} = [(234)\text{div}(1) \times 1$ .... (i) = 234
$+ (234)\text{mod}1$ .... (ii) = 0
$+ \max\{(1 \times 704)\text{div2}, 1\}$ .... (iii) = 352
$]\text{mod}704$
= 586mod704 = $\underline{586}$

Therefore, for split_PAG_CYCLE = 2, paging messages will be sent on two paging blocks, 234 and 586

Figure 7.14 Selection of paging group. This figure shows the calculations underlying the selection of paging channels shown in Figure 7.13

When the broadcast control channel carrier is in the *BCCH non-combined configuration* then there is a maximum of nine common control channels on TN0. The non-combined configuration for a BCCH carrier TN0 means that no service signalling channels called standalone dedicated control channels (SDCCH) are included on TN0. This is a common configuration for large capacity cells. It means that up to nine common control channels (CCCHs) are available on TN0.

This maximum of nine CCCHs is reduced if any of these nine are reserved as extended broadcast control channels (EBCCHs) or notification channels (NCHs).

For the example used here the full nine CCCHs are available. Of these nine CCCHs, some may be reserved exclusively as access grant channels (AGCHs). This is given by the parameter BS_PAG_BLKS_RES in the system information of the BCCH. Our example assumes that three CCCHs are reserved as access grant channels, leaving six CCCHs to be used as paging channels within the 51-frame multiframe.

Our example also assumes a BS_PG_MFRS period of 4 for the circuit switched paging channels. BS_PG_MFRS = 4 means that the paging cycle for CS paging channels is four 51-frame multiframes. As there are, in this case, six paging channels per 51-frame multiframe, then the total number of CS paging channels is four times this, giving 24 CS paging channels.

A split paging cycle of value 2 is assumed for the GPRS paging channels.

We can now use the equations of Figure 7.14 to calculate the two GPRS paging channels our IMSI 1234 will use.

$$\begin{aligned}
\text{PAGING\_GROUP} = &[[(\text{IMSI mod}1000) \text{ div } (KC \times N)] \times N \\
&+ (\text{IMSI mod}1000) \text{ mod}N + \max\{(m \times M) \\
&\text{divSPLIT\_PAG\_CYCLE, m}\}] \text{ mod}M
\end{aligned}$$

where

$m = 0, \ldots,$ min $(M, \text{SPLIT\_PAG\_CYCLE}) - 1$

KC = number of physical channels carrying PCCCH

M = number of paging blocks available on the one PCCCH

$\quad = (12 - \text{BS\_PAG\_BLKS\_RES} - \text{BS\_PBCCH\_BLKS}) \, 64$ for PCCCH

$\quad = (9 - \text{BS\_AG\_BLKS\_RES}) \, 64$ for non-combined CCCH

$\quad = (3 - \text{BS\_AG\_BLKS\_RES}) \, 64$ for combined CCCH

$N = 1$ for PCCCH

$N = (9 - \text{BS\_AG\_BLKS\_RES}) \, \text{BS\_PG\_MFRMS}$ for non-combined CCCH

$N = (3 - \text{BS\_AG\_BLKS\_RES}) \, \text{BS\_PG\_MFRMS}$ for combined CCCH

for our example:

$KC = 1$

$M = (9 - \text{BS\_AG\_BLKS\_RES})64 = (6)64 = 384$

$N = (9 - 3)4 = 24$

$\quad m$ has the range from 0 to the limit

$\quad$ min $(M, \text{SPLIT_PAG_CYCLE}) - 1 = $ min $(384, 2) - 1 = 1$

For $m = 0$

$$\text{PAGING\_GROUP} = [(234)\,\text{div}(1 \times 24) \times 24 \qquad \text{(i)}$$
$$= 216$$
$$+ (234)\,\text{mod}24 \qquad \text{(ii)}$$
$$= 18$$
$$+ \max\{(0 \times 384)\,\text{div}2,\, 0\} \qquad \text{(iii)}$$
$$= 0$$
$$]\text{mod}384$$
$$= 234\,\text{mod}384 = 234$$

For $m = 1$

$$\text{PAGING\_GROUP} = [(234)\,\text{div}(1 \times 24) \times 24 \qquad \text{(i)}$$
$$= 216$$
$$+ (234)\,\text{mod}\ 24 \qquad \text{(ii)}$$
$$= 18$$
$$+ \max\{(1 \times 384)\,\text{div}2,\, 1\} \qquad \text{(iii)}$$
$$= 192$$
$$]\text{mod}384$$
$$= 426\,\text{mod}384 = 42$$

These results are illustrated in Figure 7.15 which also shows the paging channel for circuit switched operation. This is given by:

Circuit switched paging channel $= (\text{IMSI mod}1000)\,\text{mod}(\text{BS\_CCH\_CHs} \times N)\,\text{mod}N$

where

BS_CCH_CHs $=$ the number of physical channels carrying CCCHs.

$$N = \text{number of paging channels in a 51-frame multiframe}$$
$$\times \text{ the multiframe paging channel repetition period.}$$

In our example:

Circuit switched paging channel $= (\text{IMSI mod}1000)\,\text{mod}(\text{BS\_CCH\_CHs} \times N)\,\text{mod}N$
$$= (234)\,\text{mod}(24)\,\text{mod}24$$
$$= 18\,\text{mod}24$$
$$= 18$$

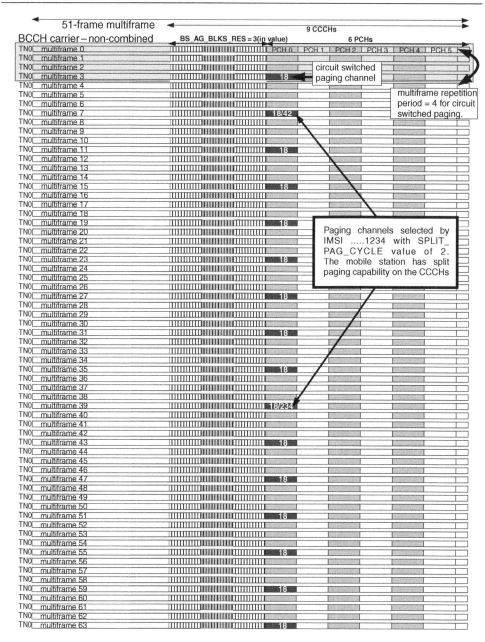

(a)

Figure 7.15 (a) Packet paging channels and circuit switched paging channels on the common control channels for IMSI...1234 with split paging cycle = 2. This corresponds to NMO2

(b)

Figure 7.15 (b) this diagram shows four repetitions of the 51-frame multiframe structure on TN0 of the BCCH carrier. The configuration of this TN is non-combined, giving nine CCCHs. The mobile station is told that three of these CCCHs are reserved exclusively for AGCHs (parameter BS_AG_BLKS_RES), leaving six CCCHs for paging, PCH0–5 illustrated. The mobile station is told that, for CS paging, these six paging channels repeat after four multiframes (parameter BS_PG_MFRS). As each of these four multiframes contain six paging channels, the total number of CS paging channels is 24. Our mobile station uses its IMSI to calculate which one of these paging channels it should listen to. The calculations result in paging channel 18 for this mobile station. For packet paging, the multiframe repetition period is sixty-four multiframes. This is illustrated in (a)

Figure 7.15 shows that the paging channels for packet and circuit switched operations are periodically coincident, which means that, at the time of coincidence, the mobile station may receive packet paging and circuit switched paging or packet paging or circuit switched paging. This arrangement minimises the battery drain of the mobile station.

- A class A mobile station is fully capable of monitoring the common control channel (CCCH) circuit switched paging channel whilst in packet transfer condition.
- A class A mobile station is fully capable of monitoring the CCCH packet paging channel whilst in circuit switched dedicated condition.
- A class B mobile station in packet transfer mode is not required to monitor the circuit switched paging channel. Many handset designs do in fact monitor the CCCH for circuit switched paging whilst in GPRS packet transfer.
- A class B mobile station in circuit switched dedicated mode is not required to monitor the packet paging channel.
- A class B mobile station working to a cell in GPRS sub-network mode of operation (NMO) 2 will receive both packet paging messages and circuit switched paging messages satisfactorily. NMO2 operation does not have a packet broadcast control channel (PBCCH) and does not have a Gs interface (the interface which allows the mobile's switching centre (MSC) to communicate with the serving GPRS support node (SGSN)).
- A class B mobile station working to a cell in GPRS sub-network mode of operation (NMO) 1 will receive both packet paging messages and circuit switched paging messages satisfactorily. NMO1 operation has a PBCCH and a Gs interface, allowing circuit switched paging messages to be sent on any paging channel. Additionally, the mobile station in packet transfer mode can receive circuit switched paging messages on the packet associated control channel (PACCH).
- A class C mobile station is blind to the other service if in circuit switched dedicated or packet transfer mode.

7.5.5 Monitoring PCCCHs and CCCHs for paging (the cell has a PBCCH but no Gs interface) (GSM 45.002 section 6)

When a class A or B mobile station is camped on a cell which offers GPRS with a PBCCH but without the Gs interface (this is called network mode of operation (NMO) 3) then it must monitor two sets of paging channels on two quite separate physical channels whilst in the circuit switched idle/GPRS GMM standby mode.

Examining Figure 7.15 for our mobile station of IMSI 1234, the circuit switched paging channel repeats at intervals of 204 frames (on a physical channel which has a 51-frame multiframe structure).

Examining Figure 7.13 for our mobile station of IMSI 1234, without split paging the packet paging channel repeats at intervals of 3328 frames (on a physical channel which has a 52-frame multiframe structure).

There is no correlation between the two sets of paging channels and the mobile station must monitor both sets on separate physical channels.

However, there is a one-frame slippage between the 51- and 52-frame multiframe structures, and at some stage they will coincide (the two sets of paging channels on separate physical channels will occur on the same frames). This coincidence means that one of the paging channels may be missed by the mobile station. If a message is present on the missed channel then the paging message may be lost. Figure 7.16 illustrates this trend.

The coincidence of the paging channels in our example will occur within 3328×204 frames $= 678\,912$ frames and may be more frequent than this.

Whether a paging channel is missed under these circumstances depends upon the location of the physical channel carrying the PCCCHs relative to the BCCH physical channel. If the PBCCH is on the same radio carrier as the BCCH then the mobile station should be capable of reading both paging channels. If the PBCCH is on a different frequency to the BCCH, then the mobile station must have time to change frequency to read the PCCCH and back again to read the CCCH. This is evidently impossible if the PBCCH is on TN0, TN1 or TN7, and depends upon the agility of its synthesiser if the PBCCH is on TNs within this range.

7.5.6 Network mode of operation (NMO) and paging (GSM 44.060 section 6)

The network mode of operation defines how the GPRS sub-network manages its paging. The GPRS sub-network may have the following configurations:

- **NMO1**. The cells have a PBCCH and there is full interconnection (Gs interface) between the CS MSC and the GPRS SGSN. This means that, for a mobile station in CS idle *and* GPRS packet idle (GMM standby), although there are two separate physical channels, one for CS paging and the other for packet paging, it needs to monitor only one of these physical channels for paging, CS paging messages are passed from the MSC to the SGSN for delivery on the packet paging channels.

  When a mobile station is active in a packet transfer, it is alerted to an incoming CS call by the MSC passing a CS paging message to the SGSN which delivers it to the mobile station on the PACCH.

  NMO1 is illustrated in Figure 7.17.

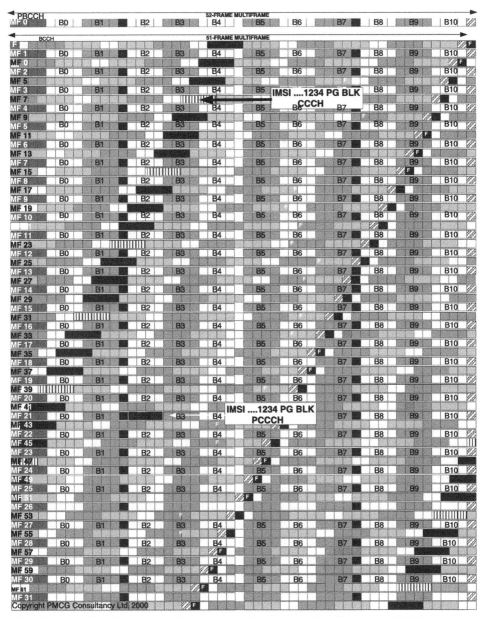

(a)

Figure 7.16 Continued

(b)

Figure 7.16   (a) Paging channels on uncoordinated CCCHs and PCCCHs; (b) The diagram shows two   separate physical channels interleaved, one following a 51-frame multiframe structure, and the other a 52-frame multiframe structure. It shows a *packet paging channel* within the 52-frame multiframe for IMSI . . . 1234 and a *circuit switched paging channel* within a 51-frame multiframe for the same IMSI. This situation occurs if the network is in NMO3. The mobile station must then monitor both sets of paging channels in the idle condition on separate physical channels. A one-frame slippage is shown between each 51-frame multiframe and 52-frame multiframe. This means that the channels sometimes coincide on the same frames and, depending upon the physical channels each occupies, may mean that one paging channel is missed this diagram illustrates the situation that the mobile station faces when the GPRS sub-network has a PBCCH, but no Gs interface exists between the MSC and SGSN. The CS and GPRS paging channels are on two separate physical channels, which must both be monitored for paging messages. The CS paging messages occur on a physical channel with a 51-frame multiframe (normally TN0 of the BCCH carrier). The GPRS paging channels occur on a physical channel with a 52-frame multiframe. The diagram shows that there is a one-frame slippage between the two multiframes, which will eventually cause the CS and GPRS paging channel to occur in the same four frames. This can cause difficulty in reading both paging channels. Part (a) illustrates this slippage more graphically

- **NMO2**. The cells do not have a PBCCH. Both sets of paging messages, CS and packet paging are delivered on the GSM common control channel (PCH). The mobile station must monitor only the CCCH PCHs for paging messages. When the mobile station is active in a packet transfer, it must also monitor its CCCH PCH for CS paging messages. This could cause difficulties, and it is not mandatory that the mobile station should do this.
  NMO2 is illustrated in Figure 7.17.
- **NMO3**. The cells do have a PBCCH, but there is no interconnection between the MSC and the SGSN (the Gs interface is not operational. This means that a mobile station in CS idle *and* GPRS packet idle (GMM standby) must monitor two separate physical channels, the physical channel carrying CCCH PCH for CS paging messages, and the physical channel carrying PCCCH PPCH for packet paging messages. This can lead to difficulties.
  When a mobile station is active in a packet transfer, it is alerted to an incoming CS call by monitoring the CCCH PCH, that is, it must switch from the physical channel(s) carrying the TBF to the physical channel carrying its CS paging message. This can lead to difficulties and it is not mandatory that the mobile station does this.
  NMO3 is illustrated in Figure 7.18.
  Class B mobile stations using a cell in NMO1 (PBCCH plus Gs interface) will not miss paging messages whilst in CS idle/GMM standby condition. Class B mobile stations using a cell in NMO2 (no PBCCH present) will not miss paging messages in the CS idle/GMM standby condition. Class B mobile stations using a cell in NMO3 (PBCCH but no Gs interface) *may* occasionally miss paging messages.

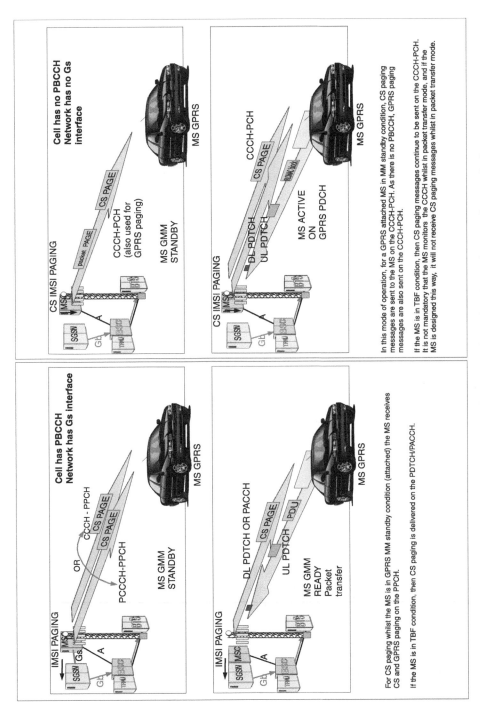

Figure 7.17 (a) Network operating mode 1; (b) Network operating mode 2

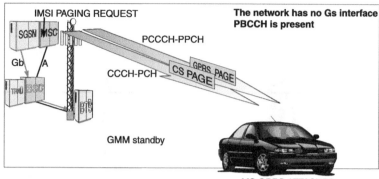

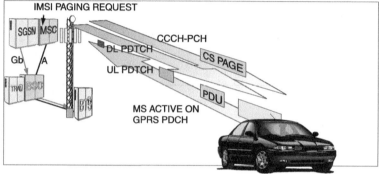

In this mode of operation, CS paging is uncoordinated to GPRS activities. As there is a PBCCH in the cell,
then the CS and GPRS paging are sent independently on their own physical channels carrying CCCH-PCH
and PCCCH-PPCH, and the MS must monitor both physical channels.

When the MS is in GPRS TBF condition it must monitor the CCCH-PCH for CS paging, but this is not
mandatory.

Figure 7.18   Network operating mode 3

7.5.6.1   Non_DRX timer

The non-DRX timer sets the time period for which a mobile station will not use
DRX after leaving the packet transfer mode. Paging messages or direct downlink
assignments for the mobile station may then be sent on *any* of the PCCCHs during
that period.

7.5.6.2   Paging procedure (GSM 44.060 section 6)

The paging procedure is illustrated in Figure 7.19 and the full paging procedure is
covered in Chapter 10.

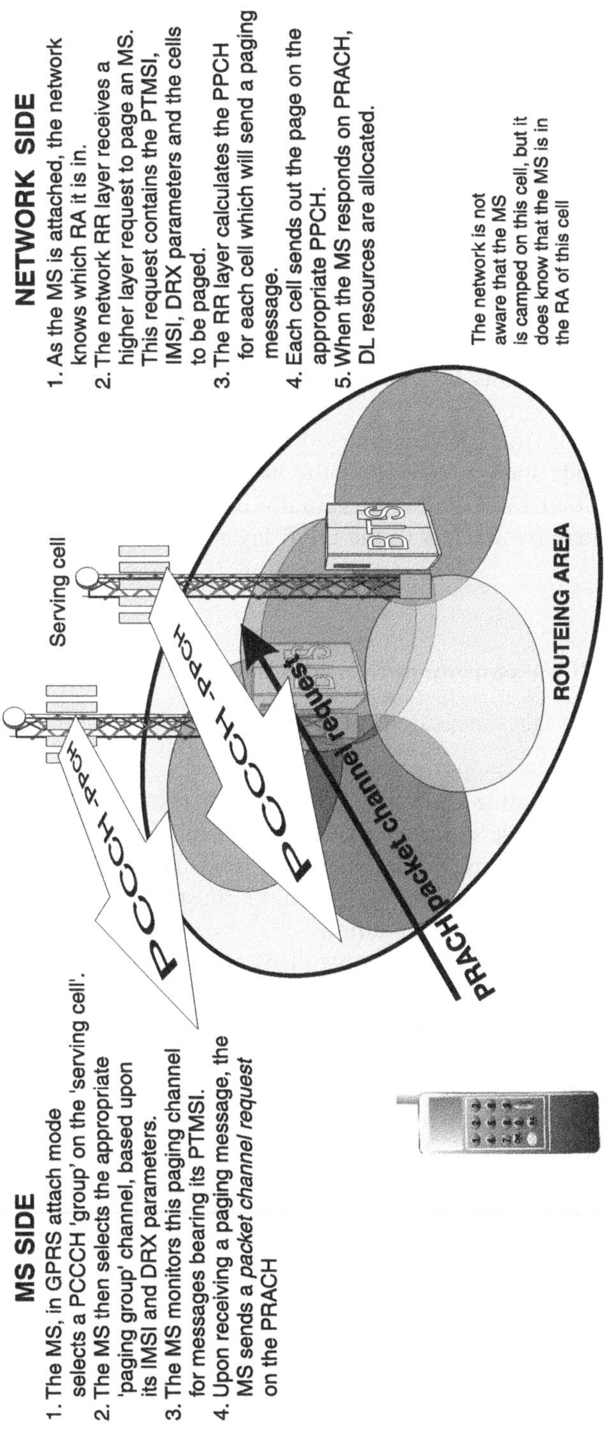

NETWORK SIDE

1. As the MS is attached, the network knows which RA it is in.
2. The network RR layer receives a higher layer request to page an MS. This request contains the PTMSI, IMSI, DRX parameters and the cells to be paged.
3. The RR layer calculates the PPCH for each cell which will send a paging message.
4. Each cell sends out the page on the appropriate PPCH.
5. When the MS responds on PRACH, DL resources are allocated.

The network is not aware that the MS is camped on this cell, but it does know that the MS is in the RA of this cell

Serving cell

PCCCH - PPCH

PCCCH - PPCH

PRACH packet channel request

BTS

ROUTEING AREA

MS SIDE

1. The MS, in GPRS attach mode selects a PCCCH 'group' on the 'serving cell'.
2. The MS then selects the appropriate 'paging group' channel, based upon its IMSI and DRX parameters.
3. The MS monitors this paging channel for messages bearing its PTMSI.
4. Upon receiving a paging message, the MS sends a *packet channel request* on the PRACH

Figure 7.19 Paging procedure on the PCCCH

7.5.6.3   DRX (ETSI 123.060 section 6, 143.064 section 6, 145.002 section 6)

Discontinuous reception (DRX) is used when the mobile station is in the packet idle condition (GMM standby and also in GMM ready if the ready timer is running and the non-DRX timer has expired); the mobile station calculates its specific paging channel or channels (depending upon the value of split paging) and when not listening to its paging channel(s), slows down its processor, saving on battery consumption. This is the 'sleep' mode. It 'wakes' at the right moment in time to listen to its allocated paging channel(s). There are a multitude of tasks a mobile station must perform whilst 'sleeping', including cell measurements and calculations for cell reselection, which reduce the effect of 'sleep' mode.

In GPRS, there are two DRX modes – normal DRX (no split paging) and split paging DRX. The calculations for the paging channel (or group) are shown in the previous section.

Figure 7.20 shows how the DRX parameters are communicated between the GPRS sub-network and mobile station. Note that the 'negotiation' of DRX parameters is a GMM function, but that the GRR layer actually uses these parameters. The GMM layer must transfer these parameters to the GRR layer.

7.6   Neighbour cell measurements

A GPRS mobile station takes measurements of its neighbour cells for one (or more) of the following reasons:

- **NCO0** Autonomous (mobile station controlled) cell reselection. Network control order zero applies to mobile stations in GMM ready (packet transfer or packet idle) and GMM standby conditions, that is when the mobile station is transferring packet data or is idle. NCO0 tells the mobile station that it must do its own cell reselection in the packet transfer or Packet idle conditions.
- **NCO1** Autonomous (mobile station controlled) cell reselection plus measurement reports to the GPRS sub-network. Network control order one tells the mobile station to do its own cell reselection in packet transfer condition or packet idle condition, but when in packet transfer to send the measurements it has made for cell reselection to the GPRS sub-network as measurement reports.
- **NCO2** Network controlled cell reselection using measurement reports sent to the GPRS sub-network. This applies to mobile stations in packet transfer condition. The measurement reports sent to the GPRS sub-network are used to command the mobile station to change cells with *cell change order*, the GPRS equivalent of *handover command*. There is nothing within the specification to stop the GPRS sub-network sending *cell change order* in NCO0 and NCO1. In the packet idle condition the mobile station does its own cell reselection.
- **Measurement reports** ordered directly by the GPRS sub-network to a particular mobile station or to all mobile stations; these are called extended measurement (EM) reports.
- **Interference reports** required by the GPRS sub-network.

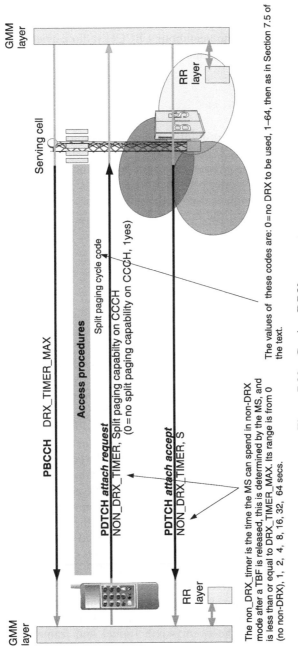

Figure 7.20 Getting DRX parameters

7.6.1   Measurements for cell reselection (GSM 45.008 section 6)

From the packet system information messages sent on the packet broadcast control channel (PBCCH) or packet associated control channel (PACCH) the mobile station receives a list of neighbour cells which it should measure. The mobile station measures the receive level of the neighbour's BCCH radio carrier, decoding the BSIC to ensure that it is the BCCH radio channel and not another radio channel of the same frequency.

7.6.1.1   Packet idle condition

If it is attached and there are no uplink or downlink radio resources allocated to the mobile station, it is in the *packet idle* condition. It must, after initially selecting a cell, take measurements of the neighbour cells specified in the BA_GPRS list, once every DRX period, or alternatively once per five seconds. Of the six strongest cells measured, the mobile station must attempt to decode their BSICs once in 14 DRX periods. A DRX period being the time between the mobile station reading its downlink packet paging channel.

7.6.1.2   Packet transfer state

During an uplink or downlink TBF, the mobile station takes measurements as illustrated in Figure 7.21, which is a simple case of an uplink transfer on one physical channel TN3. The mobile station must monitor a downlink (TN3 in this case) for PACCH control messages; this may be the same TN, as shown.

   In the illustrated case, the mobile station is monitoring the same downlink TN as for the uplink TBF, and the available measurement periods occur between the mobile station transmitting to the GPRS sub-network and monitoring the downlink TN. This period occurs every frame. During this period (four bursts, about 2 ms in Figure 7.21), the mobile station changes frequency to one of the BA_GPRS frequencies and measures the received level. If it is presynchronised to the neighbour cells, it periodically decodes the synchronisation channel (SCH) to verify that it is the correct cell. What is meant by presynchronised is this; if the mobile station has previously decoded the SCH of a neighbour cell, then it knows precisely where TN0 of the neighbour cell is located with respect to TN0 of the serving cell. Moreover, as it knows that the synchronisation channel (SCH) occurs every ten frames, it knows when the neighbour cell SCH will be present relative to the frame numbering of the serving cell. Hence, it can switch at any convenient measuring period to decode the neighbour cell's SCH. As it is likely (but not certain) that the neighbour cells are the same set of cells as measured in the packet idle condition, then the mobile station will be presynchronised.

   If the mobile station is not presynchronised to a neighbour cell it must search for the synchronisation channel (SCH). The *BSIC search period* is shown in Figure 7.21. This always occurs on frame 25 (the twenty-sixth frame) and frame 51 (the fifty-second frame) which are the idle frames. Similar empty frames may occur on the thirteenth and thirty-ninth frames if the mobile station is not using these for timing advance control purposes.

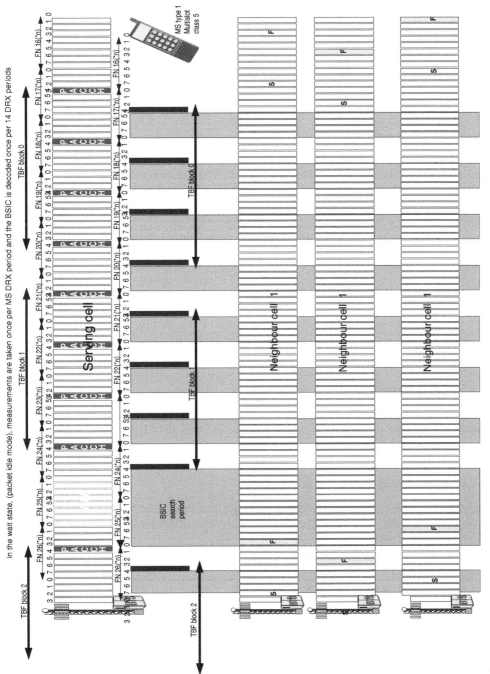

Figure 7.21   An illustration of measurements of neighbour cells during UL TBF (packet transfer state)

The combination of the 52-frame multiframe of the PDCH, and the 51-frame multiframe of the BCCH control channels and the idle frame which occurs every 26th frame, guarantees that the SCH channel carried in the 51-frame multiframe will 'slip' into a measurement period within about 0.96 s.

Once the mobile station has decoded the SCH of the neighbour BCCH radio carrier, it then has a valid set of measurements for that cell and repeats the process on the next cell in the BA_GPRS list. (The actual algorithm is left to the handset designers who will optimise it for efficiency).

7.6.1.3   Downlink multislot TBF

If the mobile station does not have the measurement periods of the example in Figure 7.21, and its multislot class does not permit measurements due to the TBF constraints, it must make the measurements after acknowledging a downlink TBF. The GPRS sub-network initiates *packet downlink ack/nack*, by setting the final block indicator (FBI) or window stall indicator with a relative reserved block period (RRBP), and it is the responsibility of the GPRS sub-network to ensure that such a mobile station is given adequate measurement opportunities.

7.6.1.4   Downlink/uplink fixed multislot TBF

If the mobile station is not able to make measurements in the fashion of Figure 7.21, due to its multislot constraints, then it must make the measurements during gaps in the TBF. The GPRS sub-network will assign inactive periods in the TBF.

7.6.1.5   BSIC decoding multislot uplink TBF

If a mobile station in this mode cannot decode the BSIC due to its multislot class constraints, it must decode the synchronisation channel (SCH) between uplink TBF transmissions.

7.6.1.6   BSIC decoding multislot downlink TBF

If a mobile station in this mode cannot decode the BSIC due to its multislot class constraints, it must decode between active periods. The mobile station requests inactive periods from the GPRS sub-network for this purpose.

The mobile station must report the measurements of neighbour cells to the GPRS sub-network if NCO1 or NCO2 is in use. Figure 7.22 shows the procedure for sending measurement reports to the GPRS sub-network

7.6.2   *Extended measurements (GSM 45.008 section 10)*

Extended measurement reports may be required from a mobile station. These reports have nothing to do with cell reselection but may be used by the GPRS sub-network, for example, to assess interference distributions. The mobile is commanded to make these measurements by the PBCCH – packet system information type 5, or by a *packet*

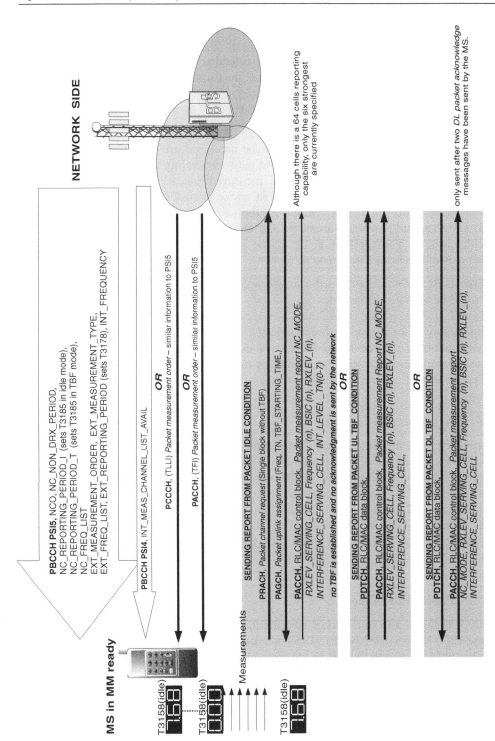

Figure 7.22 NC measurement reporting

measurement order on the packet associated control channel (PACCH) or PCCCH. The requirement is indicated by the parameter EXT_MEASUREMENT _ORDER:

- **EM0**, the mobile station shall not perform extended measurements.
- **EM1**, the mobile station shall send extended measurement reports to the GPRS sub-network.
- **RESET**, (sent only on PACCH or PCCCH), the mobile station shall return to monitoring the PBCCH parameters.

Parameters describing the frequencies to be measured are sent with the parameter EXT_FREQUENCY_LIST, and the period between sending reports by the parameter EXT_REPORTING_PERIOD. Extended measurements are always made in packet idle condition, and if a mobile station is in packet transfer mode when an extended measurement report is due, it will wait until the TBF is completed, return to packet idle mode and then make the measurements and report them.

With the EXT_FREQUENCY_LIST, the mobile station is asked to make one of three types of extended measurement report:

1. **Type 1**. BCCH carriers from the six strongest in the list, regardless of BSIC, which should be included if possible.
2. **Type 2**. BCCH carriers from the six strongest in the list that have the BSIC decoded and belong to an allowed network colour code NCC. (The BSIC broadcast by each cell consists of a three-digit NCC plus a three-digit BCC. The serving cell advises mobile stations which NCCs are valid for measurement purposes).
3. **Type 3**. BCCH carriers from the list that will be reported without BSIC decoding.

7.6.3   Interference measurements

Interference measurements reported by mobile stations will be of high value to GPRS sub-network operators. (They would be of greater value if the GPRS sub-network operator knew the location of the mobile station reporting them, and that may transpire in the future when mobile location systems become operational).

Broadcast packet system information 1 contains the parameter INT_MEAS_CHAN-NEL_ LIST_AVAIL, and this tells mobile stations that interference frequency lists are contained in PSI4. The mobile station then goes to PSI4 to get the list of frequencies, which it must measure, average and report.

7.6.3.1   Packet transfer mode

Figure 7.23 shows how interference measurements are made. Interference measurements on the INT_MEAS_CHANNEL_LIST are not made in packet transfer mode. If the mobile station is ordered to send NC measurement reports then, whilst in packet transfer mode, it measures the receive level on the packet timing advance control channel (PTCCH) and idle frames of the physical channel it is using for packet transfer. These are sent to the GPRS sub-network with the *packet measurement report* message. Thus the mobile station directly measures, and reports, the interference that it is experiencing on that physical channel. This is shown (for packet transfer mode) in Figure 7.23.

7.6.3.2 Packet idle mode measurements

The mobile station measures the packet timing advance control channel PTCCH and idle frames of the frequencies in the INT_MEAS_CHANNEL_LIST. It also measures the PTCCH burst and idle burst of the TN it is using to monitor the paging channel on the serving cell. These measurements are averaged and reported to the GPRS sub-network. The process is illustrated in Figure 7.23.

7.6.4 Measurement reports

Network controlled (NC) measurement reports may be sent by the mobile station to the GPRS sub-network from packet idle condition or whilst in a TBF – packet transfer mode. This means that the mobile station makes the measurements in both packet idle and packet transfer conditions.

 Extended measurement (EM) reports are always measured in the *packet idle condition*, and sent to the GPRS sub-network by the mobile station requesting (*packet channel request*) the radio resources for the specific purpose of transferring the measurement report.

7.6.5 NC measurement reporting procedures

Figure 7.22 shows the conditions for sending NC measurement reports. The mobile station receives from the PBCCH the data on which cells to measure and report. This data can also be received on the packet associated control channel (PACCH) or PCCCH and includes:

- **NCO**, the network control order,

 —NCO0: the mobile station will make measurements in both packet idle and active conditions for autonomous cell reselection without reporting to the GPRS sub-network.
 —NCO1: the mobile station will make measurements in both packet idle and active conditions for autonomous cell reselection and send measurement reports to the GPRS sub-network.
 —NCO2: the mobile station will make measurements for cell reselection in packet idle condition, but in active condition the GPRS sub-network will order cell reselection (*cell change order*) based upon the received mobile station measurement reports.

- **NC_NON_DRX_PERIOD** is the minimum time a mobile station must stay in non-DRX mode after an NC measurement report has been sent. In the non-DRX mode the mobile station remains fully 'awake', monitoring all the DL PCCCHs. When a mobile station is in the non-DRX mode as commanded by the GPRS sub-network, then the GPRS sub-network can take advantage of this by putting messages (such as paging, or downlink assignment) for that mobile station on any of the PCCCHs, considerably speeding up communication compared to waiting until the mobile station reads its own PCCCH paging channel.

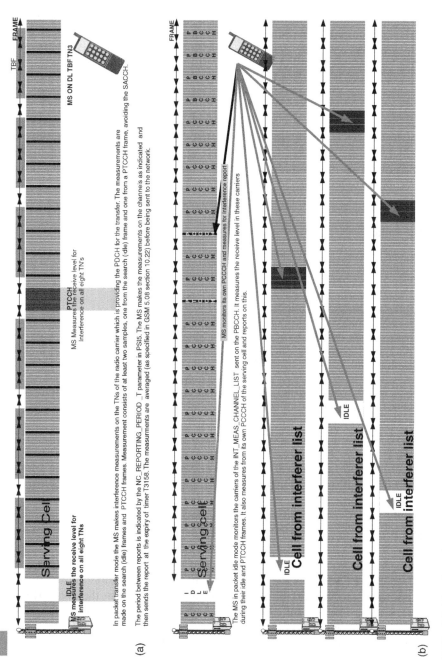

Figure 7.23 Making interference measurements. (a) Packet transfer mode NC measurement; (b) packet idle mode NC measurement

- **NC_REPORTING_ PERIOD_I** sets a timer T3158 when the mobile station is in packet idle mode and is the periodicity between reports.
- **NC_REPORTING_PERIOD_T** sets the timer T3158 when the mobile station is in packet transfer mode and is the periodicity between reports.
- **NC_FREQUENCY_LIST**, a list of the frequencies to be measured for NC reporting.
- **INT_FREQUENCY**. This parameter indicates the frequency upon which an interference measurement will be made and reported. The interference measurements are made on the idle frames and PTCCHs of the PDCHs to be measured.
- **INT_MEAS_CHANNEL_AVAIL**, this information is optionally sent on the PBCCH PSI4 or, if there is no PBCCH, it may be broadcast on the BCCH. It contains a list of frequencies, hopping or static, upon which the mobile station must perform interference measurements when in the packet idle mode. If hopping, mobile allocation (MA) and hopping sequence code (HSC) are included.

The mobile station sets the timer T3158 and when it expires, compiles the measurement report or makes the measurements and sends the report to the GPRS sub-network. Whether it waits until the timer expires (as the specification states) before making the measurements would seem to depend upon a number of factors. If the mobile station is in the packet idle mode, and its DRX period is very much greater than the reporting period, then it seems it must make the measurements upon the timer expiry. If, however, the reporting period is much greater than the DRX period, then the mobile station will have had the opportunity near to the occurrence of its paging block to measure neighbour cells. If the mobile station is in packet transfer mode, it has only limited opportunities to make neighbour cell measurements and must take full advantage of those periods. The mobile station, if in the packet idle mode, sends the report after making the access procedures, requesting a single uplink block in the *packet channel request*. The GPRS sub-network responds by giving an uplink block, and the measurement report is sent on that block.

7.6.5.1   TBF NC measurement report

The measurement report in packet transfer mode includes:

- **NC_MODE**, the network control mode currently in use by the mobile station.
- **RX_LEVEL_SERVING_CELL**, the measured received level from the serving cell.
- **FREQUENCY_n**, the frequency of a measured cell 'n' (neighbour) reported in the strongest six.
- **BSIC_n**, the base station identity code of a measured cell 'n' reported in the strongest six.
- **RXLEV (n)** the receive level of the neighbour cell.

7.6.5.2   Packet idle NC measurement report

The measurement report in packet idle mode includes:

- **NC_MODE**, the network control mode currently in use by the mobile station.
- **RX_LEVEL_SERVING_CELL**, the measured received level from the serving cell.
- **FREQUENCY_n**, the frequency of a measured cell 'n' reported in the strongest six.

- **BSIC_n**, the base station identity code of a measured neighbour cell 'n' reported in the strongest six.
- **RXLEV (n)** the receive level of the neighbour cell.
- **INTERFERENCE_SERVING_CELL**, the mobile station monitors the idle and PTCCH slots of the physical channel (TN) on the radio carrier it is using for monitoring PCCCH. It processes the results with a running average and sends the report to the GPRS sub-network, as indicated in the serving cell of Figure 7.23.

7.6.6   EM reporting

The GPRS sub-network indicates that extended measurement reporting is required (as indicated by EM1 – extended measurement report required).

The EXT_FREQ_LIST and EXT_REPORTING_PERIOD are sent on PBCCH, PACCH or PCCCH. The mobile station takes the measurements in the packet idle condition, and sends the averaged measured levels in the *packet measurement report*.

Whether BSIC is measured or reported is dependent upon the Type 1, Type 2, or Type 3 designation (see section 7.6.2) for each carrier.

An extended measurement report is only sent in the packet transfer condition if there is a set of measurements waiting to be sent when the mobile station changes from packet idle to the packet transfer condition.

7.6.7   Interference reports

If the mobile station is requested by the GPRS sub-network to send interference reports, indicated by the GPRS sub-network parameters INT_MEAS_CHANNEL_LIST, it will measure the receive levels of the listed frequencies, making the measurements in the neighbour cell idle and PTCCH slots as indicated in Figure 7.23.

The measurements are averaged and reported to the GPRS sub-network in the packet measurement report.

7.6.7.1   Packet transfer mode measurements

In this mode the mobile station measures the receive levels in the PTCCH and idle frames of the radio carrier it is using, as indicated in Figure 7.23. These are reported to the GPRS sub-network with the neighbour cell measurements only if the mobile station is asked to send NC measurement reports.

7.6.7.2   Packet idle mode measurements

The mobile station makes RXLEV measurements on the neighbour cells in the idle and PTCCH frames as specified in INT_MEAS_CHANNEL_LIST, averages them and sends the results in the *packet measurement report*. It also measures its own TN during the PCCCH and idle period and reports to the GPRS sub-network on the interference measured there. This is indicated in Figure 7.23.

7.7 Mobile station transmitter power control (GSM 45.008 section 10)

In GSM CS operations, the transmitted power of the mobile station is controlled from the uplink level and quality of the mobile station transmissions received at the BTS. As the mobile station is continually transmitting when a dedicated channel is allocated (SACCH transmission takes place four times per 480 ms irrespective of DTX) there is not a problem in controlling the mobile station's transmitted power. (Power control instructions are sent on the downlink SACCH with the same periodicity, about once per 0.5 seconds if necessary).

In GPRS, when the mobile station is allocated radio resources for a TBF, there may be no uplink transmissions for as long as 1.2 seconds if the TBF is unidirectional downlink.

Using one physical channel for a DL TBF with an RLC/MAC window size of 64 blocks, then a TBF will last for a period of 18 ms (the time taken to transfer one RLC/MAC block) $\times 64 = 1.15$ s. As, in a 52-frame multiframe, only 48 frames are used to transfer RLC/MAC blocks, then the total period for 64 blocks is 1.15 s $\times 52/48 \cong 1.2$ s. For a prolonged DL TBF the mobile station may have to transmit acknowledgments once every 1.2 s.

This control period is not tight enough and another method is used. For GPRS the mobile station adjusts its transmitted power based upon:

1. The received power measured on the downlink.
2. The dynamic control parameters sent by the GPRS sub-network on the PACCH. These dynamic control parameters are based upon the uplink received level of the mobile station at the BTS.

Figure 7.24 shows the parameters that are sent from the GPRS sub-network to the mobile station on the BCCH, PBCCH or PACCH. Another, dynamic set of parameters is sent on the PCCCH or PACCH to control dynamically the mobile's transmitted power. The parameters are:

gamma. $\Gamma_0 = 39$ dBm for GSM900 (36 dBm for GSM1800).
Γ_{ch} is a parameter sent in an RLC control message, range $0, 2, 4, \ldots, 64$ dB.
alpha. $\alpha = $ a parameter sent on PBCCH, PCCCH, PACCH, range $0, 0.1, 0.2, \ldots, 1.0$.
GPRS_MS_TXPWR_CCH is the maximum mobile station transmit power when sending *packet channel requests*, range 5–33 dBm.
PC_MEAS_CHAN is the downlink channel the mobile station will measure to assist in calculating the required transmit level.
T_AVG_W is a parameter used by the mobile station in the packet idle condition in calculating the transmit power to be used when moving from packet idle to packet transfer mode.
T_AVG_T is a parameter used by the mobile station in the packet transfer mode to calculate the transmit power used in the TBF.
N_AVG_I is an interference parameter used by the mobile station in the packet transfer mode to calculate the transmit power used in the TBF. The required transmit power is given by:

$$P_{CH} = \min\{\Gamma_0 - \Gamma_{CH} - a(C + 48), \text{GPRS\_MS\_TXPWR\_CCH}\}$$

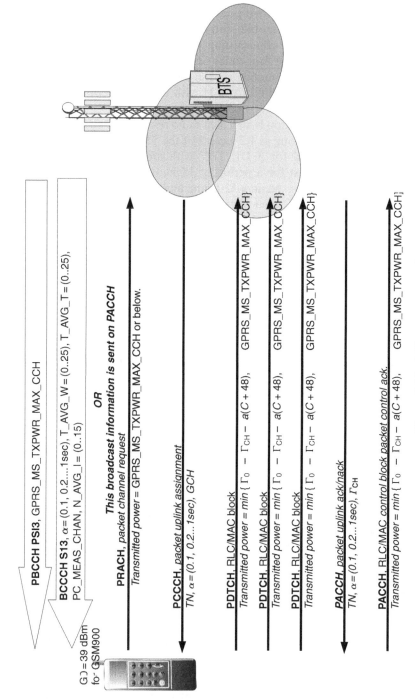

Figure 7.24 MS power control

The value of C is calculated for packet idle mode and transfer mode.

Illustrations of these calculations are given in Appendix 4 and Figure 7.24 illustrates power control in action.

7.8   Timing advance (TA) (GSM 43.064 section 6)

Timing advance is necessary in the GSM system to synchronise the arrival time of radio bursts from mobile stations to the receive time base of the BTS. The receive time base at the BTS is running in a fixed sequence, but the arrival time of radio bursts from mobile stations is variable. This variability is due to the mobile station's distance from the BTS antenna which causes a two-way propagation delay. This delay is two-way because the mobile station must synchronise to the received TN0 transmission from the network. This is delayed by the downlink propagation delay. There is an identical uplink delay for mobile station transmissions. Timing advance brings forward the mobile station's transmission by a period equal to the two-way propagation delay. This is illustrated in Figure 7.25.

However, the GPRS system imposes more stringent constraints upon TA than GSM. For GSM, when a dedicated channel is in use, there are always transmissions from the mobile station to the network and from these it can calculate the required TA instruction necessary to keep the mobile station synchronised to the network.

Even if the dedicated channel is not transmitting customer data, the SACCH channel associated with the traffic channel is transmitting four bursts within 480 ms (four times the traffic channel multiframe period), and the network can use these to calculate the TA.

In GPRS, for a prolonged downlink TBF, the mobile station may have to send uplink acknowledgments once for every 64 downlink RLC/MAC blocks (64 blocks is the window size of the RLC point-to-point data communications link). As the minimum transmission time for 64 RLC/MAC blocks is about 1.2 seconds when using one downlink physical channel, this determines the periodicity of acknowledgments sent by the mobile station. This is not considered adequate for tight TA control and an alternative system described below is used.

Initial timing advance is based upon the short access bursts which carry the *packet channel request*. The GPRS sub-network measures the offset of these bursts from the proper position, calculates the TA and sends this in the *packet resource assignment* message.

However, if it is necessary to queue the mobile station, it may subsequently be necessary to recalculate the TA. This is done by sending a *packet polling* message which forces the mobile station to respond with four short access bursts, the TA is recalculated and then resources are allocated.

Figure 7.26 shows the two methods used to control TA in GPRS operations, initial timing advance, which is identical to the GSM initial timing advance, and thereafter continuous timing advance, which differs from GSM. The first arrow in Figure 7.26 shows the mobile station making a *packet channel request* using the short access burst, which may carry 8 or 11 bits of information. The second arrow shows the GPRS sub-network responding with a *packet uplink assignment*. This carries the TA value in the range 0–63 bits, which the mobile station will use to advance its timing.

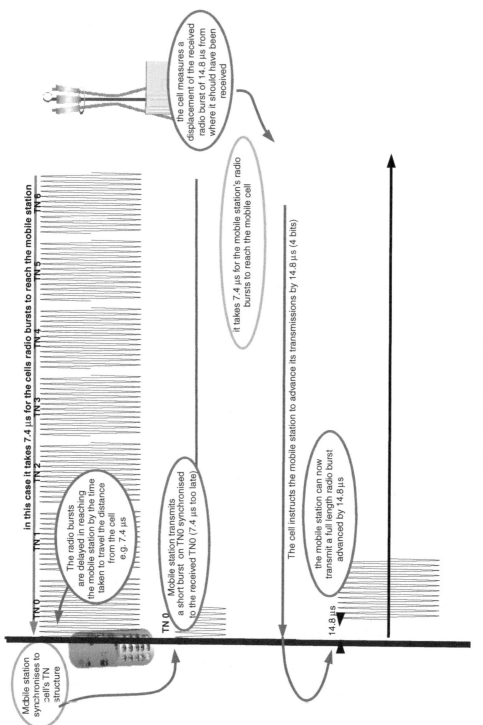

Figure 7.25 Initial timing advance

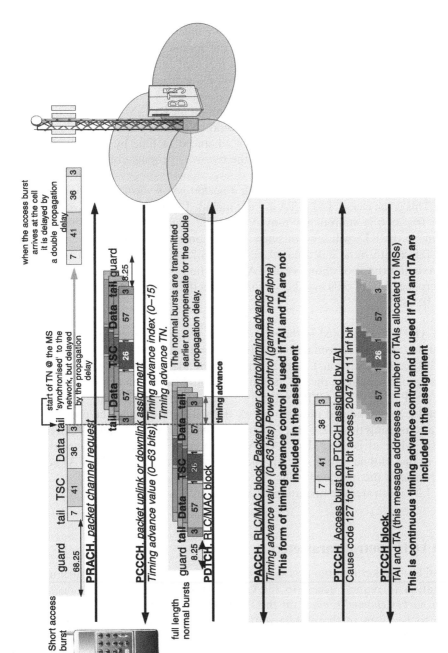

Figure 7.26   MS timing advance

Timing advance is incremented or decremented by an air interface one-bit period of about 3.7 ms. A one-bit adjustment corresponds roughly to a radial distance of 0.5 km from the BTS.

This message may also contain a *timing advance index* (TAI) coupled with the timing advance TN, which is the physical channel the mobile station must use for timing advance control. If these two parameters are included then continuous TA control is used. If they are not included then subsequent TA instructions are sent in a message *packet power control/TA*.

The third arrow shows the mobile station sending an RLC/MAC data block on the PDTCH. This is transmitted earlier than it would be in the absence of TA instructions; this is shown on the arrow with the bursts advanced in time by the TA instruction received from the BTS, compensating for the two-way propagation delay between the BTS and mobile station.

The fourth arrow shows the BTS sending a *packet power control/TA* message on the PACCH. This is the case if TAI + TA are not included with *packet uplink assignment*. In this case, the mobile station has moved by more than 0.5 km since the initial TA instruction, so the BTS will send this new instruction on the PACCH.

As the full length burst has a guard period of 8.25 bits then, theoretically, for cells with a maximum range of about 4 km, TA instruction may not be necessary. In urban areas where the radii of cells may be very much less than 4 km, TA instructions may, or may not, be used.

The fifth and sixth arrows show the process for continuous TA control, when TAI and the TAI TN are included in the *packet uplink assignment* message. The TAI has a range of 0–15 (allowing sixteen mobile stations to use one GPRS PDCH for continuous timing advance transmissions), and assigns a particular PTCCH uplink burst exclusively to a mobile station on the TN indicated. The mobile station transmits short access bursts to this PTCCH. The GPRS sub-network uses four PTCCH bursts on the downlink to send a TA command message for all the sixteen TAIs and the possible sixteen mobile stations using the PTCCHs. The message includes the TAI and required timing advance for whoever is using that TAI. As a particular TAI on a physical channel is exclusive to one mobile station, the mobile station with that particular TAI adjusts its timing advance to the indicated value.

Figure 7.27 illustrates the operation of continuous TA procedures.

7.9 PRACH control parameters

The PRACH – packet random access channel – is the logical uplink channel used by a mobile station to request access to the GPRS sub-network services. It does this by sending a *packet channel request* on the PRACH. This request asks for uplink radio resources to be given to the mobile station for message transfer. The PRACH is a common channel, common to all mobile stations on that cell requesting access to the GPRS sub-network services. It is also random access; this means there is quite a high probability of two or more mobile stations attempting to use the same channel simultaneously. A number of control mechanisms are in place to impose order upon what could be disordered randomness. These include:

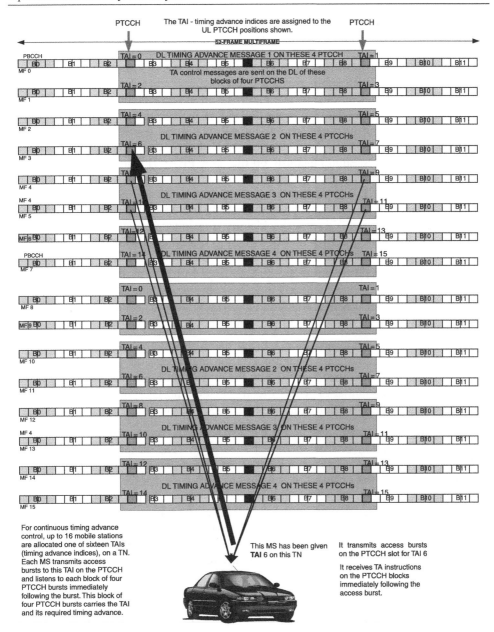

Figure 7.27   GPRS continuous timing advance control

- Controlling the *transmit power level* that a mobile station will use when it requests GPRS sub-network services.
- Regulating the *number of access attempts* made by mobile stations. (If the GPRS sub-network is too busy to respond to access requests, the mobile station must not continue sending requests *ad infinitum*).

- *Regulation based upon subscription* of the access attempts made by a mobile station. This uses the *radio priority* parameter. A radio priority is given to the mobile station by the GPRS sub-network when a PDP context is successfully established. Subscribers who pay more will be given a superior radio priority and are more likely to gain access to the GPRS sub-network services.

  For TBFs requested specifically for higher layer service signalling (from the GMM layer), and not customer data (from the SNDCP layer), the mobile station always uses the highest radio priority when sending the *packet channel request*.

- *Spreading of access requests* so that if two mobile stations do initially clash in sending *packet channel request* simultaneously on the same PRACH channel, then the probability of subsequent clashes is reduced.

- *Forced queuing* of a mobile station by responding to its *packet channel request* with a *packet queuing notification*. This gives the mobile station making the request a *temporary queuing identity* (TQI) and this identity is included in a later downlink *packet polling request* which causes the mobile station to recommence sending *packet channel requests*.

- *Forced waiting*. The GPRS sub-network may respond to the mobile station's request with *packet access reject*, which may include a *wait period*, and if it does, the mobile station must wait for this period before sending another *packet channel request*.

- *Barring* certain classes of mobile station from sending *packet channel requests*.

Figure 7.28 shows the information sent by the GPRS sub-network on PBCCH or PACCH, which determines how the mobile station shall access the GPRS sub-network. It also shows briefly the procedure used by the mobile station in accessing the GPRS sub-network and the number of access attempts made. This is expanded in Figure 7.29.

Summarizing the information given to the mobile station:

- **ACCESS_BURST_TYPE**. The access burst (*packet channel request*) may carry eight or eleven information bits as shown in Figure 3.6. The eleven-bit format allows information on radio priority to be sent and, in some cases, a longer random reference, reducing the problem of contention, as discussed below.

- **ACCESS_CONTROL_CLASS** (range 0–15). This is used by the GPRS sub-network operator to control congestion in a cell. All subscribers have a classification in the range 0–9 programmed into the SIM and if the ACCESS_CONTROL_CLASS parameter on the PBCCH indicates that one of these classes is barred from using the cell then a subscriber of that class can make no access attempt.

  There are also classes 10–15:

  —Class 10 indicates whether the cell will allow emergency calls to be made.

  —Classes 11–15 are reserved for services such as police, ambulance, security and network operator use.

- **MAX_RETRANS** (range 1, 2, 4, 7). This tells the mobile station how many repeat *packet channel requests* are allowed after the initial attempt. The total number of *packet channel requests* allowed is MAX_RETRANS + 1.

- **S**, spreading (range 12, 15, 20, 30, 41, 55, 76, 109, 163, 217). It is desirable to 'spread' access bursts after the first attempt to reduce congestion and allow the GPRS

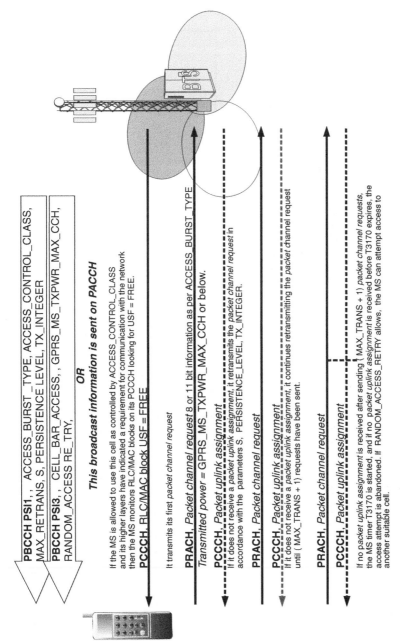

PBCCH PSI1, , ACCESS_BURST_TYPE, ACCESS_CONTROL_CLASS, MAX_RETRANS, S, PERSISTENCE LEVEL, TX_INTEGER

PBCCH PSI3, , CELL_BAR_ACCESS, , GPRS_MS_TXPWR_MAX_CCH, RANDOM_ACCESS RE_TRY,

OR

This broadcast information is sent on PACCH

If the MS is allowed to use this cell as controlled by ACCESS_CONTROL_CLASS and its higher layers have indicated a requirement for communication with the network then the MS monitors RLC/MAC blocks on its PCCCH looking for USF = FREE.

PCCCH, RLC/MAC block USF = FREE

It transmits its first *packet channel request*

PRACH, *Packet channel request* 8 or 11 bit information as per ACCESS_BURST_TYPE

Transmitted power = GPRS_MS_TXPWR_MAX_CCH or below.

PCCCH, *Packet uplink assignment*
If it does not receive a *packet uplink assignment*, it retransmits the *packet channel request* in accordance with the parameters S, PERSISTENCE_LEVEL, TX_INTEGER.

PRACH, *Packet channel request*

PCCCH, *Packet uplink assignment*
If it does not receive a *packet uplink assignment*, it continues retransmitting the *packet channel request* until (MAX_TRANS + 1) requests have been sent.

PRACH, *Packet channel request*

PCCCH, *Packet uplink assignment*

If no *packet uplink assignment* is received after sending (MAX_TRANS + 1) *packet channel requests*, the MS timer T3170 is started, and if no *packet uplink assignment* is received before T3170 expires, the access attempt is abandoned. If RANDOM_ACCESS_RETRY allows, the MS can attempt access to another suitable cell.

Figure 7.28 GPRS access control

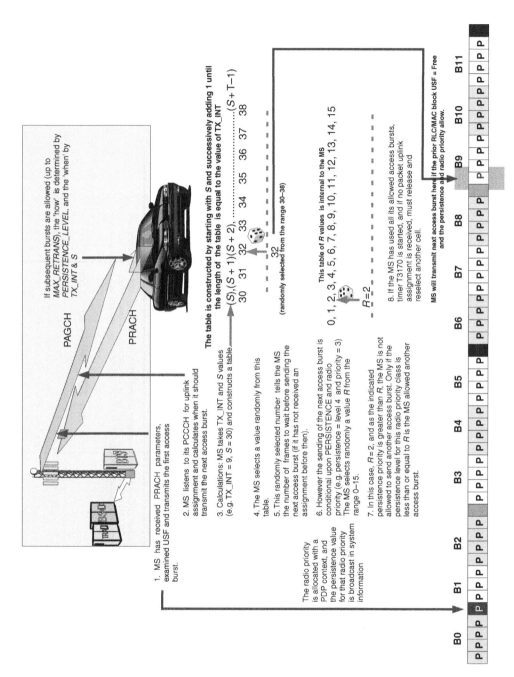

Figure 7.29 Using the GPRS PRACH parameters

sub-network to deal with requests in an orderly fashion. The value of S indicates the minimum number of frames a mobile station must wait before making a second access attempt.

The mobile station constructs a look up table with values S, $S+1$, $S+2,\ldots$ until the length of the table is equal to TX_INTEGER.

After sending its first *packet channel request*, the mobile station then selects a value randomly from this table, and this value determines how many frames it must wait before sending its next *packet channel request*, conditional *upon being allowed to send a second packet channel request*.

- **TX_INTEGER** (range 2–10, 12, 14, 16, 20, 25, 32, 50). The TX_INTEGER value determines the length of the look up table constructed upon the S value. The higher the value of TX_INTEGER, the lower the probability of subsequent clashes if a clash occurs at the first *packet channel request*.

  A note on S and TX_INTEGER. The value of S determines the minimum wait period (frames) between access attempts. A GPRS sub-network operator may choose a high value to help in congestion control. If all mobile stations must wait for a longer period before sending subsequent *packet channel requests*, then this gives the GPRS sub-network radio resources a breathing space to clear current requests. The value of TX_INTEGER is used by the GPRS sub-network operator to both reduce the probability of clashes and to give a 'spread' suited to their requirement; the higher the value of TX_INTEGER, the greater the spread of subsequent *packet channel requests* and the easier it is for the GPRS sub-network to manage the situation.

- **PERSISTENCE_LEVEL** (range 0–15 for each radio priority 1–4). This parameter is new to GPRS operations and helps to prioritise access attempts. Persistence level and radio priority are broadcast as a pair in the system information. A typical pairing set might be:

  (Radio priority 1, Persistence 0)

  (Radio priority 2, Persistence 4)

  (Radio priority 3, Persistence 9)

  (Radio priority 4, Persistence 12)

  Subscribers with a low radio priority will receive less favourable treatment than those with a high priority. The operation is described in Figure 7.29.

  The mobile station is given its radio priority when a PDP context is established. From this broadcast information, it reads the persistence value given for its radio priority.

- **GPRS_MS_TXPWR_MAX_CCH** (range see Appendix 1). The maximum power the mobile station may use for its access bursts.

- **RANDOM_ACCESS_RETRY** (range 0, 1). If set to '1', the mobile station is allowed to attempt access on a neighbour cell should the access attempt on this cell fail.

Figure 7.29 shows these parameters in action. In this example, the parameter values are:

> TX_INTEGER = 9
>
> $S = 30$
>
> Persistence level = 4 for radio priority = 3
>
> The radio priority of the mobile station in this example is 3
>
> The MAX_RETRANS is not specified for this illustration.

Figure 7.29 shows a mobile station which has been given a radio priority of 3 when a PDP context was established. It has read the corresponding persistence value of 4 applicable to radio priority 3. It has read a TX_INTEGER value of 9, and an S value of 30 from the system information.

It makes its first access attempt on frame 4, sending a *packet channel request* on the PRACH. It constructs the S table by taking the value of S from system information and incrementing it by one until the length of the table is equal to the value of TX_INTEGER. It then selects a value randomly from the S table (32 in our example). This means it must wait for 32 frames before sending its second *packet channel request* if the first is not answered. However, the mobile station must test to see if it is allowed to send a second *packet channel request* if the first is not answered.

It tests whether it is allowed to send a second request by selecting randomly from the R table. All mobile stations have the R table. In this case the R table returns $R = 2$. The mobile station compares the R value with its persistence value. If $R \geq$ persistence value, it may transmit a second *packet channel request*. In our example, $R <$ persistence value, and our mobile station may not transmit a second *packet channel request* if the first is not answered.

The higher the persistence value for a radio priority class, the lower is the probability that a subscriber with that radio priority is able to transmit a further *packet channel request*. At one limit, if persistence value = 15, for a given radio priority class then subscribers of that radio priority class will have a probability of 1 : 15 of transmitting a further *packet channel request*. In the other limit, if the allocated persistence value = 0, then subscribers with that radio priority will have a probability 15 : 1 of transmitting a further *packet channel request*. This continues until MAX_RETRANS + 1 is reached.

7.10   Contention resolution (GSM 43.064 section 6)

The GPRS sub-network identifies a mobile station during a TBF using the *temporary logical link identifier* (TLLI) after the mobile station has attached. The GPRS sub-network is addressing many mobile stations simultaneously and internally discriminates between mobile stations by the TLLI.

On the air interface, the mobile station and GPRS sub-network do not always need the TLLI, as the mobile station is defined by the cell and the physical and logical channels it is using, in addition to a temporary flow identifier (TFI)

The GPRS sub-network indexes TLLI to the cell and physical/logical connection of a mobile station, so that packets (with an IP address indexed to IMSI, which is indexed to

PTMSI which is indexed to TLLI) are routed to the correct physical/logical channel. So, although the TLLI is not often used across the air interface, it is always used at the start of a TBF to identify a mobile station.

The use of TLLI on the air interface at the start of a TBF identifies a mobile station and resolves the situation when two mobile stations attempt to use the PDTCH/ PACCH at the same time. TMSI in circuit switched operations has a similar use.

This use of TLLI is necessary because of the random access system used. TLLI is sent from the mobile station to the GPRS sub-network as an identity when it sends the TBF packets on the assigned uplink (single-phase access), or on a single block specifically given to convey this identity (two-phase access).

There are three types of TLLI:

- **Local TLLI** – the TLLI is derived from a PTMSI which was allocated in the same routeing area as that in which the current access attempt is being made.
- **Foreign TLLI** – the TLLI is derived from a PTMSI which was allocated in a different routeing area to the current routeing area in which the access attempt is made.
- **Random TLLI** – the mobile station has not been allocated a PTMSI.

The process of calculating the TLLI is shown in Chapter 10.

Contention resolution is the removal of the inherent ambiguity in *packet uplink assignment* in response to a *packet channel request*. As there are two modes of access in the GPRS system, one-phase access and two-phase access, then there are two methods of contention resolution.

The ambiguity arises in *packet uplink assignment* because, at the time of the assignment, the GPRS sub-network does not know the identity of the mobile station. A mobile station 'guesses' that the assignment is addressed to it because the assignment returns a (rather limited in size) random number, which the mobile station included in the *packet channel request*. However, two or more mobile stations may be waiting to receive *packet uplink assignment*, and if they have both used the same random number whilst sending *packet channel request* on the same frame, they will all think the uplink assignment belongs to them. More than one mobile station may try to access the data channel assigned by the GPRS sub-network.

7.10.1   One-phase access contention resolution

Figure 7.30 illustrates the principle of one-phase access contention resolution. Two mobile stations, mobile station A and mobile station B, both send *packet channel requests* in the same uplink frame, and have included the same random number in their requests. (The probability of two mobile stations using the same random number is quite high as the average length of this number is three bits and there is a 1 : 8 probability of two mobiles using the same number).

With both mobile stations sending their *packet channel request* in the same frame, the GPRS sub-network *may* not be able to decode either of the received messages. However, in large radius cells when a request from a mobile station that is close to the cell's antennae arrives early with a larger power compared to a mobile station sending a request at extreme range from the antennae, the cell may well decode the first message to arrive.

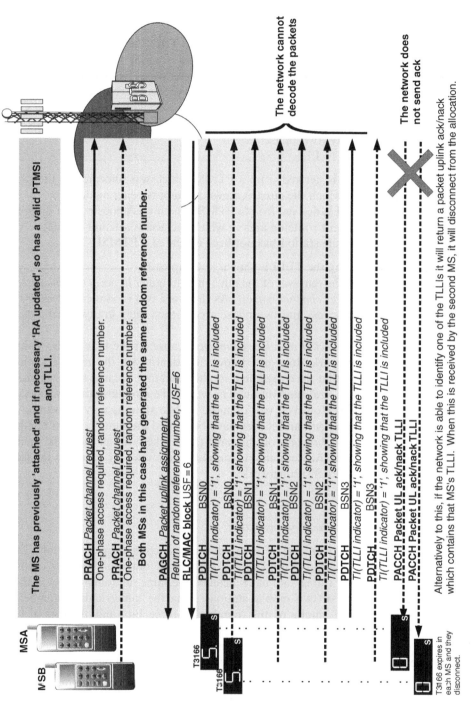

Figure 7.30   One-phase access contention resolution

Both mobile stations then keep their receivers open listening to all the DL PCCCH – PAGCH for a *packet uplink assignment*. The GPRS sub-network sends a downlink *packet uplink assignment* message containing the random reference number and the frame in which the message was received, and each mobile station identifies the message as its own and starts a TBF when the USF allocation (in this case the assignment is dynamic) appears on the downlink. They both include their TLLI in the RLC/MAC data block message. Timer T3166 in each mobile station is started when the first block is sent.

Both mobile stations continue sending blocks until the USF granularity is reached. They then start another timer (T3180 which sets the limit of tolerance of waiting for the USF to reappear) and wait for the reappearance of their USF identity on the downlink RLC/MAC blocks.

The GPRS sub-network cannot decode these two simultaneous receptions at the same time and therefore does not respond, and the two mobile stations do not receive a *packet uplink ack/nack*. T3166 expires and both mobiles disconnect from the GPRS sub-network.

Now, for example, if only one of the mobile stations has started a TBF on the first USF allocation, then the GPRS sub-network will decode the message, identify the mobile station from the TLLI, and is obliged to send a *packet uplink ack* as soon as possible. This acknowledgment is sent solely to return the mobile station's TLLI (and stop T3166). Both mobile stations receive this message. The mobile station receiving the wrong TLLI must disconnect from the cell.

For one-phase access, the exchange of TLLI between the mobile station and the GPRS sub-network can take quite a few RLC/MAC blocks.

For fixed allocation, if the mobile station does not receive TLLI from the GPRS sub-network it may continue with the uplink block transfer. If there were a lot of uplink blocks to transfer on a fixed allocation, this could waste a lot of the GPRS sub-network resources. Because of this, one-phase access is used for short TBFs only. The transmissions are also limited by T3166 and a counter.

If the radio conditions are good, then one-phase access is efficient in the use of radio resources. If radio conditions are poor, it can be wasteful of radio resources. The mobile station decides whether to request one- or two-phase access. The GPRS sub-network can, at any time, change a one-phase access request into a two-phase access by allocating only a single uplink block in response to the *packet channel request*.

7.10.2 *Two-phase access contention resolution*

If the RLC/MAC layer has a large number of RLC/MAC blocks to send, it indicates on the *packet channel request* that it requires a two-phase access.

The GPRS sub-network replies to this *packet channel request* with a *packet uplink assignment*, but assigns only *one* RLC/MAC uplink block.

The mobile station accesses the assigned PDCH at the time indicated and sends a *packet resource request, which includes the TLLI of the mobile station.* The mobile station then listens to the PACCH for a *packet UL ack + packet uplink assignment* message.

The GPRS sub-network identifies the mobile station from the TLLI sent in the *packet resource request* message, and returns the TLLI in the *packet UL ack + packet uplink assignment*. When both the GPRS sub-network and the mobile station have received the TLLI, any ambiguity is removed and the contention resolution procedure is complete.

Figure 7.31 illustrates the concept of two-phase access contention resolution. The timer, which was previously T3166 for one-phase access, now becomes T3168. For successful two-phase access contention resolution it is not necessary to send the TLLI on RLC/MAC data blocks.

If two mobile stations react to the initial *packet uplink assignment* and access the single block allocated, the GPRS sub-network may be able to decode one of the *packet resource request* messages (if one mobile station is much nearer to the BTS than the other), and the *packet UL ack + packet uplink assignment* will return the TLLI belonging to the nearest mobile station. The mobile station that receives the wrong TLLI must disconnect from the air interface.

7.11   Channel encoding (GSM 45.003 and 43.064)

Channel encoding allows error detection and correction across the air interface. This is known as forward error correction (FEC), and GPRS uses two functions, block encoding and convolutional encoding. There are four types of encoding, CS1, CS2, CS3 and CS4 used in GPRS as Table 7.1 shows.

All mobile stations are able to use CS1 to CS4 on command from the GPRS sub-network. RLC/MAC control blocks always use CS1 – coding system 1, and RLC/MAC data blocks use coding systems CS1 to CS4 as commanded by the GPRS sub-network.

The commands for setting channel encoding are illustrated in Figure 7.32. This shows that the encoding is commanded in the *packet uplink assignment* message with the parameter *Channel_Coding_Command*.

There is also a parameter *TLLI_BLOCK_CHANNEL_CODING* that tells the mobile station how to encode the TLLI if it is included in RLC/MAC data blocks. TLLI is always included in the RLC/MAC data blocks of a single-phase access TBF. If this parameter is set to '0' then the TLLI is encoded using CS1; if the parameter is set to '1' then the TLLI is encoded as per the data blocks.

The RLC/MAC *control* blocks are always encoded with CS1.

Figure 7.33 shows the process of channel encoding and interleaving. This illustrates the encoding from the GPRS sub-network viewpoint.

Table 7.1   The four types of encoding used in GPRS

Coding	Service bits	Service data rate	Air interface data rate
CS1	181	9.05 kb/s	22.8 kb/s
CS2	268	13.4 kb/s	22.8 kb/s
CS3	312	15.6 kb/s	22.8 kb/s
CS4	428	21.4 kb/s	22.8 kb/s

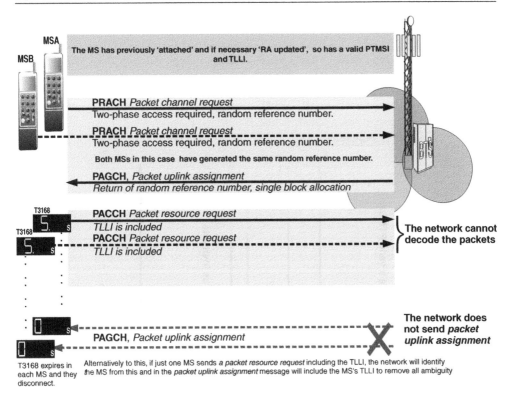

The MS has previously 'attached' and if necessary 'RA updated', so has a valid PTMSI and TLLI.

MSA

MSB

PRACH *Packet channel request*
Two-phase access required, random reference number.

PRACH *Packet channel request*
Two-phase access required, random reference number.

Both MSs in this case have generated the same random reference number.

PAGCH, *Packet uplink assignment*
Return of random reference number, single block allocation

T3168

PACCH *Packet resource request*
TLLI is included

T3168

PACCH *Packet resource request*
TLLI is included

The network cannot decode the packets

The network does not send *packet uplink assignment*

PAGCH, *Packet uplink assignment*

T3168 expires in each MS and they disconnect.

Alternatively to this, if just one MS sends *a packet resource request* including the TLLI, the network will identify the MS from this and in the *packet uplink assignment* message will include the MS's TLLI to remove all ambiguity

Figure 7.31   Two-phase access contention resolution

Shown at the BTS are four sets of three 20 ms blocks. These 20 ms block periods are the standard GSM framing periods for data.

In the first set of three, each 20 ms block has 181 bits within the period plus three bits for USF. The RLC layer puts 181 bits into each 20 ms block because it knows the physical layer is using CS1. However, the 181 bits include at least two octets – 16 bits of header information, so the customer data rate is only in the order of 8 kb/s.

In the second set of three, each 20 ms block contains 268 bits plus an encoded USF of six bits, as the physical layer is using CS2. As the physical layer is using CS2, there is not as much protective redundancy given to the data. However, the USF indicator must be given full protection, therefore it is pre-encoded.

In the third set of three, each 20 ms block contains 312 bits plus an encoded USF of six bits, as the physical layer is using CS3.

In the fourth set of three, each 20 ms block contains 428 bits plus an encoded USF of 12 bits, as the physical layer is using CS4.

These four sets of 20 ms blocks are shown entering a FIRE encoder. The Fire block encoder arranges the data into blocks and adds parity bits and the convolutional encoder adds redundancy in the form of forward error correction (FEC). The Fire and convolutional encoders are set by command of the GPRS sub-network to operate in modes CS1, CS2, CS3 or CS4.

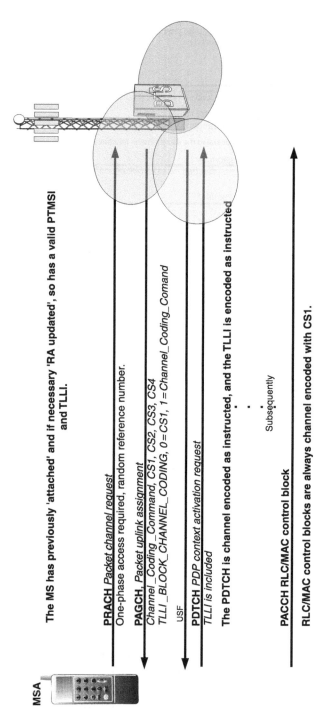

Figure 7.32   Channel coding set-up

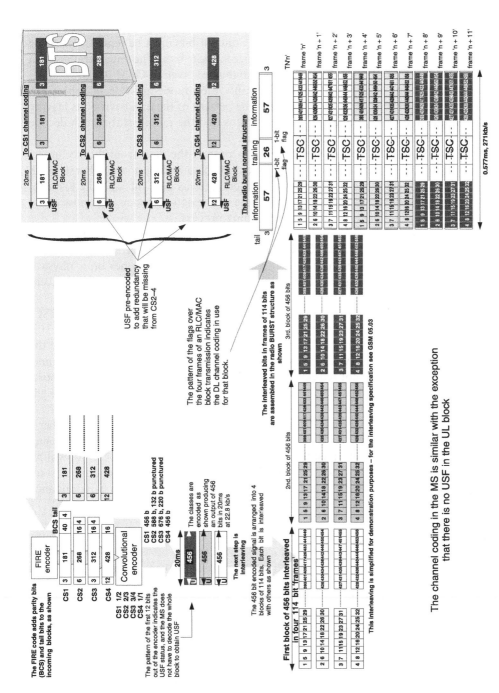

Figure 7.33 GPRS channel encoding

The RLC layer places the required number of user data bits shown in each RLC block in accordance with the channel encoding system used.

The output from the convolutional encoder is always a block of 456 bits. CS2 and CS3 must be 'punctured', that is, some of the redundant bits must be removed after the convolutional encoding to result in a block size of 456 bits. The puncturing and a complete specification for encoding and interleaving are given in GSM 05.03.

All mobile stations on a PDCH in dynamic allocation mode must decode the USF. The USF is encoded separately for CS2–CS4. The pattern of the first twelve bits leaving the convolutional encoder indicates the USF status and the mobile station can decode the USF separately from the rest of the message.

The blocks of 456 bits are next interleaved to protect the data from 'bursty' interference, which tends to destroy complete blocks of data within a radio burst. With interleaving, an air interface burst contains non-contiguous data bits and the loss of a burst results in a loss of bits 'scattered' across the whole 456-bit block. It is possible to recover these scattered data bits from the decoding process.

Figure 7.33 shows simple interleaving. GSM 05.03 specifies the actual interleaving used. The interleaved data is assembled into four sub-blocks, each of 114 bits. Each of these four sub-blocks is fitted into four radio bursts over four frames on a physical channel as shown in the diagram. The encoding for a mobile station is similar except that USF is not included.

In GSM, two flag bits are used, one on each side of the training sequence code (TSC) in the centre of the burst. These flag bits indicate whether each half of the radio burst is carrying customer data (on the traffic channel TCH) or L3 service signalling where the TCH is changed to an L3 service signalling channel, the fast associated control channel (FACCH). These flags are not required in GPRS and are used for a different purpose.

For an RLC/MAC block with a radio burst in each of four frames, there are eight flags available.

As each mobile station using a GPRS physical channel (PDCH) must know whether a particular RLC/MAC block on that physical channel is addressing data to it (using TFI), then the mobile stations must know the channel encoding used on each downlink RLC/MAC block. The eight bits are encoded to indicate the channel encoding used on each RLC/MAC block. Eight bits for four states (CS1–4), give some redundancy against errors introduced by the air interface.

7.12 Frequency hopping

The final topic in this chapter on the physical layer is frequency hopping, which is common to GSM circuit switched and packet switched operations. Frequency hopping is very important to GSM900 network operators. GSM900 has only one hundred and twenty four radio channels and the maximum number of these channels that a European operator can expect is half of these – sixty-two.

Now sixty-two radio channels is not a lot to serve millions of customers. If demand for GPRS services increases in the same way as fixed-line Internet access, then the already hard-pressed GSM900 system may not have the capacity to redouble its customer base (although using half-rate circuit switched channels will help). GSM1800 and GSM1900 have much bigger radio channel allocations and GSM1800 could hold the key to GPRS success! Because of their increased number of radio channels GSM1800 and GSM1900 are not so reliant upon frequency hopping.

The number of radio channels available to an operator is effectively increased through frequency reuse. In the early days of GSM, the frequency reuse factor was 4/12, that is, nearly all the radio channels were placed into a group of twelve cells, if each BTS site controls three cells, then four BTSs are required – hence 4/12. This group of twelve cells was then repeated throughout the PLMN service area. This was fine until demand started to soar, then more capacity (radio channels) was required, but of course there wasn't any!

A solution was to make the cell grouping less than twelve – the smaller the grouping of cells the bigger the slice of radio channels (from a fixed size radio channel pie!) that could be placed in each cell.

If an operator has, for example, 60 radio channels, then the distributions will be:

4/12 cluster size, five radio channels per cell.

3/9 cluster size, six radio channels per cell (plus spare radio channels).

7-cell cluster size, eight radio channels per cell (plus spare radio channels).

But this brings cells that are using the same frequencies nearer to one another and the problem of interference becomes paramount.

Into this picture of increasing capacity demand leading to increased interference steps the knight in white armour – frequency hopping.

Frequency hopping has the ability to spread interference across a number of physical channels. Customers on a physical channel that is experiencing very poor quality due to interference can be returned to acceptability when frequency hopping is used (at the cost of customers who were on very high quality circuits with very little interference having their quality degraded).

This effect of returning the previously unacceptable to acceptability effectively increases the network capacity! And that is the main reason why frequency hopping is now used – to redistribute interference and increase capacity.

Frequency reuse today has reached 1/3 (one BTS, three cells), and in some cases 1/1. Astonishing levels of customer demand can be met with the limited GSM900 radio channel allocations. Some system suppliers claim that they can increase the capacity of a network operator by 100% without any extra BTS sites through the use of sophisticated frequency hopping systems.

However, frequency hopping was not included in the GSM specification to allow increased capacity. It was included to counteract the effects of Rayleigh fading.

7.12.1   Rayleigh fading

Rayleigh fading occurs when there are multiple reflections of a transmitted signal in a reception area and there is no predominant line-of-sight signal.

Figures 7.34 and 7.35 show a mobile station experiencing peaks and troughs in receive level when moving over a small reception area. This variation is due to multiple

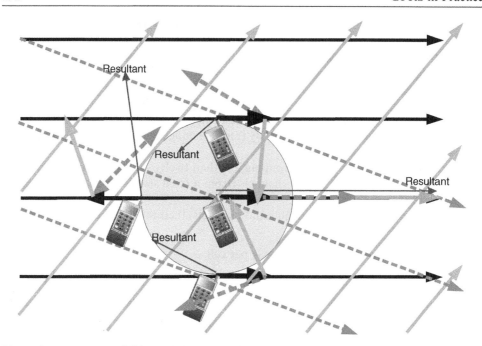

Figure 7.34 Vector addition of the same signal with multiple reflections. This diagram shows a mobile station receiving a reflected signal three times, each signal represented by the thick black, grey and dashed lines in the direction indicated. The black signal is taken as a reference phasor. The grey circle in the centre shows the mobile in four positions. In the centre of the circle the thick black, grey and dashed signals are in phase and their different amplitude signals add as shown. As the mobile station moves to the north, the phase of the solid black signal is unchanged, whereas the phases of the other signals change as shown, resulting in the reduced signal (it is assumed that the amplitudes are constant over a short distance). The other two positions, west and south cause the relative phases and resultant amplitude to change as shown. For multiple reflected signals the distribution of peaks and troughs follows a Rayleigh distribution

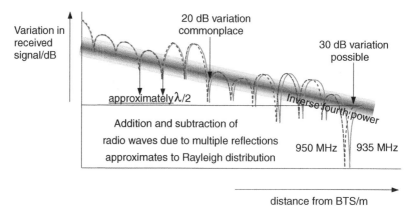

Figure 7.35 Rayleigh fading

reflections of the same signal and Figure 7.35 shows the Rayleigh distribution of the resultant fading. This is superimposed upon a log-normal distribution.

The nulls due to Rayleigh fading occur about every half-wavelength (about 15 cm for GSM900). The depths of fade commonly reach 20 dB and sometimes 30 dB. Such fading has little effect upon fast moving subscribers as a very small time is spent in a signal null and the number of radio frames lost are compensated for by the forward error correction and interleaving used in GSM.

For a slow moving or stationary subscriber, losing many GSM radio frames whilst in the signal null area can lead to severe degradation or total loss of the signal.

One of the advantages of frequency hopping is that the radio field is moved rapidly through a stationary subscriber as indicated in Figure 7.35.

GSM uses frequency hopping with a change of frequency in every radio frame of 4.615 ms. The hopping rate is the inverse of this period, giving 217 hops per second. This is called slow frequency hopping.

Figure 7.35 shows two frequencies, 935 and 950 MHz, corresponding to wavelengths of 32.08 and 31.5 cm respectively. Although the amplitude envelopes of the two signals will differ a little, Figure 7.35 shows that there is a shift in the positions of the nulls from one frequency to the other.

The frequency hopping effect of moving the radio field through a stationary subscriber reduces the effect of Rayleigh fading and the efficacy depends upon the number of frequencies hopped through.

Pseudorandom hopping, on average, does not have the same range of hopping frequencies as cyclic hopping and is not as effective in countering Rayleigh fading.

7.12.2   Rician fading

For a signal field which includes a dominant line-of-sight signal and multiple reflections of the same signal (such as in a micro cell) then the distribution changes from Rayleigh to an approximate Rician distribution. The depth of fading in a Rician distribution is less than in a Rayleigh distribution and the fading problem for slow-moving subscribers is reduced.

A model of Rician fading is illustrated in Figure 7.36. At <1km, the slope of the mean loss is about equal to the free space loss d^2 and changes $\sim$ > 1km to about d^4. If there is no predominant signal, the fading changes to a Rayleigh-like distribution.

7.12.3   Interference spreading

Frequency hopping 'spreads' interference between customers on different frequencies and mitigates to some extent the effects of heavy interference between cells. This is illustrated in Figure 7.37.

The figure shows groups of eight subscribers occupying successive frames of frequencies F1 to F7. (The diagram assumes that subscribers use the eight physical channels on each radio channel as TCHs). One of these frequencies, F4, is experiencing severe interference and the eight subscribers using F4 lose their service.

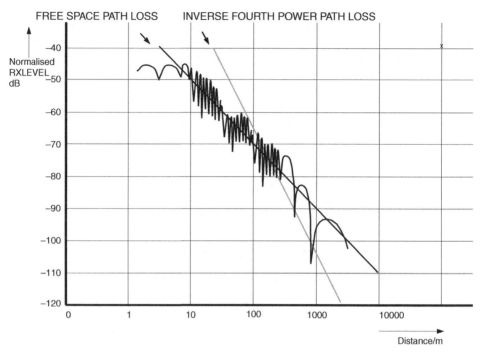

Figure 7.36   A model of rician distribution

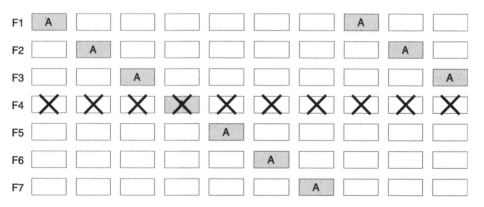

Figure 7.37 Spreading of interference through frequency hopping. This diagram illustrates eight subscribers 'A', hopping through seven frequencies. One of these frequencies, F4, is suffering severe interference. As the subscribers hop into this frequency, communication is lost, but only one seventh of the data of each subscriber is lost. By suitable encoding and interleaving, such as the block encoding, FEC and interleaving used in GSM, this lost data can be recovered. It is evident that without hopping the eight subscribers on F4 would not have service. By hopping all subscribers hop through F4, all are degraded but usable, and the system capacity is increased, in this case by eight subscribers

When the groups of eight subscribers on each frequency are hopped through the seven frequencies, then the eight subscribers on F4 who, without hopping, had lost their services, with hopping, recover their services.

The group of subscribers labelled 'A' are shown hopping F1–F2–F3–F4–F5–F6–F7, with all the other subscribers hopping in a similar (offset) pattern. Each group loses a radio frame at frequency F4, but the loss of one in seven frames is fully recoverable due to the channel encoding and interleaving used in GSM.

Before frequency hopping, the interference was restricted to the group of subscribers using frequency F4, and after frequency hopping the interference is shared or spread between all the subscribers.

The overall system capacity may be increased from 48 to 56 subscribers!

7.12.4 Frequency hopping

GSM cellular communications, like all radio communications, are subject to fading and interference. Fading is frequency and path dependent and can be overcome by using both path diversity and frequency diversity.

Path diversity at the BTS is provided by using two receive aerials, separated by a number of wavelengths so that the paths taken from a mobile station to the BTS antennas are substantially different (in terms of the wavelength of the 900 MHz radio wave – about 30 cm). Changing the frequency on each frame of a traffic channel in the GSM system provides frequency diversity.

Interference at a particular frequency or set of frequencies in the GSM system is normally due to the frequency reuse that is necessary to all cellular systems. What is meant by frequency reuse is that the basic set of absolute radio frequency channel numbers (ARFCNs) that are allocated to a network operator by the government agency are reused from one cell cluster to other cell clusters in order to provide capacity which meets customer demand. A common cell pattern used in GSM is the 4/12 cluster, which uses four sites, each with three cells covering a 120-degree sector, giving a total of twelve cells. These twelve cells commonly use all the ARFCNs allocated to a network operator, and the 4/12 cluster is repeated to provide geographical coverage. The ARFCNs are reused.

This reuse causes downlink interference between cells sharing the same frequency and uplink interference between mobile stations using the same frequency to different cells. This type of interference is called *co-channel interference*.

Another common type of interference is *adjacent channel interference*. ARFCNs that are adjacent to one another actually overlap into each other's frequency band.

GSM networks should be designed to keep these interference problems at a minimum but it is never possible to totally eliminate them. If a traffic channel is using an ARFCN which suffers interference, communication can be made unusable.

It is possible, as we have seen, to 'spread' the interference received by one or more communication channels across many communication channels by 'hopping' each communication channel through all the frequencies used by the cell. This degrades all the communications channels, but channels unusable prior to hopping may be recovered. The effectiveness of frequency hopping in spreading interference is dependent to some extent upon the number of frequencies that are hopped through.

Let us now turn our attention to the organisation of GSM frequency hopping.

There are generally two methods of frequency hopping given to network operators by infrastructure suppliers, *synthesiser hopping* and *baseband hopping*. From the GSM specifications there is no difference in operation between the two, and a mobile station commanded to hop by the network is indifferent to, and does not know how, the network implements hopping.

There are operational advantages and disadvantages to the two methods and both methods will be considered here.

Figure 7.38 illustrates the two hopping methods. The top diagram shows a standard no-hopping configuration. There are four TRXs in a cell and each TRX receives every 4.615 ms eight blocks of 260 bits of speech from eight subscribers. (These are shown as 16 kb/s as there is generally a 3 kb/s signalling overhead between the TRAU – transcoder rate adapter and BSC/BTS). These 260 bit blocks are placed on air bursts, each burst corresponding to a timeslot number. This diagram shows the eight subscribers on each TRX transferred to the air interface on eight physical channels numbered TN0 to TN7. Each TRX stays on a fixed frequency ARFCN.

The next diagram in Figure 7.38 shows the baseband hopping method of achieving frequency hopping. The incoming blocks of 260 bits, one block from each of thirty-two subscribers go into a baseband distribution module. This switches groups of eight subscribers between TRXs, frame by frame every 4.615 ms in a cyclic fashion (it could also be a pseudorandom fashion). Each TRX remains on a fixed frequency but the subscriber traffic channels are switched from TRX to TRX for each frame period of 4.615 ms. The effect is illustrated on the air interface where the 'white' frame of channels is first on F1 – the ARFCN of TRX1, then on F4 – the ARFCN of TRX4, then F3 (TRX3), then F2 (TRX2).

An advantage of baseband hopping to the network operator is that the transmitters remain on fixed frequencies and can be coupled together through low loss filters increasing the available transmitted power at the antenna.

A disadvantage of this method is that the number of hopping frequencies is limited to the number of TRXs in a cell.

The third diagram in Figure 7.38 shows the *synthesiser frequency hopping* method. This is the method used by mobile stations and, because of its advantages, tends to be used by the network operators. Eight blocks of 260 bits enter each TRX once per 4.615 ms frame period. Each block is one of the eight subscriber traffic channels at each TRX (control channels will be considered later). Each TRX changes the eight subscriber traffic channels into eight air interface radio bursts, numbered TN0–TN7, the eight bursts last for a frame period, 4.615 ms. This frame of eight bursts is transmitted at a frequency F1, F2, F3 or F4 and we shall take F1 as an example. On the next frame the frequency of each TRX is changed. Looking at the output of TRX1, this changes from frequency F1 to F2. The other TRXs are hopped in the same sequence shown, but sequentially 'offset' by one frequency.

The transmitter and receiver synthesiser is changed in frequency every 4.615 ms following the required hopping sequence.

The BTS, with many more traffic channels than a mobile station, is more complex in its organisation to achieve the same end.

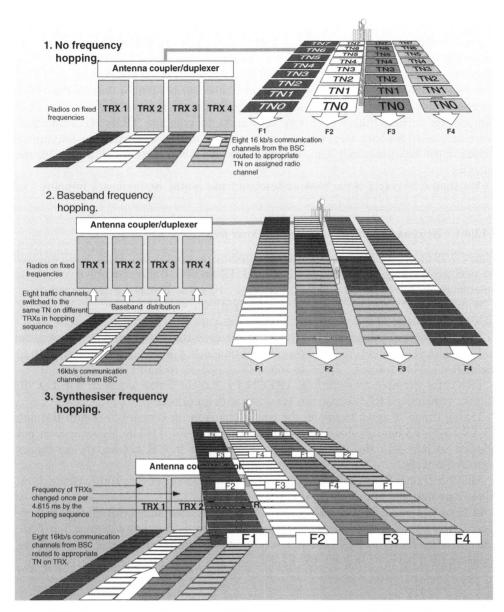

Figure 7.38   Illustrations of no frequency hopping, and baseband and synthesiser hopping

An advantage of synthesiser hopping is that the range of hopping frequencies is not restricted to the number of TRXs in use. This can be of great assistance in the frequency organisation of GSM cells when frequency interference problems arise. However, the organisation of the BCCH carrier physical channels within the hopping sequence must be treated with care.

In general, the BCCH carrier physical channels (TN1–7 in general) must only be included in the frequency hopping sequence when the number of frequencies used for hopping is the same as the number of TRXs. *TN0 of the BCCH carrier never hops in frequency, nor do CCCHs if they are on timeslots other than TN0.* These points are considered in detail later.

A disadvantage to the network operator of synthesiser hopping is that, as the TRXs are continually changing frequency, it is not possible to use low loss frequency sensitive components such as filters or cavity combiners to combine TRXs on a common antenna. Resistive combiners, with a loss of about 3 dB must be used, wasting power (which must be dissipated) and reducing the available transmitted power at the antenna.

We shall now take a closer look at baseband and synthesise frequency hopping.

7.12.4.1 Baseband frequency hopping – a closer look

Figure 7.39 examines baseband frequency hopping in more detail. Here, there are five transceivers working on ARFCNs F1, F2, F3, F4 and F5, transmitted by a cell whose antenna is on the left of the diagram. Frequency F5 is the BCCH carrier frequency.

With five TRXs there is a maximum capability for 40 traffic channels, but as the BCCH carrier uses TN0 for control channels there are only 39 traffic channels.

The air interface frames are labelled A–E for each transmitted frequency, and the traffic channels are numbered as shown, traffic channels 1–8 belong to frame A, 9–16 to frame B, 17–24 to frame C, 25–32 to frame D and 33–39 to frame E.

TRX1, for example, transmitting ARFCN F1, has the frame sequence A-E-D-C-B, and the remaining TRXs follow this sequence with an offset.

Taking frame A as an example, the traffic channels on timeslots 1–7 hop through F1-F2-F3-F4-F5 but timeslot 0 hops through F1-F2-F3-F4. TN0 in this sequence is not allowed to hop into the BCCH carrier, TN0 in the BCCH carrier always carries control channels.

All other frames follow this same sequence but with an offset. The traffic channels on TN1–7 are hopping into the BCCH carrier.

In baseband frequency hopping, there is always the same number of carrier frequencies as there are TRXs, and the BCCH carrier can be in the hopping sequence, but TN0, and any other TN of the BCCH carrier, is left out of the hopping sequence. The hopping sequence for TN0 is F1–F2–F3–F4–F1–F2 . . . etc.

7.12.4.2 Synthesiser frequency hopping – a closer look

Figure 7.40 shows five transceivers employing synthesiser frequency hopping. Although this diagram looks the same as that of baseband hopping, and externally (to a mobile station) they are identical, the difference is that each set of eight subscribers stays on the same TRX which changes frequency every frame. The traffic channels in TN0 in each case hop over only four frequencies; as before TN0 of the BCCH frequency is reserved for the BCCH control channels and takes no part in the hopping sequence.

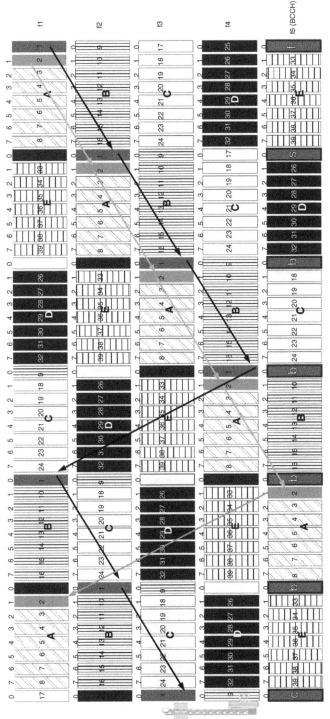

Figure 7.39 Frequency hopping over five frequencies using 'baseband hopping'

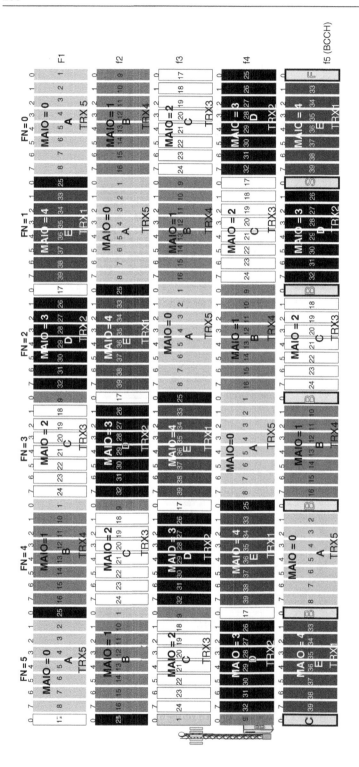

The MAIO is used in the following way:
To access the correct frequency within the hopping sequence ARFCN = (frame number + MAIO)mod 'n' where
n is the number of frequencies in the hopping sequence (MA).

For example for TN2 on frame 0, ARFCN = (0+0)mod 5 = 0, the MS will access the lowest frequency in the sequence.
for TN2 on frame 0, ARFCN = (0+1)mod 5 = 1, the MS accesses the next highest frequency
for TN2 on frame 1, ARFCN = (1+1)mod 5 = 2, the MS accesses the next highest frequency

Figure 7.40   Frequency hopping over five frequencies using 'synthesiser hopping'

This figure is very similar to the baseband hopping sequence of the previous diagram because the number of frequencies hopped through is equal to the number of TRXs. In that case, synthesiser hopping has no advantage over baseband hopping. (In fact it has a disadvantage in that the aerial couplers must be resistive and high loss).

A network operator normally makes the synthesisers hop over more frequencies than there are TRXs in a cell. This is very useful for low capacity cells, which have only two or three transceivers – the hopping is not constrained to two or three frequencies. However, if more frequencies are hopped through than there are TRXs (for example five frequencies are hopped through in a cell which has only three TRXs), it is not possible to hop the BCCH frequency if the BCCH frequency is carrying traffic channels (on TN1 to TN7 for example).

If a network operator wishes to hop through more frequencies than there are TRXs, and carry traffic on all the TRXs, then *the TRX that is reserved for the BCCH frequency does not hop*. In that way the BCCH frequency may carry non-hopping traffic channels.

The BCCH carrier may only be hopped through if it does not carry traffic channels. If the BCCH TRX does carry traffic channels then it must transmit at all times on the BCCH frequency – without hopping.

Figure 7.41 shows the effect of hopping through five frequencies with only three TRXs, with all TRXs carrying traffic channels. With only three TRXs, then only three frequencies can be transmitted simultaneously. If the traffic channels of the BCCH carrier are hopped, this results in a hopping sequence for one of the TRXs as shown (F5–F5–F5–F1–F2–F5) and such irregular sequences are not allowed in the GSM system.

In GSM, the sequences of hops are limited to cyclic codes, which step from the lowest frequency progressively to the highest frequencies, or one of a set of 63 pseudorandom codes. If the BCCH TNs are hopped in this way then they cannot carry traffic channels. If the BCCH carrier TNs are not carrying TCHs then no mobile station will be commanded to follow the irregular sequence. Each of the empty BCCH TN bursts must, however, be transmitted, and this is possible only by transmitting dummy bursts where the traffic channels would have been. Valuable spectrum is used for no benefit.

The way that hopping is arranged in the case where synthesiser hopping is used with more hopping frequencies than there are TRXs is this:

- The BCCH carrier is totally left out of the hopping sequence. The free TNs on the BCCH carrier can then be loaded with TCHs.
- The TRXs which are not reserved for the BCCH frequency hop over as many frequencies as the network operator has available.

To allow frequency hopping, the mobile station is given the following parameters by the network when radio resources are assigned.

Mobile allocation (MA)

The mobile allocation (MA) is a list of frequencies sent in the *immediate assign* or *packet uplink/downlink assignment* message which gives the mobile station a sub-set of

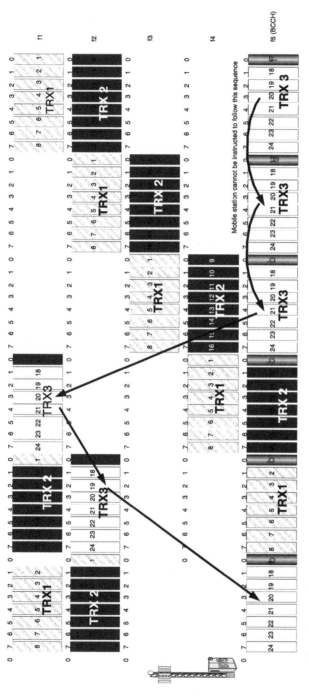

Figure 7.41 Attempting to synthesiser hop over five frequencies with only three TRXs – all TRXs carrying traffic. Attempting to frequency hop over five frequencies (or even four) with only three TRXs is not possible if one of the frequencies hopped through is the BCCH frequency and the BCCH frequency is carrying traffic channels. This can be seen above, where TRX1 and TRX2 are hopping cyclically through the BCCH, but as they go through the BCCH frequency, TRX3 must change frequency to one of the other five frequencies. The hopping sequence for TRX3 is irregular and not allowed in GSM. If TRX3 is not carrying traffic channels then this irregular hopping is acceptable as mobile stations will not have to follow it. Another alternative is to have TRX1 and TRX2 hopping over only four frequencies, not including the BCCH frequency and TRX3 carrying traffic channels in non-hopping mode

frequencies of the set used by the cell (the full set of frequencies used by a cell is called the cell allocation (CA) on the system information of the BCCH) The MA subset is used for frequency hopping.

The MA list is the set of frequencies over which the mobile station must hop. In the case of Figure 7.39 the MA will have one of two values:

- If the mobile station is allocated to a physical channel on TN0 then the MA will contain F1, F2, F3 and F4.
- If the mobile station is allocated to a physical channel in any other timeslot the MA will contain F1, F2, F3, F4 and F5.

Hopping sequence number (HSN)

This information is given to the mobile station in the *immediate assign* or *packet uplink/downlink assignment* message and tells the mobile station which one of the 64 possible hopping sequences is used by the cell, where;

- HSN0 = cyclic hopping which starts from the lowest frequency in the MA group of frequencies and hops progressively through higher frequencies.
- HSN1–63 = GSM pseudorandom hopping sequences which hop in a pseudorandom fashion through the MA list.

Mobile allocation index offset (MAIO)

This information is provided to the mobile station in the *assignment* messages. It allows the mobile station to calculate the start frequency used when joining the hopping sequence at a particular frame number.

For example, in Figure 7.40, one of the MA lists is F1–F5 (for mobile stations allocated TN1–TN7) and the cyclic sequences below could be taken for TN1 joining the sequence at various frames:

Frame 0	F1	F2	F3	F4	F5
Frame 1	F2	F3	F4	F5	F1
Frame 2	F3	F4	F5	F1	F2
Frame 3	F4	F5	F1	F2	F3
Frame 4	F5	F1	F2	F3	F4

The MAIO tells the mobile station at which frequency for which frame number the MA sequence will be joined. For this example, the MA list is indexed 0–4, where 0 is the lowest frequency in the MA list and 4 the highest frequency.

For frame number FN:

$$\text{ARFCN index} = (\text{FN} + \text{MAIO}) \bmod n$$

where n is the number of frequencies in the MA.

Example 1:

A mobile station accesses the sequence on frame number FN=0.

It is given in the assignment message MAIO=0, MA = 12 frequencies.

The frequency index for accessing the sequence on FN0 = (0 + 0) mod12 = 0

The mobile starts at the lowest frequency in the allocation.

Example 2:

A mobile station accesses the sequence on frame number FN = 1.

It is given in the assignment message MAIO = 0, MA = 12 frequencies.

The frequency index for accessing the sequence on FN1 = (1 + 0) mod12 = 1

The mobile starts at the next to lowest frequency in the allocation.

Example 3:

A mobile station accesses the sequence on frame number FN = 12.

It is given in the assignment message MAIO = 0, MA = 12 frequencies.

The frequency index for accessing the sequence on FN12 = (12 + 0) mod12 = 0

The mobile starts at the lowest frequency in the allocation.

The MAIO gives the network operator the flexibility to use the same frequencies in different cells controlled by the same BTS. By giving different MAIOs to each cell the operator can arrange the hopping so that each cell in the sector is transmitting on a different frequency at any instant of time. This requires careful planning of the MAIO.

If the HSNs applied to cells belonging to the same BTS are orthogonal, it is possible to arrange that all the cells belonging to that BTS use the same frequency set. This is only possible using 'synthesiser hopping' using more hopping frequencies than there are TRXs.

7.12.4.3 Pseudorandom hopping

Figure 7.42 shows a simplified algorithm to generate the hopping sequence used for pseudorandom hopping. This algorithm is used at the BTS and the mobile station. This type of hopping is more resilient to interference than cyclic hopping.

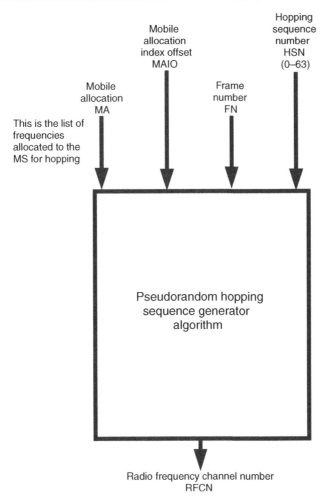

Figure 7.42 The inputs to the GSM frequency hopping algorithm. This algorithm is used at the TRXs and at the MS to determine which frequency is used for a TCH or SDCCH at a particular frame of a burst (TN)

Figure 7.43 shows that each of the three cells under the control of the BTS is using the same frequency set. Because each cell is allocated a different set of MAIOs, it is guaranteed that a frame transmitted by one cell will be at a different frequency to the same frame transmitted by the other cells. For this to be possible, there must be as many hopping frequencies are there are TRXs in use. The total number of TRXs in this case is six hopping TRXs, therefore there must be six hopping frequencies in each cell. In addition, there are three non-hopping BCCH frequencies.

This is an example of 1/1 frequency reuse. However, it only works well if each cell is in synchronism with the others, that is TN0 and the frame numbering for each cell are

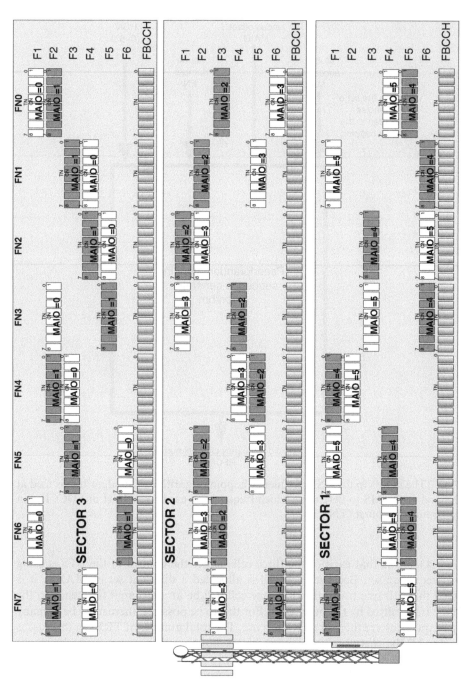

Figure 7.43 Illustrating frequency hopping using the same HSN (pseudorandom) and frequency set in each sector of a BTS through the allocation of separate MAIOs

transmitted at the same time. In the case of cells controlled by a single BTS site, this is always the case, but in the GSM system, different BTS sites are rarely synchronised, and hence the application is limited.

7.13 Abbreviations used in this chapter

AGCH	Access grant channel
ARFCN	Absolute radio frequency channel number
BA	BCCH allocation
BCC	Base station colour code
BCCH	Broadcast control channel
BSIC	Base station identity code
BTS	Base transceiver station
CA	Cell allocation
CCCH	Common control channel
CS	Circuit switched
CS	Coding system
DL	Downlink
DRX	Discontinuous reception
EBCCH	Extended broadcast control channel
EM	Extended measurement
FACCH	Fast associated control channel
FBI	Final block indicator
FEC	Forward error correction
FN	Frame number
GMM	GPRS mobility management
GMSK	Gaussian minimum shift keying
GPRS	General packet radio service
GRR	GPRS radio resources
GSM	Global system for mobile communication
HCS	Hierarchical cell structure
HPLMN	Home public land mobile network
HSC	Hopping sequence code
HSN	Hopping sequence number
IMSI	International mobile subscriber identity
IP	Internet protocol
KC	Number of physical channels carrying PCCCHs
LA	Location area
LAI	Location area identifier
LLC	Logical link control
MA	Mobile allocation
MAC	Medium access control
MAIO	Mobile allocation index offset
MS	Mobile station
MSC	Mobile switching centre
MSPC	Mobile station power class

NC   Network controlled
NCC   Network colour code
NCH   Notification channel
NCO   Network control order
NMO   Network mode of operation
PACCH   Packet associated control channel
PAGCH   Packet access grant channel
PBCCH   Packet broadcast control channel
PCH   Paging channel
PCCCH   Packet common control channel
PDCH   Packet data channel
PDP   Packet data protocol
PDTCH   Packet data traffic channel
PDU   Protocol data unit
PLMN   Public land mobile network
PPCH   Packet paging channel
PRACH   Packet random access channel
PSI   Packet system information
PTCCH   Packet timing advance control channel
PTMSI   Packet temporary mobile subscriber identity
RA   Routing area
RACH   Random access channel
RF   Radio frequency
RLA   Receive level average
RLC   Radio link control
RR   Radio resources
RRBP   Relative reserved block period
RXLEV   Receive level
S   Spreading for PRACH
SACCH   Slow associated control channel
SCH   Synchronisation channel
SDCCH   Standalone dedicated control channel
SGSN   Serving GPRS support node
SIM   Subscriber identity module
SM   Session management
SMS   Short message service
SNDCP   Sub-network dependent convergence protocol
TA   Timing advance
TAI   Timing advance index
TBF   Temporary block flow
TCH   Traffic channel
TFI   Temporary flow identifier
TLLI   Temporary logical link identifier
TN   Timeslot number
TRAU   Transcoder rate adapter unit
TRX   Transceiver

TSC	Training sequence code
TX	Transmit
UL	Uplink
UMTS	Universal mobile telecommunications system
USF	Uplink status flag

8

RLC/MAC Layer Procedures
(GSM 44.060)

The RLC layer provides a layer two data communications link that guarantees (in acknowledged mode) the delivery of logical link control (LLC) PDUs across the air interface. For this guarantee, the RLC layer must be operating in RLC acknowledged mode, but it can also operate in RLC unacknowledged mode, where no guarantee is given for the delivery of logical link control (LLC) PDUs.

The RLC layer processes the LLC data blocks into standard sub-blocks, which can be conveniently transferred across the air interface. Each sub-block is called an RLC/MAC block, which, we will recall, requires a radio burst in each of four consecutive air interface frames to transfer. Within these four frames, the number of blocks transferred depends upon the number of physical channels allocated to the mobile station; if it is given just one TN then one RLC/MAC block is transferred, if it is given two TNs, then two RLC/MAC blocks are transferred, and so on. At the start of the RLC/MAC block is the header information which, in conjunction with RLC registers at both ends of the air interface, controls the orderly transfer of data.

This chapter is about the radio link control layer; the functions of the medium access control layer have been covered in previous chapters. This chapter covers:

- The inputs and outputs of the RLC layer, the signalling to and from the logical link control (LLC) and MAC layers.
- RLC/MAC control and data block structures.
- Segmentation of logical link control (LLC) PDUs and the construction of the RLC/MAC headers.
- RLC registers and control fields.
- Operation of the RLC/MAC window.
- Arrow diagrams illustrating RLC procedures for acknowledged and unacknowledged modes.
- Half and full duplex operation of the RLC/MAC layer.
- RLC/MAC counters and timers.

GPRS in Practice: A Companion to the Specifications   Peter McGuiggan
© 2004 John Wiley & Sons, Ltd   ISBN: 0-470-09507-5

8.1   Introduction

Figure 8.1 shows a simplified picture of the functions of the RLC and MAC layers. These include the segmentation of LLC PDUs and RLC control PDUs. It provides acknowledged or unacknowledged operation for LLC PDUs and unacknowledged operation for the RLC control PDUs. ARQ is operational in acknowledged mode.

The conditions for RLC acknowledged mode transmission are:

- Sub-network PDUs belonging to a PDP context with a QoS profile which includes reliability classes 1, 2 or 3.
- GMM (higher layer service signalling) PDUs are also acknowledged.

The conditions for RLC unacknowledged transmission are:

- Sub-network PDUs belonging to a PDP context with a QoS profile which includes reliability classes 4 or 5.
- RLC/MAC control PDUs are unacknowledged.

The functions of the MAC layer are also indicated in Figure 8.1. The MAC layer is responsible for the management of the logical channels and physical channels to get the correct information onto the right logical channel and over the air interface at the correct time on the correct physical channel.

Measurements, measurement reports, DRX and cell reselection are the responsibility of the radio resource physical link layer, and they have been described in Chapter 7.

Figure 8.1 also shows all the message types flowing over the air interface and the logical channels that carry them. For example, the PTCCH shown at the bottom of the physical link/MAC interface carries access bursts from the mobile station and receives timing advance messages from the GPRS sub-network. (The figure is drawn from the mobile station's point of view.)

8.2   RLC procedures

The purpose of the RLC component of the RLC/MAC layer is:

- To provide an interface for the logical link control (LLC) primitives, allowing transfer of LLC protocol data units between the MAC and LLC.
- To segment LLC PDUs into RLC data blocks, and reassemble received RLC data blocks into LLC PDUs for transfer to the LLC layer.
- To segment and reassemble RLC/MAC control messages into and out of RLC/MAC control blocks.
- To use, where appropriate, ARQ for retransmission of RLC/MAC data blocks received with errors (RLC acknowledged mode).

This section looks at the procedures used by the RLC/MAC layer to achieve these purposes.

Figure 8.2 illustrates the structure of RLC/MAC blocks. An RLC control block is transmitted to a mobile station on the PACCH on physical channel TN5. It takes four

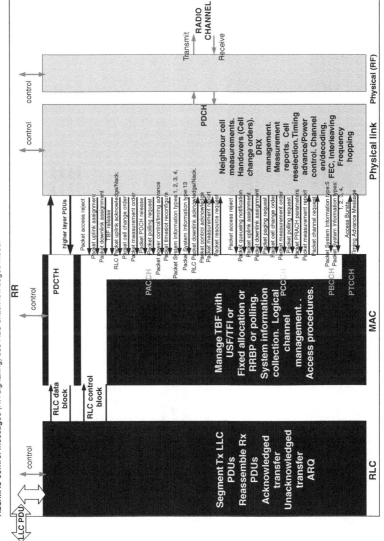

Figure 8.1 The RLC/MAC layer in context

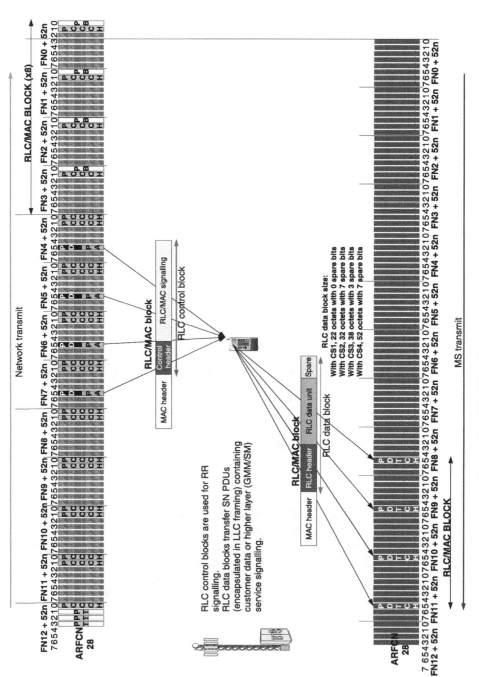

Figure 8.2 RLC/MAC control and data blocks

bursts of this TN to carry an RLC/MAC block. The mobile station is shown sending an RLC data block to the GPRS sub-network on the same (uplink) physical channel. RLC control blocks are always sent on either the PACCH or PCCCH. RLC control blocks carry radio resource signalling. RLC data blocks are always sent on the PDTCH. RLC data blocks carry either customer data or higher layer (GMM or SM) service signalling.

Of course, the GPRS sub-network may also send RLC data blocks and the mobile station RLC control blocks.

Figure 8.2 also shows the number of octets sent in an RLC data block for each of the four channel encoding systems available. CS1 is the commonly used coding system as this is currently the only mandatory system for the GPRS sub-network (all four coding systems are mandatory for the mobile station). CS1 allows 22 octets per RLC/MAC data block. Of these 22 octets, at least two are used as RLC headers; one or more extra may be used for other purposes, giving customer data block lengths of 20, 19 or 18 octets (in some cases even less than 18). As it takes $4.615\,\mathrm{ms} \times 4 =$ (frame period $\times$ four frames) $= 18.46\,\mathrm{ms}$ to transmit an RLC/MAC block, then the customer data rate on the air interface using $CS1 = (20 \times 8\,\mathrm{bits})(18.46\,\mathrm{ms})^{-1} = 8.7\,\mathrm{kb/s}$ maximum to $7.8\,\mathrm{kb/s}$ minimum. This disregards overheads in the sub-network dependent conversion protocol and LLC layers and also disregards the inclusion of TLLI (four octets) when one-phase access is used.

Figure 8.3 shows the overall flow for transferring N PDUs (network protocol data units) between peer entities. The first primitive SN-DATA-REQ carries a network PDU from the mobile station's user layer (more correctly the network layer) to the SNDCP layer. This is compressed and changed to an SN PDU (sub-network PDU) by the SNDCP layer and carried by the primitive LL-DATA-REQ to the LLC layer.

The maximum allowable number of octets in an SN PDU is 1520. The LLC layer frames the SN PDU within a data link protocol, which may guarantee delivery to the LLC peer entity; it is ciphered and passed on as an LLC PDU carried by the GRR-DATA-REQ primitive. The maximum size of an LLC PDU is 1563 octets, including headers etc.

The LLC layer guarantees delivery (LLC acknowledged mode) of SN PDUs which have a quality of service reliability class one or two. All other SN PDUs are carried by the LLC layers between the mobile station and GPRS sub-network in unacknowledged mode.

An LLC PDU arriving at the RLC layer initiates a TBF and the RLC /MAC layer processes this LLC PDU for a TBF.

The RLC layer segments the larger LLC PDU into smaller RLC data blocks, adds an RLC header to each block and passes the RLC blocks to the MAC sub-layer, which adds a MAC header to each of the blocks and arranges for the RLC/MAC blocks to be transmitted as a TBF at the correct time on the correct air interface logical channel.

For the case illustrated in Figure 8.3, the LLC PDU has 1563 octets and must be segmented into RLC data blocks, which carry a maximum of 20 octets (CS1) of the LLC PDU. Each RLC/MAC block is then encoded and interleaved before modulation and transmission.

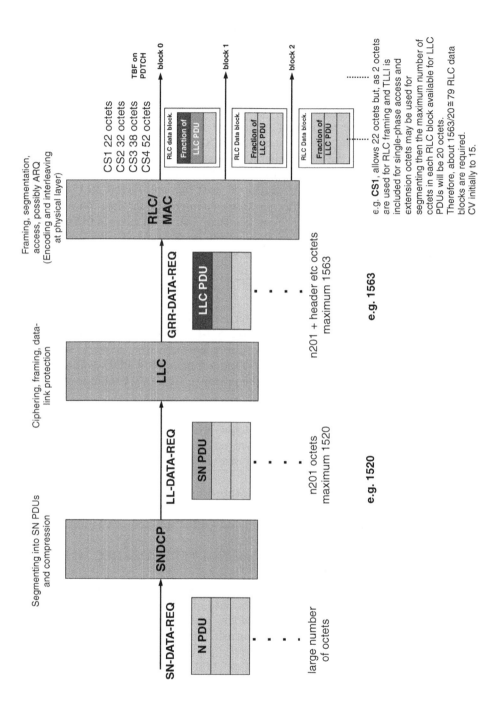

Figure 8.3   A simplified view of RLC segmenting

8.2.1   Segmenting LLC PDUs into RLC data blocks (GSM 44.060 section 9)

Three RLC parameters are used to indicate segmentation of LLC PDUs. These parameters are transmitted in the RLC headers of RLC data blocks when necessary. The parameters are:

1. **LI (the length indicator)**. This is only used if:

   —An RLC block is not totally filled with octets from an LLC PDU, or
   —An RLC block contains octets from two LLC PDUs. Then the LI indicates how many octets of the first LLC PDU are carried. The length of the second LLC is implicit as long as it totally fills the remaining space of the RLC block.
       In theory, an RLC block could contain part of one LLC PDU, followed by a complete second LLC PDU, followed by part of a third LLC PDU. If this were to be the (highly unlikely) case, then two or three length indicators would be required, one to signify the end of the first LLC PDU, and another to signify the end of the second LLC PDU, and another if the third PDU did not fill the block.

2. **M (more)**. This indicates if the end of the first LLC PDU within the RLC block is followed by part of a second (more) LLC PDU.
3. **E (extension octet)**. This indicates whether an LI octet is included in the RLC block; this will only be necessary if an LLC PDU does not totally fill the RLC/MAC block, or if the RLC/MAC block is carrying more than (parts of) two LLC PDUs.

Figure 8.4 shows the principles of the segmentation of logical link control (LLC) PDUs into RLC data blocks. Three LLC PDUs, PDU1, PDU2 and PDU3, are shown reaching the RLC/MAC layer.

The RLC/MAC layer can send multiple LLC PDUs as one TBF if the sum total of the octets contained within all the LLC PDUs is less than the storage capacity of the RLC buffer store. The RLC buffer store must be capable of storing up to 64 RLC/MAC blocks. As the largest size RLC/MAC block for CS4 carries 52 octets, then the storage capacity of the buffer must be 3328 octets.

The illustrated LLC PDUs contain 256, 165 and 180 octets of data respectively. These are processed by the RLC layer and sent in one TBF.

Shown leaving the RLC/MAC layer are arrows representing RLC data blocks. One arrow represents one RLC data block, occupying one burst per frame in four consecutive frames on the air interface. One RLC data block can transfer only 22 octets in these four bursts if channel encoding CS1 is used and at least two of these octets are used in the RLC header. The structure of the RLC/MAC block is shown at the bottom of Figure 8.4.

For segmenting, we are interested in the parameters block sequence number (BSN), E, M and LI. Note that octets 3, 4, 5 and 6 are carrying the extension octets. These are only included if necessary; this depends upon the E bit; if $E =$ '0', then an extension octet is used. So, looking at octet 2, if the E bit there is set to '0', then octet 3 is used as an extension octet. This happens when a new PDU starts in the RLC data block. If octet 3 is used as an extension octet, then it contains another E bit; if this is set to '0', then another octet – octet 4 – is used as a further extension octet. This second extension octet gives the number of octets the new LLC PDU occupies in the RLC data block. This process is shown to the top of Figure 8.4.

The maximum number of octets in an LLC-PDU I frame is 1520 data octets (N201) +37 Control /Address octets + 3 FCS octets = 1560 octets. The minimum number of octets in an LLC-PDU (carrying an SN-PDU) is 140 data octets + 37 Control/Address octets + 3 FCS octets =180 octets.

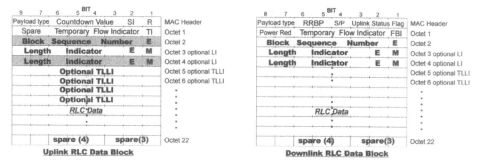

E	M	Function
0	0	Reserved, if received by the MS, the MS will ignore all fields except for the MAC header.
1	0	There is no More LLC data (from another LLC PDU) after this LLC PDU and (hence) no more Extension octets.
0	1	There is a new (More) LLC PDU within this RLC data block after the current LLC PDU, and (hence) another Extension octet to delimit the new LLC PDU (show where it starts within this RLC data block)
1	1	There is a new (More) LLC PDU within this RLC data block after the current LLC PDU, and it continues until the end of the RLC information field octets ; there are no more Extension octets.

The LI - Length Indicator field is used to delimit (mark the beginning and end) of LLC PDU's within an RLC data block. The first LI header octet included in a BSN indicates the number of octets in that BSN belonging to the first LLC PDU of the TBF; if a second LI is present it indicates the number of octets of a second LLC PDU The *final* RLC data block of a TBF has an LI field unless the final LLC PDU has precisely 20 octets *remaining* and completely fills the information field of the RLC data block

Uplink RLC Data Block

	BIT							
8	7	6	5	4	3	2	1	
Payload type		Countdown Value				SI	R	MAC Header
Spare		Temporary Flow Indicator					TI	Octet 1
Block	Sequence	Number					E	Octet 2
Length	Indicator					E	M	Octet 3 optional LI
Length	Indicator					E	M	Octet 4 optional LI
Optional TLLI								Octet 5 optional TLLI
Optional TLLI								Octet 6 optional TLLI
Optional TLLI								
Optional TLLI								
RLC Data								
spare (4)				spare(3)				Octet 22

Downlink RLC Data Block

	BIT							
8	7	6	5	4	3	2	1	
Payload type		RRBP		S/P		Uplink Status Flag		MAC Header
Power Red		Temporary Flow Indicator					FBI	Octet 1
Block	Sequence	Number					E	Octet 2
Length	Indicator					E	M	Octet 3 optional LI
Length	Indicator					E	M	Octet 4 optional LI
								Octet 5 optional TLLI
								Octet 6 optional TLLI
RLC Data								
spare (4)				spare(3)				Octet 22

Figure 8.4   Segmenting of LLC PDUs into RLC blocks

In general, when $E = 0$ this means that the current LLC PDU comes to an end in this RLC/MAC block and the extension block gives information on the remaining length of this current LLC PDU.

When $M = 1$ in an extension octet this means that there is a second (or third or fourth) LLC PDU starting in this RLC/MAC block. If this RLC/MAC block is totally filled by the second (third, fourth) LLC PDU then there is no need for another extension block to indicate its length; if the following LLC PDU does not completely fill the RLC/MAC block, an extension octet is needed to indicate its length.

Looking at the RLC/MAC block arrows in Figure 8.4, after the RLC layer has processed the LLC PDUs into RLC blocks, the access procedure requests uplink radio resources to transfer the blocks across the air interface. Upon access procedures completion we see the first RLC data block with BSN0 sent.

This access procedure is two-phase. This means that the RLC/MAC blocks carrying the LLC PDUs do not include the TLLI as this is exchanged separately as part of the two-phase access procedure. If the access had used the single-phase procedure then the RLC/MAC blocks would carry the TLLI.

- **BSN0**. The first octet of all BSNs contains the TFI allocated by the GPRS subnetwork, and TI (TLLI indicator), which tells the GPRS sub-network whether the mobile station's TLLI is included in this RLC/MAC block. The E bit of the second octet of BSN0 is set to one, indicating that octet 3 is not used as an extension octet. Octet 3, which would contain the length indicator, is not used as the LLC PDU1 completely fills BSN0. There are 20 octets of LLC PDU1 in BSN0; two of the available 22 octets are used as headers.
- **BSN1–11**. As these RLC data blocks are totally filled with octets from LLC PDU1, no extension blocks are necessary and each BSN carries its full complement of 20 octets of LLC PDU1. The total number of LLC PDU1 octets carried by BSN0–11 is 240 and only 16 octets remain to be carried by BSN12.
- **BSN12**. As this contains 16 octets of LLC PDU1, a new PDU, PDU2, will also start in BSN12. This is indicated by octet 2 having $E = $ '0', and octet 3 becomes the first extension octet. Octet 3 indicates with LI that, for PDU1, 16 octets are carried in this BSN. It also indicates ($E = 1$, $M = 1$), that the new LLC PDU fills up the remaining space in the RLC block. BSN12 therefore carries 16 octets of LLC PDU1 and three octets of LLC PDU2, with three octets required for the RLC header.
- **BSN13–21**. The sequence continues as indicated in Figure 8.3, until BSN21 is reached. This BSN carries the final two octets of LLC PDU2 and 17 octets of LLC PDU3, with three octets required for RLC headers.
- **BSN22–30**. BSN22–29 each carry 20 octets of LLC PDU3. The total carried in BSN21–29 is 177 octets, therefore there are three remaining octets of LLC PDU3 to be carried in BSN30. In the second octet of BSN30, $E = 0$ indicates an extension octet. This extension octet length indicator indicates $LI = 3$, the quantity of LLC PDU3 octets carried in this RLC block. $E = 1$, $M = 0$ indicates that there are no more LLC PDUs after this RLC block. The LLC PDU transfer is completed.

The MAC header which is placed on top of the RLC block has a countdown value (CV). When this reaches zero, the uplink LLC PDU transfer is completed. The CV counter has a maximum (and default) value of fifteen and for TBFs of more than sixteen blocks, the counter does not start decrementing until there are fifteen blocks remaining to be sent.

8.3   RLC/MAC block headers and parameters (GSM 44.060 section 10)

Figure 8.4 shows the RLC/MAC block headers and parameters transmitted in the uplink and downlink directions for the *RLC data block* and Figure 8.5 shows the structures of the UL and DL *RLC control blocks*. Figure 8.6 shows the same parameters plus registers in use at each end of the radio link with a brief description of the function of each field. The parameters indicated on this diagram will be used when RLC block flow is considered.

8.3.1   RLC data block (uplink) parameters

8.3.1.1   In the MAC header

- **MAC R bit**. This is used by the mobile station on the uplink only to indicate how many attempts – retry – were made with *packet channel request* messages – one or more than one.
- **MAC SI (stall indicator) bit**. This indicates whether the mobile station's window can be advanced or not. If it cannot be advanced it is 'stalled'. The window size is 64 blocks and unless all the transmitted blocks within this window have been acknowledged, the

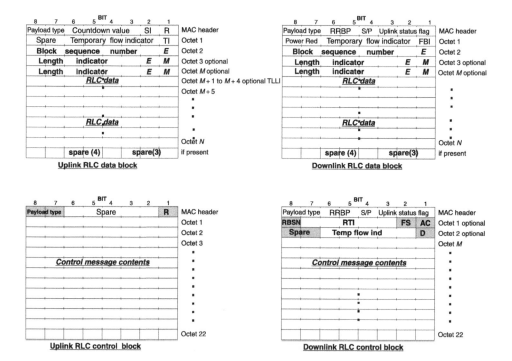

Figure 8.5   RLC/MAC block structures

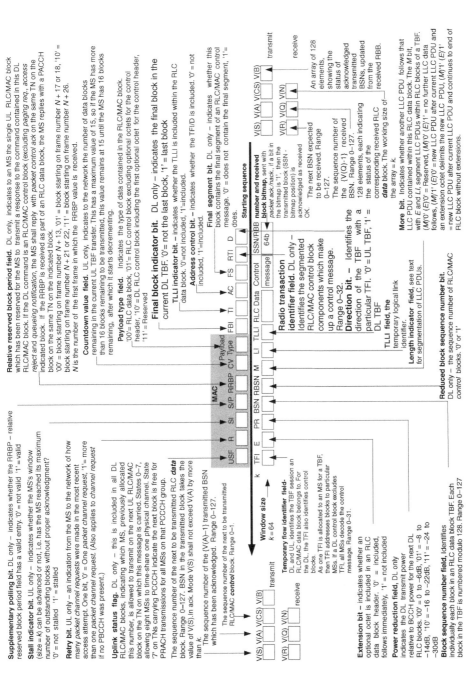

Figure 8.6 RLC/MAC registers, headers and parameters

mobile station cannot transmit more blocks. As blocks are acknowledged, the window advances. This is illustrated in Figure 8.7.

- **MAC payload type**. This indicates whether the payload is an RLC data block ('0'), or an RLC control block ('1'), and defines the air interface logical channel, PDTCH or PACCH respectively.
- **MAC countdown value**. This starts at (default) BSN15 and then decrements. When it reaches zero it indicates to the GPRS sub-network that this is the last block in this TBF.

8.3.1.2   The RLC data block header and static parameters

The RLC block buffer stores all transmitted blocks until they have been acknowledged.

- **V(S)** indicates the BSN of the next to be sent block. When a block is transmitted with the BSN taken from V(S), V(S) increments to indicate the next to be transmitted block. The block being sent by the first arrow in Figure 8.8 is BSN '0', taken from V(S). Note that V(S) increments to '1' when this block is transmitted.
- **V(A)** indicates the BSN of the oldest transmitted block not yet acknowledged. Note that it is set to '0' on the first arrow in Figure 8.8 because no blocks have been acknowledged; the oldest possible block is BSN '0'.
- **V(B)** is a 128-bit register which holds the acknowledgment status (ack or nack) of up to 128 transmitted blocks. It receives this status from the far end of the data communication link (in our case in Figure 8.8 the GPRS sub-network), when a control *packet ack/nack* message is received from the far end.
- **V(R)** indicates the next BSN expected to be received from the far end of the data communication link. As, in this case, the mobile station is the receiver, the other side of the link (the GPRS sub-network) will be considered where this receive array is active. Here in Figure 8.8 it is seen that V(R) is set to '0', the next BSN it is expecting; upon receiving this BSN it increments to the next expected BSN.
- **V(Q)** indicates the oldest BSN which has not been received. On the right of the first arrow in Figure 8.8 it is set to '0'. If, for any reason, blocks transmitted by the mobile station are not received by the GPRS sub-network, then the first the GPRS sub-network will know of this is the reception of a block whose BSN is greater than V(R). In that case, V(R) will increment from the BSN of the block received, but V(Q) will remain at the next block that was expected to be received, and will remain on that oldest block until it is retransmitted by the mobile station and received by the GPRS sub-network. On the first arrow in Figure 8.8, V(Q) is shown incrementing when a block is correctly received.
- **V(N)** is a 128-bit register which records that BSNs are received. A representation of the array is shown to the right in Figure 8.8. This has 128 bits, each bit position representing a BSN. In this diagram, BSN '0' has been received OK so the bit representing BSN '0' is set to '1'; all other bits are set to '0' as they have not been received. The contents of this register are sent to the mobile station (in the start sequence number/ received block bitmap SSN/RBB) with the *packet uplink ack/nack* message and update V(B) in the mobile station. MAC headers are shown for completeness.
- **TI (TLLI Indicator)**, this indicates whether the TLLI is included within the RLC data block.

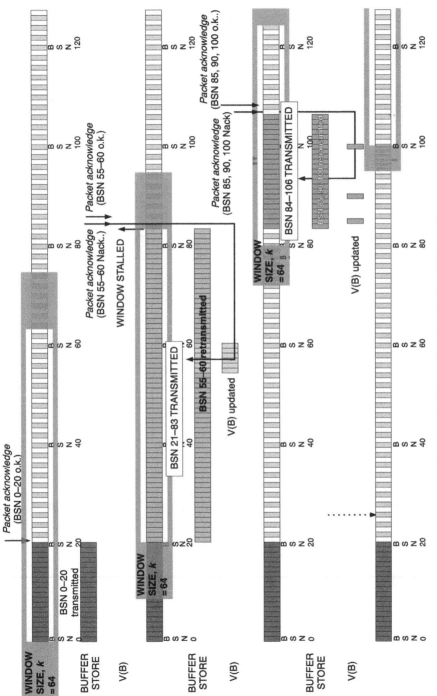

Figure 8.7 Illustrating the window (*k*) in action

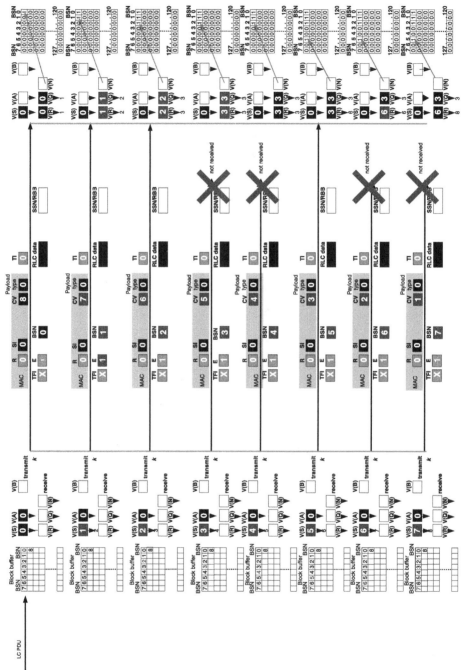

Figure 8.8 RLC acknowledged UL TBF procedure (includes MAC header)

- **TFI**, a number in the range 0–31, this identity is given to the mobile station by the GPRS sub-network to assist in identifying the mobile station during a TBF. It is also used on the downlink to identify RLC/MAC blocks which are addressed to this mobile station. It identifies a TBF.
- *E* bit, extension bit – see section 8.1.1.
- **BSN**, a number in the range 0–127 which identifies a block within the TBF.
- **SSN/RBB**, start sequence number/received block bitmap; this field is included in the *packet ack/nack* and indicates those blocks which have been received. Its action is described later.
- **k**, window size, this is 64, and is the maximum number of blocks allowed to be transmitted without acknowledgment.

The easiest way to visualise what all these parameters do is to watch them in operation. This is what the next section does, using appropriate diagrams.

8.3.2   RLC/MAC parameters in action

This section examines the procedures for an acknowledged and unacknowledged TBF across the air interface using the RLC/MAC procedures.

8.3.2.1   Acknowledged uplink TBF

This section uses Figure 8.6 in conjunction with Figures 8.8 to 8.10.

The acknowledged TBF of Figures 8.8 and 8.9 are examined block by block, and the changes between successive blocks highlighted. A single arrow on these diagrams represents one RLC/MAC block.

BSN0 transfer

MAC header

R is '0', therefore only one *packet channel request* was made for this TBF.
SI is '0', therefore the window is not stalled.
CV is at eight, indicating that there are eight more blocks after BSN0 to complete the TBF.
Payload type is '0', indicating an RLC data block.

Mobile station side and RLC data block

BSN0 takes the value of V(S) and is sent from the mobile station to the GPRS sub-network.
BSN0 is stored in the buffer. (In fact all RLC blocks, after processing by the RLC layer, are stored in the buffer as sequential BSNs).
V(S) increments from 0 to 1.
V(A) is at zero – this is the BSN of the oldest transmitted block not yet acknowledged.
TI is at zero, indicating that the TLLI is not included in BSN0.
TFI shows X, representing whatever the GPRS sub-network previously allocated to the mobile station.

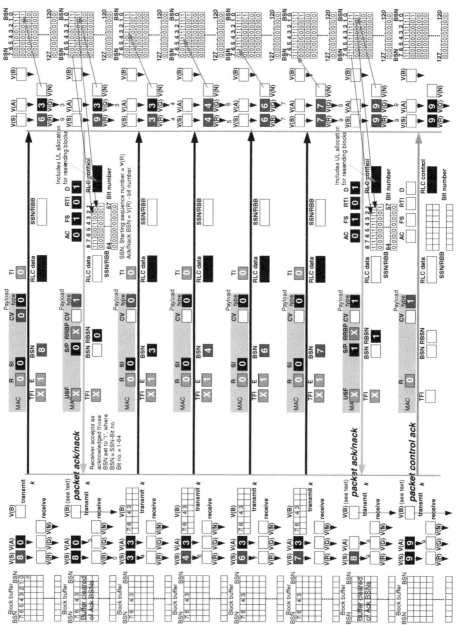

Figure 8.9 Continuation of Figure 8.8

E is at 1, indicating that there is no extension octet within the block.
The unshaded (white) blocks are not used

GPRS sub-network side

V(R), the next BSN expected to be received, is initially set to 0 and increments to 1 when BSN0 is received.
V(Q) registers the oldest BSN, which has not been received. It is initially set to 0 and increments to 1 when BSN0 is received.
V(N) which tracks received BSNs sets the bit corresponding to BSN0 to '1', indicating that BSN0 is received.

BSN1 transfer

MAC header CV is at seven, indicating that there are seven more blocks after BSN1 to complete the TBF.

Mobile station side and RLC data block

BSN1 takes the incremented value of V(S) and is transmitted.
BSN1 is stored in the buffer.
V(S) increments from 1 to 2.
E is at 1, indicating that there is no extension octet within the block used to identify the first LLC PDU.

GPRS sub-network side

V(R) increments to 2 when BSN1 is received.
V(Q) increments to 2 when BSN1 is received.
V(N) sets the bit corresponding to BSN1 to '1', indicating received.

BSN2 transfer

MAC header CV changes to six, indicating that there are six more blocks after BSN2 to complete the TBF.

Mobile station side and RLC data block

BSN2 takes the incremented value of V(S) and is transmitted.
BSN2 is stored in the buffer.
V(S) increments from 2 to 3.

GPRS sub-network side

V(R) increments to 3 when BSN2 is received.
V(Q) increments to 3 when BSN2 is received.
V(N) sets the bit corresponding to BSN2 to '1', indicating received.

BSN3 transfer

MAC header CV changes to five, indicating that there are five more blocks after BSN3 to complete the TBF.

Mobile station side and RLC data block

BSN3 takes the incremented value of V(S) and is transmitted.
BSN3 is stored in the buffer.
V(S) increments from 3 to 4.

GPRS sub-network side

BSN3 is not received and there are no changes in the registers.

BSN4 transfer

MAC header CV changes to four, indicating that there are four more blocks after BSN4 to complete the TBF.

Mobile station side and RLC data block

BSN4 takes the incremented value of V(S) and is transmitted.
BSN4 is stored in the buffer.
V(S) increments from 4 to 5.

GPRS sub-network side

BSN4 is not received – no change.

BSN5 transfer

MAC header CV changes to three, indicating that there are three more blocks after BSN5 to complete the TBF.

Mobile station side and RLC data block

BSN5 takes the incremented value of V(S) and is transmitted.
BSN5 is stored in the buffer.
V(S) increments from 5 to 6.

GPRS sub-network side

V(R) increments to 6 when BSN5 is received.
V(Q) remains at 3, the oldest BSN not received.
V(N) sets the bit corresponding to BSN5 to '1', indicating received.

BSN6 transfer

MAC header CV changes to two, indicating that there are two more blocks after BSN6 to complete the TBF.

Mobile station side and RLC data block

BSN6 takes the incremented value of V(S) and is transmitted.
BSN6 is stored in the buffer.
V(S) increments from 6 to 7.

GPRS sub-network side

BSN6 is not received – no change.

BSN7 transfer

MAC header CV changes to one, indicating that there is one more block after BSN6 to complete the TBF.

Mobile station side and RLC data block

> BSN7 takes the incremented value of V(S) and is transmitted.
> BSN7 is stored in the buffer.
> V(S) increments from 7 to 8.

GPRS sub-network side

> BSN7 is not received – no change.

BSN8 transfer

From Figure 8.9. MAC header CV changes to 0, indicating that there are no more blocks after BSN8 to complete the TBF. This is the last block of the TBF.

Mobile station side and RLC data block

> BSN8 takes the incremented value of V(S) and is transmitted.
> BSN8 is stored in the buffer.
> V(S) has reached the total TBF length.

GPRS sub-network side

> V(R) increments to 9 when BSN8 is received.
> V(Q) remains at 3, the oldest BSN not received.
> V(N) sets the bit corresponding to BSN8 to '1', indicating it is received.
> The GPRS sub-network detects that the TBF is completed (CV = 0) and sends an RLC control block with a *packet acknowledge/nacknowledge* message.

> Before examining the processes in this message, the downlink parameters are discussed.

8.3.3   RLC control block (downlink) parameters

8.3.3.1   In the MAC header

- **Uplink status flag (USF)** identifies which of eight mobile stations with this identity should use the next uplink block for transmission.
- **Supplementary polling bit (S/P)** indicates whether the relative reserved block period (RRBP) is valid.
- **RRBP** indicates a block is reserved for the mobile station to allow it to reply to a downlink control message. The uplink block allocated to the mobile station is numbered relative to the frame of this control message. The mobile station will use this uplink block to send a control message, normally *packet control acknowledge*.

8.3.3.2   The RLC control block header and static parameters

- **Address control bit (AC)** indicates whether the downlink TFI is included in this block.
- **Final segment bit (FS)** indicates whether this block contains the final segment of the control block message, '0' means it does not, '1' means it does.

- **Radio transaction identifier (RTI)** identifies individual downlink control blocks in a range 0–32. This may be used if a control message is segmented into a number of RLC/MAC blocks.
- **Direction bit (D bit)** used with TFI, '0' = uplink, '1' = downlink.
- **Reduced block sequence number (RBSN)** '0' or '1' for downlink control blocks.
- **Start sequence number/received block bitmap (SSN/RBB)** is not a part of the header, but part of the information content of the control block. SSN is the start sequence number, the value of V(R) when this message is sent. RBB is the received block bitmap, and indicates which of the blocks sent by the transmit end (mobile station in our case) have been received (in our case by the GPRS sub-network).

Figure 8.9 indicates how RBB is assembled from the V(N) array. The mobile station changes the entries in RBB to an acknowledged or unacknowledged BSN by the equation BSN = V(R) – bit number, a 0 or 1 in the bit number position indicating ack and nack respectively. RBB can send the status of 64 received blocks. The static arrays have similar functions to those described for RLC data blocks and where these functions differ, they will be described below.

Continuing the TBF from reception of BSN8.

GPRS sub-network response RLC control block message

uplink packet acknowledge

MAC header

> USF indicates which mobile station may use the next uplink block.
> S/P is 0, indicating that the RRBP field is not valid.
> RRBP not used for this transaction.
> CV, only used in the uplink direction for RLC data blocks.
> Payload type, set to 1, indicating an RLC control block.

GPRS sub-network side

> RLC control block RBSN is set to 0. This alternates between 0 and 1.
> RTI set to 0, the first block for this particular message.
> FS set to 1, indicating the block contains the final segment of this control message.
> AC is set to 0, indicating that the downlink TFI is not included. (In fact the TFI would normally be included).
> TFI, the temporary flow identifier previously allocated to the mobile station. Not included in this block, although it would normally be included.
> *D*, the direction bit, set to 1, indicates downlink.
> SSN/RBB is loaded as shown in Figure 8.9.

Mobile station side

> Block buffer, SSN/RBB is indicating BSNs 0, 1, 2, 5 and 8 are received OK, so these blocks are erased from the buffer.
> V(A), the oldest BSN that has been sent but is not acknowledged is BSN3, and V(A) is loaded with this BSN.
> V(B) is loaded with the unacknowledged BSNs. The mobile station now resends blocks, with V(S) taking the BSNs of V(B).
> As these are received by the GPRS sub-network, so V(R) and V(Q) at the GPRS sub-network are updated.

Continuing the TBF from the successful GPRS sub-network reception of retransmitted BSN7 (blocks 3, 4 and 6 have been sent).

Downlink RLC control block message

uplink packet acknowledge

MAC header

USF indicates which mobile station may use the next uplink block.
S/P is 1, indicating that the RRBP field is valid.
RRBP allocates to the mobile station an uplink block for the reply to this message.
CV, only used in the uplink direction for RLC data blocks.
Payload type, set to 1, indicating an RLC control block.

GPRS sub-network side and RLC control block

RBSN set to 1. This alternates between 0 and 1.
RTI set to 0, the first block for this particular message.
FS set to 1, indicating the block contains the final segment of this control message.
AC is set to 0, indicating that the downlink TFI is not included.
TFI, the temporary flow identifier previously allocated to the mobile station. Not included in this block.
D, the direction bit, set to 1 indicates downlink.
SSN/RBB is loaded as shown in Figure 8.9.
V(R) increments, but this is the end of the TBF.
V(Q) is set to 9, the oldest BSN in the window size 64 not yet received. (This signifies all eight blocks are received OK.)
V(N) all BSNs recorded as received.

Downlink on the mobile station side

Block buffer, SSN/RBB is indicating all of the outstanding BSNs are received OK, so these blocks are erased from the buffer.
V(A), the oldest BSN that has been sent but is not acknowledged, is BSN9, and V(A) is loaded with this BSN.
V(B) is cleared.

Uplink RLC control block message

packet control acknowledge

MAC header

R, retry bit, one previous *packet channel request*.
Payload type, set to 1, indicating an RLC control block.

Uplink mobile station side and RLC control block

V(B) is cleared.

GPRS sub-network side

V(N) is cleared. The TBF is completed.

8.3.3.3 Acknowledged downlink TBF

The differences between uplink and downlink parameters for RLC data blocks can be seen in Figure 8.10 and only those differences will be listed here.

Downlink MAC header

- **MAC RRBP, the relative reserved block period**. This allocates an uplink block for the mobile station to send a control message. It is validated by the supplementary polling (S/P) bit and, as we shall see, is normally included in the final block of a downlink TBF (indicated by the FBI status in the RLC header).
- **MAC S/P, supplementary polling bit**. '0' indicates the RRBP is not valid, '1' that it is valid.
- **MAC USF, uplink status flag**. This indicates which mobile station, previously allocated the value of this field, is allowed to use the next uplink block on this PDCH.

Downlink RLC header

- **RLC PR, power reduction**. This 2-bit field indicates the reduction of the current PDCH transmit power below the BCCH transmit power, $0 = 0$–6 dB, $1 = 8$–14 dB, $2 = 16$–22 dB, $3 = 24$–30 dB.RLC
- **FBI, final block indicator**. '1' = final block. An RRBP is normally included with the final block to allow the mobile station to send downlink *packet acknowledge*.
- **RLC static parameters**. These are the same as those listed in 8.3.3.2.

Now the acknowledged downlink TBF of Figure 8.10 will be examined block by block, and the changes between successive blocks highlighted.

BSN9

Figure 8.10 enters the downlink TBF when BSN9 is being transferred. Downlink MAC header payload type $= 0$, indicating an RLC data block. $S/P = 1$, indicating that the RRBP applies, and the mobile station is allocated an uplink block. USF has a meaning to other mobile stations.

GPRS sub-network side and RLC data block header

V(S) $= 9$, the BSN currently being transmitted, after transmission, V(S) changes to 10, the next BSN to be transmitted.
Block buffer stores BSN data blocks 0–9.TFI will identify the mobile station that this block is addressed to.
$E = 1$, indicating that no extension octets are used in this block.
PR indicates how far below the BCCH transmit power the PDCH is.
FBI, the final block identifier, set to '1', indicates this is the final block in the TBF.

Mobile station side

V(N). It can be seen for the V(N) array that BSNs 6 and 7 have not previously been received, when BSN9 is received, the bit position for BSN9 is set to '1', indicating that it is received.

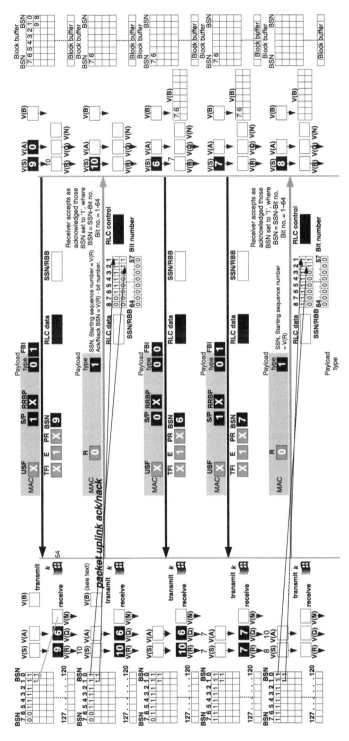

Figure 8.10 RLC acknowledged DL TBF procedure

V(R) is expecting BSN9 and when it is received increments to 10.
V(Q) is indicating 6, of the BSNs not received, this is the oldest.
With FBI = 1, the mobile station is forced to respond on the allocated uplink RRBP.

Uplink packet downlink acknowledge/nack

Mobile station side RLC control block header

SSN/RBB, the start sequence number and received block bitmap are loaded with the
V(R) value and V(N) values as shown, indicating that BSNs 6 and 7 have not been
received.
Payload type '1' indicates an RLC control block.
R indicates only one *packet channel request* was needed to establish this TBF.

Uplink GPRS sub-network side

V(B) is loaded with the receive array of RBB.
Buffer is cleared of acknowledged blocks.

Downlink Retransmission of blocks

BSN6

MAC header

S/P = 0 indicating that the RRBP does not apply.

GPRS sub-network side and RLC data block header

V(S) = 6, the BSN currently being retransmitted.
Block buffer stores BSN data blocks 6 and 7.
FBI, the final block identifier, set to '0' indicates this is not the final block in the
TBF.

Mobile station side

V(N). It can be seen for the V(N) array that BSNs 6 and 7 have not been received and
when BSN6 is received, the bit position for BSN6 is set to '1', indicating that it is
received.
V(R) is expecting BSN9 and when BSN6 is received increments to BSN7.
V(Q) indicating 6, now changes to 7 which, of the BSNs not received, is the oldest.

BSN7

MAC header

S/P = 1 indicating that the RRBP does not apply.

GPRS sub-network side and RLC data block header

V(S) = 7, the BSN currently being retransmitted.
Block buffer stores BSN data blocks 6 and 7.
FBI, the final block identifier, set to '1' indicates this is the final block in the TBF.

Mobile station side

V(N). It can be seen for the V(N) array that BSN7 has not been received, and when BSN7 is received, the bit position for BSN7 is set to '1', indicating that it is received. V(R) is expecting BSN7 and when BSN7 is received, increments to BSN8. V(Q) indicating 7, now changes to 10 which, of the BSNs not received, is the oldest. With FBI = 1, the mobile station is forced to respond on the allocated uplink RRBP.

Uplink packet downlink acknowledge/nack

Mobile station side and RLC control block header

SSN/RBB, the start sequence number and received block bitmap are loaded with the V(R) value and V(N) values as shown, indicating that BSNs 6 and 7 have been received.

Uplink GPRS sub-network side

V(B) is loaded with the receive array of RBB.
Buffer is cleared of acknowledged blocks.

The downlink TBF is completed.

8.4 RLC unacknowledged mode

Unacknowledged mode does not retransmit blocks reported as missing. Block sequence numbers (BSNs) are still applied to RLC blocks and are used to reassemble the received blocks.

Packet acknowledgment messages are used, but the information for retransmission (SSN/RBB) is disregarded. If the number of blocks transmitted is equal to the window size k, then the mobile station starts a timer (T3182) and if no *packet acknowledgment* is received before this timer expires, the mobile station releases the TBF and attempts to re-establish a new TBF with *packet channel request* messages.

Acknowledgments are used in the unacknowledged mode of RLC transmission because acknowledgments have a dual purpose:

1. To regulate the reception of RLC/MAC blocks when in acknowledged mode.
2. In both acknowledged and unacknowledged modes to verify that the radio link remains good.

8.5 RLC/MAC timers and counters (GSM 44.060 section 13)

Various counters and timers are used on both the network and mobile station sides to protect the flow of RLC/MAC blocks and to ensure an orderly and predictable close down of services, if, for example, an irrecoverable loss of radio communication occurs.

The proliferation of timers and counters in comparison to the GSM circuit switched operations is necessary because GPRS does not have the GSM *slow associated control channel (SACCH)* which is always transmitted with a circuit switched dedicated channel, and whose presence or absence indicates that the radio connection between the mobile station and network is good or bad.

Figures 8.11–8.14 show these counters and timers in action and although these diagrams are largely self-explanatory, additional notes are included here, and descriptive names (not included in the specifications) are given to them.

8.5.1   Mobile station side RLC/MAC counters and timers

8.5.1.1   Window stall timeout for dynamic allocation (T3182 and N3102)

When the RLC transmit window k is filled with transmitted RLC/MAC blocks, any one of which is unacknowledged, then the TBF window 'stalls' and no more blocks can be sent. This is indicated by setting the stall indicator (SI) to '1'. T3182 is then started. N3102 has been previously loaded with a count PAN_MAX (Read PAN as packet acknowledge).

If T3182 expires without acknowledgment of the blocks causing the stall, then the mobile station decrements N3102 (by the amount PAN_DEC sent in packet system information), disconnects from the radio resources, and attempts to re-establish the TBF in the same cell with *packet channel request*. If this is successful, the timer T3182 is reset.

If a *packet uplink acknowledge* is received from the GPRS sub-network allowing the window to be advanced and transmissions resumed, N3102 is incremented by PAN_INC (but never to more than PAN_MAX) and T3182 stopped.

If no *packet uplink acknowledge* is received, the procedure above is repeated until N3102 reaches zero, when the mobile station will reselect another cell (if possible) and attempt to establish a TBF with that cell.

The same procedure is followed when an uplink TBF is completed with $CV = 0$ indicated on the final block.

8.5.1.2   USF assignment timeout for dynamic allocation (T3164)

When a mobile station accesses the GPRS sub-network through *packet channel request*, a *packet uplink assignment* allocating a USF is received and T3164 is started. When the allocated USF granting access to the uplink PDTCH appears on the downlink the timer T3164 is stopped.

If the USF does not appear before T3164 expires, the mobile station disconnects from the radio resources and the access procedure is repeated. After four such access attempts and four failures of the USF, further access attempts are abandoned and the mobile station higher layers are informed of the failure.

8.5.1.3   BSN acknowledgment timeout (T3198)

When an RLC/MAC block is transmitted by a mobile station it starts a timer T3198. If this timer expires without acknowledgment, then the status of the block is set to Nack. There are 64 timers T3198, one for each BSN within the window size of 64.

A possible use of this timer is when a window stall occurs on a dynamic UL TBF, the mobile station's USF may well continue to appear on the downlink. The mobile station can take advantage of this to retransmit those RLC/MAC blocks for which T3198 has expired.

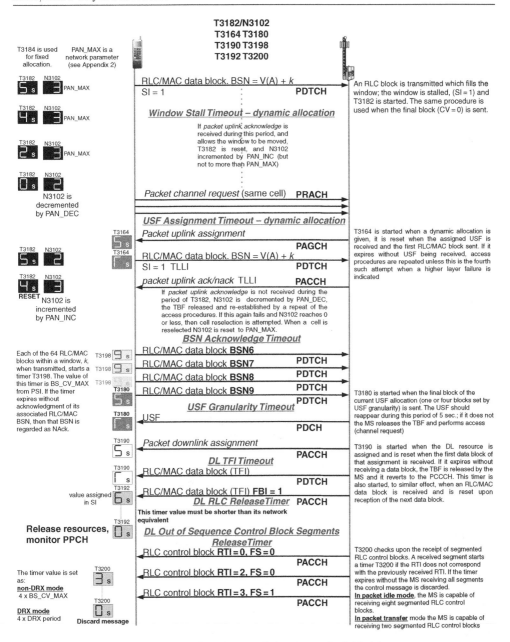

Figure 8.11   Counters and timers on the MS side

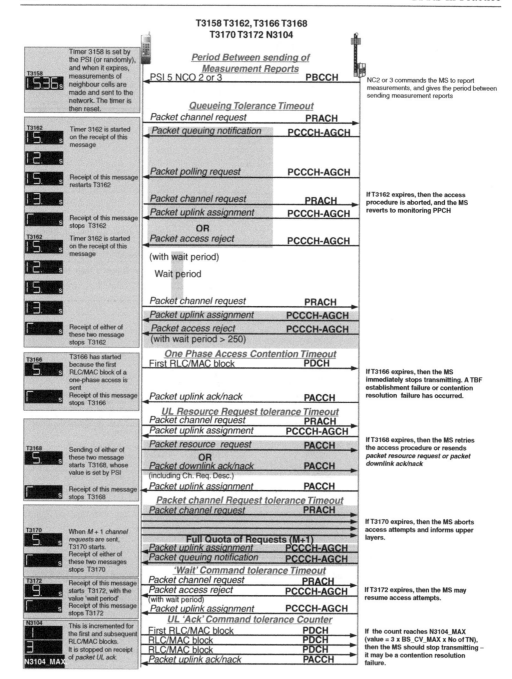

Figure 8.12   Counters and timers on the MS side

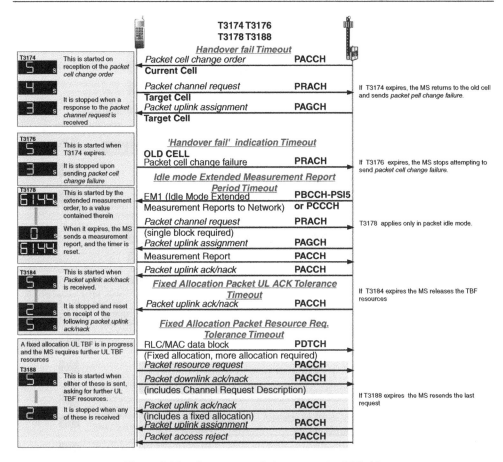

Figure 8.13   Counters and timers on the MS side

8.5.1.4   USF granularity timeout (T3180)

When a mobile station is in a dynamic allocation uplink TBF, this timer is started when the last block allowed by the USF granularity is sent. USF granularity allows one or four blocks to be sent for each appearance of the USF.

If the USF does not appear on the downlink before this timer expires, then the mobile station abandons the TBF and reattempts access procedures.

8.5.1.5   Downlink TFI timeout (T3190)

This timer is started when a *packet downlink assignment* is received by the mobile station. If a downlink data block is received carrying the allocated TFI before timer T3190 expires, the timer is reset.

If the timer expires during the downlink TBF, the TBF is released, radio resources disconnected and the mobile station monitors the packet common control channel (PCCCH).

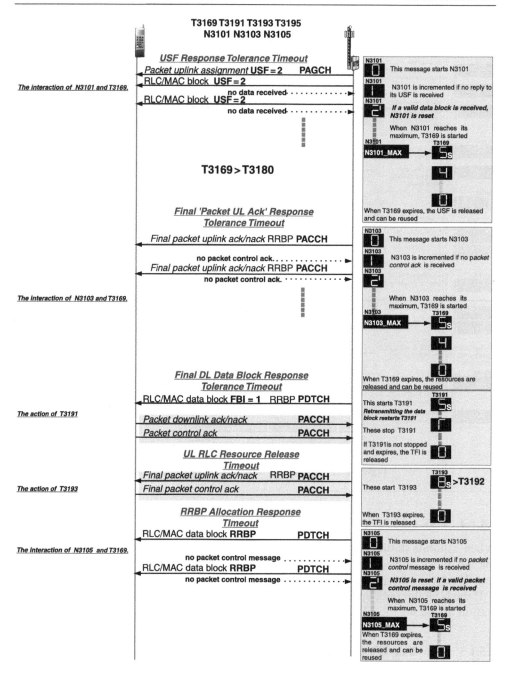

Figure 8.14   Counters and timers on the network side

8.5.1.6   Downlink RLC release timer (T3192)

This timer is started when the final block of a downlink TBF is received. At its expiry the mobile station releases radio resources.

This timer has a shorter duration than its GPRS sub-network counterpart T3193, so that the GPRS sub-network does not reassign the TFI to another mobile station before this mobile station has released it.

This is the normal radio resource release timer.

8.5.1.7   Downlink control block segments watchdog (T3200)

If a downlink RLC/MAC control message is segmented into a number of RLC/MAC blocks, the segments are identified by the combination of the radio transaction identifier (RTI) and the final segment indicator (FSI).

Timer T3200 is started if an RTI does not sequentially follow a previous RTI. If the missing RTI is received before the expiry of T3200, then T3200 is stopped.

Multiple timers T3200 can be used for one segmented control block message. If a T3200 expires, then the whole of the control block message is rejected, as a segment is missing.

8.5.1.8   Period between measurement reports (T3158)

If the packet system information (PSI) indicates NCO2 (the mobile station selects cells but sends measurement reports) or NCO3 (the GPRS sub-network controls cell selection and the mobile station sends measurement reports) then the mobile station must measure neighbour cells and send measurement reports to the GPRS sub-network. The measurement reports may be made in packet idle mode or packet transfer mode, the timer being set by NC_REPORTING_PERIOD_I (packet idle) or NC_REPORTING_PERIOD_T (transfer).

8.5.1.9   Queuing tolerance timer (T3162)

This timer is started when the mobile station receives *packet queuing notification* (a tool the GPRS sub-network operator may use to control congestion in a cell by forcing a mobile station to pause its access attempts until it is convenient for the network to handle them) carrying temporary queuing identity (TQI) after it has sent *packet channel request*. It is restarted if a *packet polling request* carrying the TQI is received. It is stopped when, in response to *packet polling request*, the mobile station has made another *packet channel request* and received *packet uplink assignment*.

If T3162 expires without an uplink assignment, then access attempts are aborted and the mobile station monitors the PPCH.

The DL *packet access reject* also affects this timer as shown in Figure 8.12.

8.5.1.10   One-phase access contention timer (T3166)

One-phase access demands the sending of TLLI in the RLC/MAC blocks sent by the mobile station.

The first *packet uplink ack/nack* sent by the GPRS sub-network will return this TLLI, and any mobile station that receives this but does not own the TLLI must disconnect. This is a part of the *contention resolution* procedure of GSM systems.

The timer T3166 is started when the first RLC/MAC block is transmitted, and is stopped when a *packet uplink ack/nack* returning TLLI is received.

If it expires, then the mobile station must stop transmitting and disconnect from the radio resources. A contention failure or other failure has occurred.

8.5.1.11   Resources request tolerance timer (T3168)

When a mobile station is in a TBF, it may require more or new uplink TBF resources (for example, to send another LLC PDU which has arrived at the RLC layer during the current TBF).

It does this by sending a *packet resource request*, or asking for uplink resources in a *packet downlink ack/nack* message. These requests start T3168.

The timer is stopped when the mobile station receives *packet uplink assignment* or a similar message.

If T3168 expires, the mobile station sends *access request* or resends the *packet resource request*.

8.5.1.12   Access request tolerance timer (T3170)

This timer determines how long the mobile station will wait for a response after sending its full quota of *packet channel requests*.

The timer is started when the full quota has been sent and is stopped on receipt of *packet uplink assignment* or *packet queuing notification*.

If it expires, the mobile station abandons access attempts and informs upper layers.

8.5.1.13   Access congestion 'wait' timer (T3172)

This timer is loaded with the 'wait' value, which may be contained in a *packet access reject* message responding to a *packet channel request*.

If the mobile station receives a *packet uplink assignment* or *packet queuing notification* message during this period, the timer is stopped.

If the timer expires, the mobile station may resume access attempts.

8.5.1.14   Count of RLC blocks before Ack (N3104)

This counts the number of RLC/MAC blocks transmitted on the uplink during a *single-phase* access TBF.

If a *packet uplink ack/nack* is received before the count reaches N3104_MAX, then the count is stopped. If the counter reaches N3104_MAX, then the mobile station must stop transmitting and disconnect from the radio link.

The purpose of this counter is to stop an uplink TBF which results from a single-phase access if no *packet uplink ack* is received within a reasonable period – the GPRS sub-network should send *packet uplink ack* as soon as it decodes the TLLI carried on one of the RLC/MAC blocks. (All RLC/MAC blocks carry TLLI for one-phase access and the network returns this on a *packet uplink ack* message to resolve contention.)

If the mobile station does not receive this acknowledgment, there is a possibility that contention has occurred – two mobile stations are on the same PDTCH and the GPRS sub-network cannot decode TLLIs, or a radio link failure has occurred.

8.5.1.15   Handover (cell change order) failure timer (T3174)

When NCO3 is in operation, the GPRS sub-network is in command of cell selection and may tell the mobile station to change to another cell with *packet cell change order*. This command starts T3174 and the mobile station changes physical channels, sending a *packet channel request* to the new cell.

When a *packet uplink assignment* is received from the new cell, T3174 is stopped.

If T3174 expires, the handover or cell change has failed and T3176 starts.

8.5.1.16   Handover (cell change order) failure signalling period (T3176)

This is the period during which the mobile station is allowed to send *packet cell change failure* messages to the original cell.

When T3176 expires, these attempts must stop (for example the original cell may not be reachable).

8.5.1.17   Packet idle mode extended measurement report period (T3178)

The value of this timer is contained in the idle mode extended measurement reports command contained in PSI5 or PCCCH. The mobile station will report measurements with this periodicity.

8.5.1.18   Fixed allocation packet uplink ack/nack tolerance timer (T3184)

Started on receipt of *packet uplink ack/nack*, this timer is reset on receipt of the next *packet uplink ack/nack*.

If it expires, the mobile station releases the TBF and radio resources.

8.5.1.19   Fixed allocation further resource request tolerance timer (T3188)

This is started on the sending of an *uplink resources request* when a fixed allocation uplink TBF is in progress.

If *packet uplink assignment, packet uplink ack/nack* is received whilst the timer is running, then the timer is stopped.

If the timer expires, the mobile station resends the request.

8.5.2   Network side RLC/MAC counters and timers

8.5.2.1   'Response to USF' tolerance timer (N3101 and T3169)

If no data is received when an assigned USF is placed on the downlink, N3101 is incremented.

If, before the count reaches N3101_MAX, data arrives on the uplink for the assigned USF, N3101 is reset.

If N3101 reaches its allowed maximum count, N3101_MAX, then the USF is no longer placed on the downlink, timer T3169 starts and runs to expiry; on expiring, the USF is released (which means it can be reassigned to another mobile station).

8.5.2.2   Response to final packet uplink ack/nack tolerance timer (N3103 and T3169)

If no response is received to a final *packet uplink ack/nack* message, N3103 is incremented and the message resent.

If N3103_MAX is reached without a response, T3169 starts and, upon expiry, releases the resources.

8.5.2.3   Response to final data block tolerance timer (T3191)

When the final RLC data block is transmitted in a downlink TBF, this timer is started. Receipt of *packet downlink ack/nack* or *packet control ack* will stop T3191.

If T3191 expires, the TFI and other resources are released.

8.5.2.4   Uplink RLC release timer (T3193)

This is the network counterpart of T3192, and releases the TBF resources on reception of *packet downlink ack/nack* or *packet control ack*. This timer must be longer than T3192.

8.5.2.5   Response to RRBP tolerance timer (N3105 and T3169)

When the network indicates a valid RRBP during a downlink TBF, it expects a response on that reserved block; if no response is forthcoming, N3105 is incremented. The network resends a block with RRBP and, if a response is obtained, N3105 is reset. If a string of no-responses occurs and N3105 reaches N3105_MAX, then no further messages are sent and T3169 is started and will expire, releasing the TBF resources.

8.6   Abbreviations used in this chapter

AC   Address control bit
ARQ   Automatic request for retransmission
BCCH   Broadcast control channel
BSN   Block sequence number
CS   Coding system
CV   Countdown value

D	Direction bit
DL	Downlink
DRX	Discontinuous reception
E	Extension octet indicator
FBI	Final block indicator
FS	Final segment bit
FSI	Final segment indicator
GMM	GPRS mobility management
GPRS	General packet radio service
GSM	Global system for mobile communication
LI	Length indicator
LLC	Logical link control
M	More
MAC	Medium access control
NCO	Network control order
N PDU	Network protocol data unit
PACCH	Packet associated control channel
PCCCH	Packet common control channel
PDCH	Packet data channel
PDP	Packet data protocol
PDTCH	Packet data traffic channel
PDU	Protocol data unit
PPCH	Packet paging channel
PR	Power reduction
PSI	Packet system information
PTCCH	Packet timing control channel
QoS	Quality of service
R	Retry bit
RBSN	Reduced block sequence number
RLC	Radio link control
RRBP	Relative reserved block period
RTI	Radio transaction identifier
SACCH	Slow associated control channel
SI	Stall indicator
SM	Session management
SNDCP	Sub-network dependent convergence protocol
SN PDU	Sub-network protocol data unit
S/P	Supplementary polling bit
SSN/RBB	Start sequence number/received blocks bitmap
TBF	Temporary block flow
TFI	Temporary flow identity
TI	TLLI indicator
TLLI	Temporary logical link identifier
TN	Timeslot number
TQI	Temporary queuing identifier
UL	Uplink

USF	Uplink status flag
V(A)	Acknowledge register
V(B)	128-bit register
V(N)	Register of received PDUs
V(Q)	Oldest PDU not received
V(R)	Next PDU expected to be received
V(S)	Next PDU to be sent

9

LLC Layer Procedures

(GSM 44.064)

This chapter covers the elementary operation of the logical link control layer, which is a layer 2 data communications link between the mobile station and the SGSN. You may wonder why it is necessary to have another data communications link sitting on top of the RLC layer 2 data communications link. There are perhaps three reasons for this:

1. As the MS does its own cell reselection during a TBF (at least it does this in NCO0 and NCO1), for a downlink TBF the RLC layer in the new cell has no idea which SN PDUs have been delivered to the MS. To keep track of these, a L2 function separated from the BTS is introduced at the BSC or SGSN.
2. *Encryption.* With the MS doing its own cell reselection during a TBF it might be inconvenient to repeat the encryption procedures each time a cell is reselected. Having a L2 function divorced from the BTS helps achieve this.
3. The air interface TBF may guarantee the delivery of RLC/MAC blocks, but there may be residual errors within the reassembled PDUs. Having a second L2 function may detect these residual errors.

The LLC layer is disposed of in 3G operations where the user equipment (UE) does not perform its own cell reselections whilst using a traffic channel.

The primary function of the LLC layer then, is to provide guaranteed delivery of packet data units between the GPRS sub-network and mobile station SNDCP layers. If this is not possible due to poor radio conditions, the LLC layers inform the higher layers (SNDCP) that guaranteed delivery cannot be provided.

The LLC L2 layer places SNDCP PDUs within LLC frames, enciphering the contents if necessary, and sends them to the RLC layer in asynchronous balanced (acknowledged) mode (ABM) if requested.

The data link protocol used by LLC is derived from the HDLC protocol. (LAPD used in GSM circuit switched operations is another derivative of HDLC.) HDLC – high speed data link control – was originally developed by IBM for data transmission.

GPRS in Practice: A Companion to the Specifications Peter McGuiggan
© 2004 John Wiley & Sons, Ltd ISBN: 0-470-09507-5

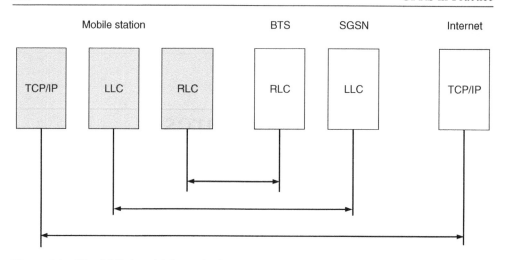

Figure 9.1   The L2 links which can be in operation from the mobile station during a TBF. The largest LLC PDU delivered from the MS LLC layer to RLC layer has 1563 octets. Using CS1, the RLC layer will segment this into 79 RLC/MAC blocks. It will take at least $79 \times 18\,\text{ms} \times 52/48 = 1.5\,\text{s}$ to deliver this LLC PDU across the air interface (under perfect radio conditions). It is evident that the retransmission timer for the LLC layer must be greater than this period, and this is easily arranged. It is also evident that the TCP L2 retransmission timer must be set to a greater value than the LLC retransmission timer. This is not so easily arranged and can lead to the problem of the TCP timer expiring during a TBF delivering the NPDU for which the timer is awaiting a response. When the TCP timer expires without this response, the N PDU is sent again to the LLC layer!

The LLC may also transmit SN PDUs in unacknowledged, connectionless mode with no guarantee of delivery. This mode is sometimes called 'send and pray' and is also known as unitdata mode; connectionless mode may use communications resources more efficiently (if radio conditions are good) than connection mode.

Applications with their own error checking and correction facilities, such as TCP/IP may not require acknowledged mode transmission from the LLC layer; applications without those facilities may require LLC acknowledged mode.

The LLC protocol is similar to LAPD. In the GSM circuit switched mode, the L2 data link layer provides a service to the radio resource (RR) layer; oddly, in GPRS operation it is the other way round – the L3 RR layer provides a service for the L2 data link layer.

The LLC data link layer is a layer 2 function in terms of the OSI 7 layer structure.

Figure 9.1 shows the L2 links which can be in operation from the mobile station during a TBF.

Many network operators force LLC to unack. mode by giving reliability class 3 to Ms's

9.1   Function of the LLC layer

Looking at the simplified diagram in Figure 9.2 (which should now be familiar from previous chapters), there are effectively three access and exit points from higher layers to the LLC layer:

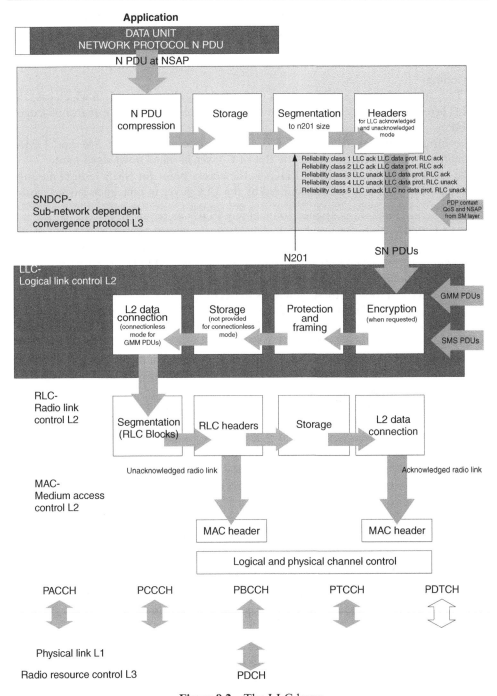

Figure 9.2 The LLC layer

1. The input arrow labelled 'SN PDUs' is the access point for PDUs from the SNDCP layer. These may belong to a PDP context with reliability class 1 or 2, in which case the LLC layer will set up a point-to-point data communications link in acknowledged mode. The LLC layer will also apply data protection in the form of block encoding.

   They may belong to a PDP context with reliability class 3, 4 or 5, in which case the LLC layer will set up a point-to-point data communications link in unacknowledged mode.
2. The input arrow labelled GMM PDUs is the access point for SM and GMM signalling. These are transmitted by the LLC layer in unacknowledged mode.
3. The input arrow labelled SMS PDUs is the access point for SMS – short message service – PDUs. These are transmitted by the LLC layer in unacknowledged mode.

The ciphering function is applied to incoming PDUs as required.

The LLC PDUs are buffered when the LLC layer is operating in acknowledged mode. They are stored until the far end acknowledges reception. The LLC layer processes the incoming higher layer PDUs and passes them to the RLC layer, requesting RLC acknowledged mode or RLC unacknowledged mode.

This chapter will look at these functions in detail.

Figure 9.3 outlines the functions of the LLC layer. The first diagram of Figure 9.3 shows connectionless communication established as a result of the LLC layer receiving the higher layer primitive UNITDATA.REQ. This causes the SN PDU (or GMM PDU) to be encapsulated in an LLC UI frame, where UI means unacknowledged information. This frame is numbered and, if it is lost, an empty frame filled with 0s is delivered to the higher layers.

When a UI frame is received by the distant LLC layer it is delivered by that LLC layer to its higher layer with the primitive UNITDATA.IND. If the numbered frames are received out of sequence, they are delivered out of sequence to the higher layer. In this example, the UNITDATA request instructs the LLC layer not to encipher.

The second diagram of Figure 9.3 again shows connectionless communication. The SNDCP requests the LLC layer to forward SNDCP XID (exchange identity) parameters to the network. This information is sent on the air interface logical channel PDTCH in unacknowledged mode. It causes an LLC XID frame (encapsulating the SNDCP XID parameters) to be sent on the LLC data link and will include, if necessary, LLC XID parameters for negotiating the L2 LLC parameters with the peer entity (the network LLC layer). LLC XID parameters include IOV, T200, N200, N201, m DL, m UL, k DL, k UL. These are discussed in a following section.

The LLC layer passes received layer 3 XID parameters to the SNDCP layer using an XID.IND primitive. The cipher function is shown switched off in this example although it would most probably be on.

The third diagram shows the establishment of acknowledged asynchronous balanced mode (ABM) operations. It is initiated by the SNDCP layer sending the primitive ESTABLISH.REQ. This primitive will normally carry a higher layer PDU, and this will be encapsulated within an LLC SABM (set asynchronous balanced mode) frame.

The exchange across the LLC data link of SABM and UA (unnumbered acknowledge) sets up the data link for ABM operation – acknowledged mode. The LLC layer

The LLC layer has the functions

Establish a data link with its corresponding layer (peer entity) – network LLC layer, using the TLLI as the MS address.
Provide a highly reliable data link between the MS and network for delivery of SN PDUs which are delivered in connection mode (acknowledged-I-ABM) or connectionless mode (unacknowledged-UI-ADM) dependent upon the higher layer requirements. This is done by framing SN PDUs in appropriate DCL-data communications link frames.
Handle multiple N PDU sessions.
Frame SN PDUs into LLC frames.
Frame signalling messages delivered by the MM layer for transmission in unacknowledged mode.
Frame SMS messages in unacknowledged transmission mode.
Cipher and decipher SN PDUs and MM and SMS messages.

Reliability class 1, invokes LLC acknowledged mode, LLC data protection, RLC acknowledged mode, GTP acknowledged mode
Reliability class 2, invokes the same as class1 for LLC and RLC, but unacknowledged mode for GTP
Reliability class 3, invokes LLC unacknowledged mode, LLC data protection, RLC acknowledged mode and GTP unacknowledged mode.
Reliability class 4, invokes LLC unacknowledged mode, LLC data protection, RLC unacknowledged mode and GTP unacknowledged mode.
Reliability class 5, invokes unacknowledged mode on LLC, RLC and GTP and unprotected data on the LLC link.
GMM PDUs use RLC acknowledged mode.
RLC/MAC control messages (RR signalling) use RLC unacknowledged mode.

(1) **UNACKNOWLEDGED MODE**
The LLC frames are numbered but their reception by
the far end is not acknowledged.

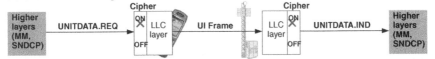

(2) **UNACKNOWLEDGED MODE**
In this case the higher layers are requesting negotiation of compression parameters;
the LLC layer may also negotiate LLC parameters.

(3) **ABM Establishment**
A higher layer has requested ABM operation; the exchange of SABM and UA sets up the acknowledged
mode data link. LLC XID parameters may be exchanged with SABM and UA.

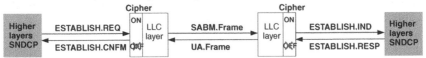

(4) **ACKNOWLEDGED MODE-ABM**
SN PDU data is now being transferred in acknowledged mode by the LLC.

Figure 9.3   The function of the LLC layer

receiving SABM waits for the higher layer to respond to a layer 3 message contained in SABM.

The receipt by an LLC layer of the SABM frame causes the indication to the higher layer that a data connection is being established with primitive ESTABLISH.IND.

The return UA frame causes the receiving LLC layer to indicate to the higher layers via the primitive ESTABLISH.CNFM that a data connection is established.

The two LLC peer entities are now set up to guarantee delivery of higher layer PDUs through the data link connection established.

The final diagram of Figure 9.3 shows SN PDUs being transferred in ABM operation; they are fully protected and delivery is guaranteed (or the higher layers are informed that communication has broken down).

The LLC layer sends an IS (information/supervisory frame) when it receives a DATA.REQ from the higher layer. It encapsulates the SN PDU in an IS frame, numbers the frame and expects acknowledgment of that frame. Transmitted IS frames are acknowledged with:

> S frames (supervisory frames);
> RR (receiver ready) frames;
> ACK (acknowledge) frames; or
> SACK (selective acknowledge) frames.

These are discussed in a later section.

The LLC layer receiving an IS frame and delivers it to the higher layer within a DATA.IND primitive. When acknowledgment has been received by the sending end, it confirms delivery to the sending higher layer with DATA.CNFM.

These procedures are discussed below.

9.2 LLC frames

Before looking at the operation of the LLC data link, this section looks at the flags, registers and other parameters used to control the data flow.

Figure 9.4 shows all the registers and flags used by the LLC data link. This diagram is used to illustrate LLC data link frame transfer in a later section.

A brief description of the flags and registers is given below.

9.2.1 Types of information/supervisory frame

Information (I) frames are normally combined with supervisory (S) frames and then called *IS frames*. For convenience, we will call all I frames IS frames. IS frames are always used for ABM-protected/guaranteed operation when layer 3 data units requiring acknowledged mode are transmitted.

Below the block labelled IS on Figure 9.4 are seen the supervisory frame functions:

- **RR, receiver ready**. This frame is used to indicate that the LLC layer is ready to receive I frames. It is also used to acknowledge receipt of I frames.
- **ACK**. This is an acknowledgment frame, acknowledging receipt of I frames.
- **SACK selective acknowledgment**. This is an acknowledgment frame which can individually indicate the satisfactory reception of up to 255 frames.

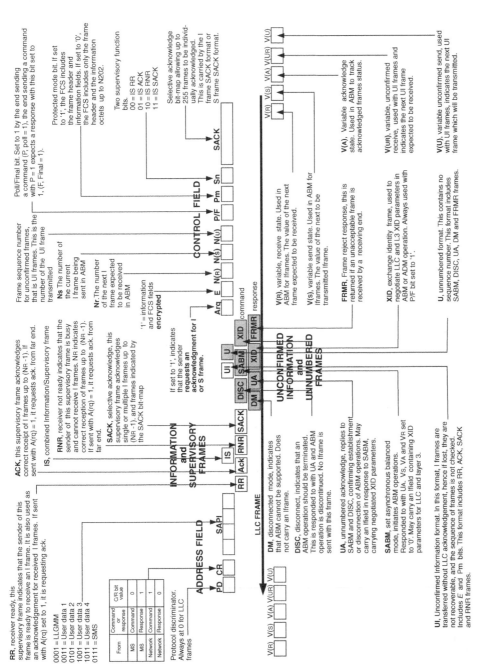

Figure 9.4 LLC registers, headers and parameters

Which of these three acknowledge frames should be used and when? RR is sent as an acknowledgment if the number of the received frame is the number which was expected. Formally:

$$\text{if } N(S) = V(R), \text{ then acknowledge with RR}$$

where N (S) is the number of the transmitted frame, V(R) is the frame number expected to be received.

ACK is sent as an acknowledgment if the received frame is one greater than the frame expected. Formally:

$$\text{if } N(S) = V(R) + 1, \text{ then acknowledge with ACK}$$

SACK is sent as an acknowledgment if neither of the above equalities apply.

9.2.2   The unacknowledged information frame

Frames labelled UI are unacknowledged information frames (carrying an SNDCP or GMM PDU) transferred in connectionless mode.

9.2.3   Types of unnumbered frame

- **SABM (set asynchronous balanced mode)**. This LLC frame indicates to the receiving end that the sending end wishes to set up a data link connection in ABM.
- **UA (unnumbered acknowledge)**. The expected response from an LLC layer which has received SABM, indicating it is now prepared for ABM operations.
- **XID (exchange identity)**. An LLC frame which carries layer 3 requested XID parameters; the response to this frame indicates the XID parameters which will be used. This LLC frame is only sent if the SNDCP requires the LLC layer to work in unacknowledged mode. This XID frame may also carry layer 2 LLC parameters for negotiation.
- **DISC (disconnect)**. An LLC frame which seeks to disconnect the data link from ABM operation. The proper response from an LLC layer receiving this command is UA, indicating that the data link connection is released.
- **DM (disconnect mode)**. An LLC frame sent by an LLC layer which receives an ABM type frame (IS) when no ABM operation has been established, indicating it cannot service the ABM type frame.
- **FRMR (frame reject)**. An LLC frame sent by an LLC layer in response to a received layer 2 frame containing severe errors.

9.2.4   Flags and registers used for frame address fields

The format of the address field is shown in Figure 9.4.

- **PD (protocol discriminator)**. This is always set to '0' for LLC frames.
- **CR (command/response)**. This bit indicates whether the frame being sent is a command or response frame. The coding is shown in Figure 9.4.
- **SAPI, the service access point identity** of the higher layers which are using the LLC layer. The coding is shown in Figure 9.4.

9.2.5   Flags and registers used for control fields

There are a number of control indicators included with LLC frames. These are shown in Figure 9.4.

- **A(rq) (acknowledge bit)**. This indicates ('1') that an acknowledgment is required for transmitted frames. The LLC layer receiving this is forced to send an acknowledgment.
- **E (encryption bit)**. If this is set to '1', it indicates to the receiver that the information and FCS fields of the received LLC frame are encrypted.
- **N(R)**. The number of the next I frame expected to be received. This, sent in each I frame transmitted, tells the far end that frame N(e)−1 has been received.
- **N(S)**. This is the number of the current I frame being transmitted.
- **N(U)**. This is the number of the current UI frame being transmitted.
- **P/F (poll/final)**. If an LLC frame with the poll bit set to '1' is sent, the sender expects a response from the receiving end with the final bit set to '1'.
- **Pm (protection mode)**. If set to '1', then the frame check sum (FCS) includes the information and frame header fields; set to '0', it indicates the frame header and information fields up to N202 are included.
- **Sn (supervisory bits)** as indicated in Figure 9.4.
- **SACK (selective acknowledge)**. A bit-map indicating which of 255 frames have been received correctly. Included with the SACK supervisory frame.

9.2.6   Registers used for variable arrays in the LLC

Both ends of an LLC data communication link have a number of registers which track the status of transmitted and received LLC frames in both the ABM and connectionless modes. Figure 9.4 shows these variable arrays.

- **V(R)**, in ABM, the next I frame expected to be received.
- **V(S)**, in ABM, the value of the next-to-be-sent I frame.
- **V(A)**, in ABM, the next I frame expected to be acknowledged, or, put another way, the oldest transmitted frame as yet unacknowledged.
- **V(UR)**, in connectionless mode, the number of the next UI frame expected to be received.
- **V(U)**, in connectionless mode, the number of the next UI frame to be transmitted.

9.3   LLC operational parameters

Figure 9.5 illustrates some of the LLC data link parameters.

- **N201**, the maximum number of octets allowed in an LLC frame information field. Default values are shown in the table of Figure 9.5. Note that the default value of N201 for SAPI 3 QoS 1 access is 1520 octets. This means for this service access point (SAP), that the SNDCP layer may not have to segment SN PDUs before sending them to the LLC layer.
- **N202**, the number of octets allowed in the layer 3 header of a UNITDATA header, (maximum 5).

- *k* **DL,** *k* **UL,** the window size for transmission of LLC frames in ABM.
- *m* **DL,** *m* **UL,** the LLC buffer sizes. Note that for SAPI 3, the *m* value is 1520×16 octets, that is, the LLC layer must be capable of storing 16 SN PDUs. As we would expect, this is also the window size of this SAP.
- **T200,** the timeout value during ABM operations, within which an acknowledgment should be received.
- **N200,** a counter used in ABM to count the number of retransmissions of unacknow- ledged frames.
- **IOV,** input offset value applied to encryption, allocated by SGSN.

9.4 LLC data link flow – ABM establishment

This section explains the operation of the LLC data communications links, using the parameters covered briefly in Section 9.2. Figures 9.6–9.8 illustrate LLC frame inter- change for the LLC data link.

Let us look at ABM establishment. Asynchronous balanced mode is the acknow- ledged transmission of LLC PDUs, and before this can happen a data communications link connection must be established. This is called establishing ABM. This section uses Figure 9.6 to explain ABM establishment.

9.4.1   Normal (successful) ABM establishment

The first LLC frame in Figure 9.6 shows unacknowledged transmission of a UI frame. The GMM layer sends the primitive LL-UNITDATA-REQ, which requests unacknow- ledged transmission of a GMM PDU. The LLC layer wraps the PDU in a UI – unacknowledged information – frame and allocates a number for this in $N(U) = 0$, the 'zeroeth', or first UI frame to be transmitted.

The V(U) register is incremented to 1 (the V(U) register always holds the next UI frame due to be transmitted).

The LLC frame is labelled UI and sent to the far end (GPRS sub-network) LLC layer. The network V(UR) register holds the number of the next UI frame it is expecting to receive, and receiving $N(U) = 0$ increments V(UR) (to the number of the next frame expected to be received).

The UI frames are numbered, and the numbering sequence is monitored, but there is no buffering of the received or transmitted frames.

The next frame shows the establishment of ABM operation. ABM operation is normally initiated (except for the case of ABM re-establishment) by the SNDCP layer and this is shown by an LL-ESTABLISH-REQ primitive sent from the SNDCP layer. This primitive causes a SABM frame to be transmitted by the LLC layer:

> Timer T200 is started,
> Counter N200 set,
> The C/R bit set to '0',
> P/F to '1',
> The register array comprising V(R), V(S) and V(A) is set to zero.

LLC frame

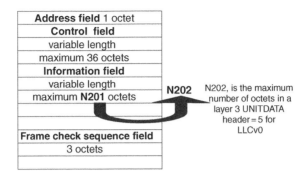

| Address field 1 octet |
| Control field |
| variable length |
| maximum 36 octets |
| Information field |
| variable length |
| maximum **N201** octets |
| |
| |
| Frame check sequence field |
| 3 octets |
| |

N202 — N202, is the maximum number of octets in a layer 3 UNITDATA header = 5 for LLCv0

Window size *k* and buffer size *m*

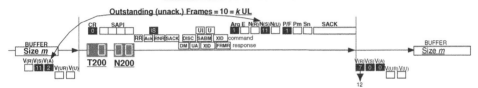

LLC parameters default values

Maximum and Minimum Values		SAPI 1 GMM	SAPI 3 Qos1	SAPI 5 Qos2	SAPI 7 SMS	SAPI 9 Qos3	SAPI 11 Qos4	
	T200	5s	5s	10s	20s	20s	40s	default
	N200	3	3	3	3	3	3	default
maximum 24320 x 16 octets	m DL	n/a to ADM	1520	760	n/a to ADM	380	190	default *units = 16 octets*
	m UL	n/a to ADM	1520	760	n/a to ADM	380	190	default *units = 16 octets*
	k DL	n/a to ADM	16	8	n/a to ADM	4	2	default
	k UL	n/a to ADM	16	8	n/a to ADM	4	2	default
minimum 140 octets	N201-U	270	500	500	270	500	500	default *units = octets*
maximum 1520 octets	N201-I	n/a to ADM	1503	1503	n/a to ADM	1503	1503	default *units = octets*
N202 maximum 5 octets	IOV-UI	0	0	0	0	0	0	default
	IOV-I	n/a to ADM	$2^{2^7} \times$ SAPI	$2^{2^7} \times$ SAPI	n/a to ADM	$2^{2^7} \times$ SAPI	$2^{2^7} \times$ SAPI	default

LLC ciphering parameters

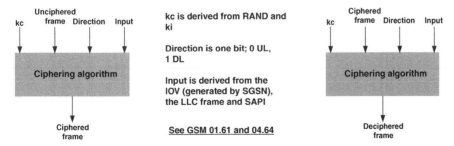

kc is derived from RAND and ki

Direction is one bit; 0 UL, 1 DL

Input is derived from the IOV (generated by SGSN), the LLC frame and SAPI

See GSM 01.61 and 04.64

Figure 9.5 LLC parameters

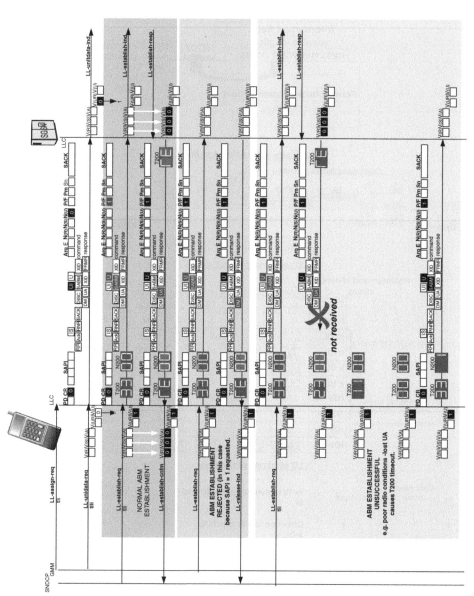

Figure 9.6 LLC data link protocol establishing ABM operation

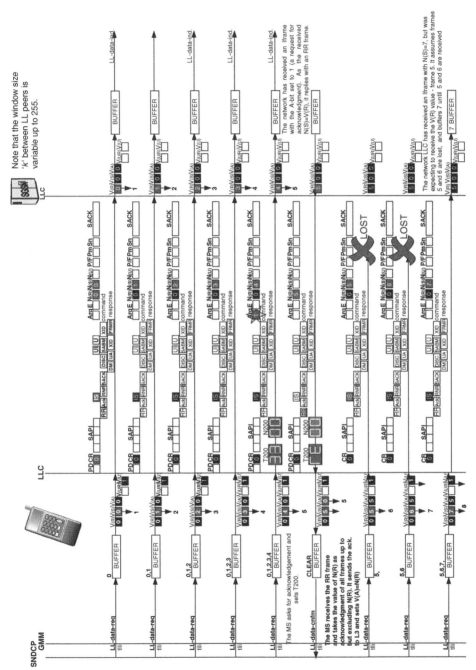

Figure 9.7 LLC data link protocol using ABM operation. Normal operation with acknowledge request and with lost frames

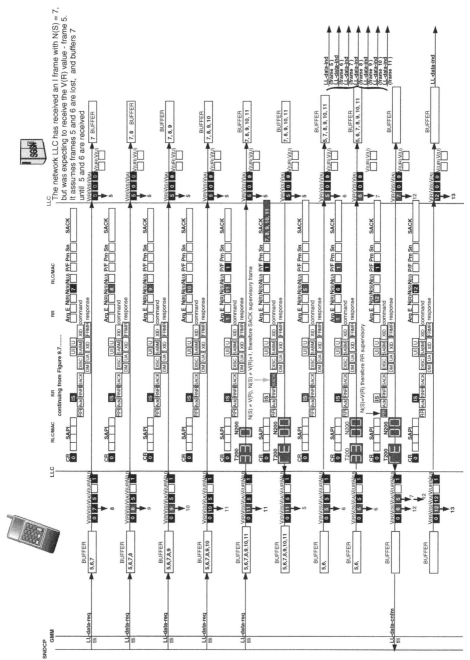

Figure 9.8 LLC data link protocol using ABM operation – normal operation with acknowledge request and with lost frames

The SABM frame received by the network indicates ABM operation is required. The SABM frame will normally carry an SNDCP PDU requesting SNDCP XID parameter settings. The sending LLC layer may add to this its own XID parameters to set the operating condition for the LLC data link connection.

When the network LLC layer receives the SABM frame, it sets V(R), V(S) and V(A) to zero, and returns to the mobile station a UA frame with the P/F bit set to '1', and the C/R to '0'. The UA frame will be delayed until the network higher layers have responded to the mobile station's SNDCP XID request. This higher layer response will then be included on the UA frame. (The layer receiving a SABM frame can respond autonomously if the SABM frame does not contain a layer 3 PDU; in this case the SABM frame does contain a layer 3 PDU and waits for a layer 3 response before sending the UA frame – the UA frame then carries the layer 3 response PDU.)

The GPRS sub-network LLC layer indicates to the SNDCP layer with LL-ESTAB-LISH-IND that ABM operation is being established.

When the mobile station LLC layer receives UA with P/F = 1, timer T200 is reset and ABM is established.

9.4.2   ABM establishment rejection (SAPI = 1)

This anomalous situation should not arise but is included as an illustration.

The mobile station LLC layer receives the LL-ESTABLISH-REQ primitive from the GMM layer. This is prohibited as all GMM communications are UI.

However, the mobile station LLC layer transmits a SABM frame with the SAPI = 1 (GMM SAPI). The receiving end cannot accept SABM operation for this SAP and responds with disconnected mode (DM). On receipt of DM, the LLC layer resets T200 and sends LL-RELEASE-IND to the L3 entity (GMM layer) requesting this ABM operation.

9.4.3   ABM establishment rejection, poor radio reception

The mobile station SNDCP layer requests ABM operation and the LLC layer sends SABM. The network LLC layer receives SABM and responds with UA. This is not received by the initiating LLC layer due to poor radio reception.

The mobile station timer T200 counts down, until it reaches timeout. The mobile station LLC layer then resends the SABM frame, incrementing N200, and resetting T200. If the mobile station LLC layer does not receive UA within the T200 period, it will again resend SABM, increment N200 and reset T200.

This continues until either UA is received or N200 reaches the maximum retry attempt count. When N200 reaches this count, the mobile station LLC layer sends LL-RELEASE-IND to the SNDCP layer and LLGMM-STATUS-IND to the GMM layer to indicate that there is no response to SABM. The LLC layer enters the ADM state. (In the ADM state, ABM operations are not possible, UI operations are possible.)

9.5   Information/supervisory (IS) frame transfer

This section is about the transfer of IS frames after establishment of ABM operation. It uses Figures 9.7 and 9.8 to explain the operation.

9.5.1 IS frame transfer

In Figure 9.7 the SNDCP layer sends an SN PDU with an LL-DATA-REQ primitive. The status of the LLC register variables array has been set to zero by the previous SABM/UA exchange (V(R), V(S), V(A) = 0).

The LLC layer encrypts the SN PDU and encapsulates it within an IS L2 frame. The N(S) and N(R) flags in the IS frame take the values of V(S) and V(R). When the IS frame is transmitted, the mobile station variable V(S) increments. N(S) in the sent IS frame indicates the sequence number of the IS frame; N(R) indicates the sequence number of the next IS frame that the mobile station expects to receive from the network.

The network LLC layer receives the numbered (N(S)) IS frame, and this is the frame it expects to receive (V(R) = N(S)). It deciphers the received frame and passes the SN PDU to the SNDCP layer with the LL-DATA-IND primitive. The network V(R) variable register is incremented to '1', the next frame it is expecting from the mobile station.

This process continues in an orderly fashion with network N(R) incrementing sequentially and the mobile station V(R) tracking the sent frames.

For the fifth frame (frame 4) transmitted by the mobile station, the mobile station requests acknowledgment by setting the A-bit to '1'. It simultaneously starts the timer T200 and the counter N200 and sets the P/F bit to '1'.

The network responds to the request for acknowledgment by sending a receiver ready (RR) frame with the N(R) bit set to 5, the number of the next frame the network is expecting to receive from the mobile station. The network responds with RR because the network V(R) = mobile station N(S) on the frame with the A-bit set to '1'.

The conditions for the type of response frame to a request for acknowledgment are

$$\text{If } V(R) = N(S) \text{ then response} = RR;$$

$$\text{If } V(R) + 1 = N(S) \text{ then response} = ACK.$$

If neither of these equalities is met then the correct response is SACK.

When the mobile station receives the RR frame with P/F = '1', it resets T200 and N200, increments V(A) to the status of the received N(R) and sends an LL-DATA-CNFM primitive to the SNDCP layer. Note that all sent frames are stored in the sending buffer store until acknowledgment, when the buffer store is cleared.

The mobile station then continues transmitting IS frames. Unbeknown to the mobile station, two of the transmitted frames are lost due to poor radio conditions. The network knows that mobile station transmitted frames are lost when it receives frame 7 when it was expecting frame 5. It stores frame 7 and does not pass it to L3.

Note that the LLC always delivers IS frames in the correct sequence to the SNDCP layer.

Figure 9.8 repeats the last mobile station transmitted frame N(S) = 7 of Figure 9.7. As the network V(R) is not equal to the received N(S), the network knows frames have been lost and holds frame 7 in its buffer store. It then receives frames 8, 9, 10 and 11, with frame 11 asking for acknowledgment with the A-bit set to '1'. Frames 7–11 in the network LLC buffer store are not passed to the SNDCP layer.

The correct response frame to the mobile station's *acknowledgment request* is SACK – selective acknowledgment. This is the frame returned to the mobile station with P/F = '1' and with the SACK bit-map acknowledging frames 7, 8, 9, 10 and 11.

When the mobile station receives this SACK frame it resets T200, clears the buffer store of the acknowledged frames and sets about resending the unacknowledged frames (frames 5 and 6). With frame 6 retransmission, the mobile station requests acknowledgment and the network responds with RR. The network has now received all the frames sent by the mobile station and sends them in sequence to the SNDCP layer.

If the response RR frame is not received from the network within the time frame of T200, frame 6 is resent and N200 incremented.

If no response is received within the time frame T200 × N200, the sending end will attempt to re-establish ABM operation by sending SABM frame(s).

When the mobile station receives RR it knows that the network is expecting to receive frame 12. This is an acknowledgment by the network that frames 5 and 6 have been received. All frames up to 11 have therefore been acknowledged. Transmission of IS frames continues from frame 12.

Further information on LLC operation can be found in the recommended reading list.

9.6 Abbreviations used in this chapter

A (bit)	Acknowledge demand
ABM	Asynchronous balanced mode
ADM	Asynchronous disconnect mode
BTS	Base transceiver station
CR	Command response
DISC	Disconnect
DL	Downlink
DM	Disconnect mode
E	Encryption bit
FRMR	Frame reject
GMM	GPRS mobility management
GPRS	General packet radio service
GSM	Global system for mobile communication
HDLC	High speed data link control
IOV	Input offset value
IS	Information/supervisory
k	Window size
LAPD	Link access procedure D channel
LLC	Logical link control
MS	Mobile station
NCO	Network control order
N(R)	The next PDU expected to be received
N(S)	The current LLC
N(U)	The number of the unacknowledged LLC
PD	Protocol discriminator
PDP	Packet data protocol

PDTCH	Packet data traffic channel
PDU	Protocol data unit
P/F	Poll final bit LLC
Pm	Protection mode
QoS	Quality of service
RLC	Radio link control
RR	Radio resources
RR	Receiver ready
SABM	Set asynchronous balanced mode
SACK	Selective acknowledge
SAP	Service access point
SAPI	Service access point identity
SGSN	Serving GPRS support node
SM	Session management
SMS	Short message service
Sn	Supervisory bits
SNDCP	Sub-network dependent convergence protocol
SN PDU	Sub-network protocol data unit
TBF	Temporary block flow
TCP/IP	Transmission control protocol/Internet protocol
UA	Unnumbered acknowledge
UE	User equipment
UI	Unacknowledged information
V(A)	Acknowledge register
V(R)	Next PDU expected to be received
V(S)	Next PDU to be sent
V(U)	Number of transmitted unacknowledged PDU LLC
V(UR)	Next expected unacknowledged PDU LLC
XID	Exchange identities

10

GMM Layer Procedures

(GSM 23.060 section 6 and 44.008 sections 4 & 9)

The GPRS mobility management layer has similar functions to the circuit switched MM layer, with some variation. These functions include:

- **Registration and deregistration**. The GMM layer registers and deregisters with the GPRS sub-network by using one of the procedures:

  - **GPRS attach**. This can, in some cases, be combined with circuit switched IMSI attach. The GPRS attach is initiated by the GMM layer when requested by the session management (SM) layer or the application (more correctly the network) layer.
    Attach informs the GPRS sub-network of the mobile station's identity, capabilities and routeing area and allows the GPRS sub-network to authenticate the mobile station and exchange encryption parameters. When a mobile station is GPRS attached it performs the GPRS cell reselection procedures (if there is a PBCCH).
  - **GPRS detach**. The mobile station must detach from the GPRS sub-network under a number of conditions, which are discussed in this chapter.

- **Location management**. In the attached condition, the mobile station GMM layer must keep track of the routeing area or cell identity, performing routeing area and cell updates to the GPRS sub-network as appropriate. The conditions for a GPRS attached mobile station to report routeing area changes to the GPRS sub-network are:

  - The RR layer indicates a change in the RA. (The RA reported to the GMM layer is continuously compared to the RA stored on the SIM card. When these differ, an RA update to the GPRS sub-network is initiated by the GMM layer). RA updates are performed in GMM standby and GMM ready states.
  - The routeing area periodic update timer expires. This timer period is set by the GPRS sub-network. When it expires, the mobile station initiates an RA update irrespective of the RA it is in. The purpose of this timer is to allow the GPRS sub-network to monitor whether mobile stations are still in its service area after long periods of mobile station inactivity. Periodic RA updates are performed from the GMM standby state only.

GPRS in Practice: A Companion to the Specifications   Peter McGuiggan
© 2004 John Wiley & Sons, Ltd   ISBN: 0-470-09507-5

The condition for a GPRS attached mobile station to report cell changes to the GPRS sub-network is:

—The RR layer reports the cell identity of the current cell to the GMM layer. The GMM layer compares this to the cell identity stored in the SIM card and, if they differ, initiates a cell update to the GPRS sub-network. Cell updates are only actioned if the mobile station is in the GMM ready condition.

- **Authentication**. The GPRS sub-network will normally authenticate the mobile station during the GPRS attach procedures. The authentication is similar to circuit switched authentication. The same parameters are used with RAND (a random number related to IMSI in the SGSN) sent to the mobile station and the mobile station responding with SRES, the signed response. The A3 algorithm is used by the mobile station and GPRS sub-network to derive SRES (the customer identification key, ki, and RAND are the input parameters to A3 and used by the GPRS sub-network AUC and mobile station SIM card). This is the same A3 as used for circuit switched operations.
- **Ciphering**. The GMM layer is not responsible for ciphering – that is the function of the LLC layer, but it is responsible for passing the RAND received from the GPRS sub-network to the SIM card that uses the A8 algorithm to produce the ciphering key, kc. The GMM layer then passes this key to the LLC layer, which uses it as an input to the GPRS encryption algorithm.

    The GPRS sub-network gives RAND with the *authentication and ciphering request* during the GPRS attach procedure. It includes with RAND a cksn – ciphering key sequence number. This cksn changes sequentially as further MM procedures occur. During the other MM procedures, the mobile station reports the cksn last used by the GPRS sub-network. The GPRS sub-network then knows which ciphering key (kc) the mobile station can use. The GPRS sub-network operator may then bypass the authentication and go directly to encrypted mode. (The GPRS sub-network, knowing the kc that the mobile station can use, uses this key itself in sending an encrypted message, indicating to the mobile station that LLC encryption is used (the *E* bit in the LLC header). The mobile station decrypts using the same kc.
- **Identification**. The mobile station GMM layer will respond to the GPRS sub-network with the following identities when requested:

    —IMSI, the international mobile subscriber identity (held on the SIM card).
    —IMEI, the international mobile equipment identity (the serial number of the mobile equipment).
    —IMEISV, the software version of the mobile equipment.

- **TLLI generation**. A successful attach to the GPRS sub-network is signified when the GPRS sub-network sends *attach accept* to the mobile station. This message may contain PTMSI (packet temporary mobile subscriber identity), an apparently random 32-bit number, which the mobile station and GPRS sub-network will use instead of IMSI, giving subscriber security. For increased security, the PTMSI is sent by the GPRS sub-network in encrypted form. It is however, sent by the mobile station over the air interface in clear (unciphered) form with the *attach request* message.

    From the PTMSI the GMM layer constructs the temporary logical link identifier. The TLLI is used by the GPRS sub-network as the mobile station's identity during a

TBF, and the mobile station identifies itself to the GPRS sub-network with TLLI during a TBF. The GMM layer passes the TLLI to the LLC layer and the RLC layer.
- **Ready state activity monitoring**. When a mobile station is in the GMM ready condition, a ready timer T3314 is started in the GMM layer. If no activity occurs (no LLC PDUs are received or transmitted during this period), the GMM layer reverts to GMM Standby condition.

Figure 10.1 shows many of these functions of the GMM in the mobile station. It also shows the information the GMM layer requires to perform these functions and the sources of the required information, for example:

- The *Identity* function may require IMSI as an input and this is provided from the SIM card.
- 'Authentication' requires ki and RAND, and these are provided respectively by the SIM and the DL PDTCH.
- 'TLLI' requires PTMSI and RA, the routeing area identifier, and these are provided by the PBCCH (or BCCH if no PBCCH is used) and the PDTCH.

This chapter will examine how these functions are used by the mobile station and between the GPRS sub-network GMM and mobile station GMM layers.

10.1 GMM states (GSM 23.060 section 6)

The GMM layer can be in one of three states:

1. **GMM idle state** *(The mobile station is in packet idle condition in this state)*. In this state the mobile station is not registered with the GPRS sub-network GMM layer. The GPRS sub-network is not aware of the mobile station's existence and will not attempt to page the mobile station. The mobile station does not monitor the PCCCH-PPCH for paging messages. It may monitor the BCCH and PBCCH of cells but it does not perform GPRS cell reselection. It does not perform routeing area updates or cell updates. No GMM attach context exists. It may receive point to multipoint multicast (PTM-M) data from the GPRS sub-network. The mobile station cannot activate a PDP context from this state.

2. **GMM standby state** *(The mobile station is in packet idle condition in this state)*. In this state the mobile station has performed 'GPRS attach' and the GPRS sub-network is aware of the mobile station's routeing area. A PTMSI is allocated by the GPRS sub-network, and both the GPRS sub-network and mobile station derive a TLLI from this. The TLLI is used to identify any subsequent air interface communication.

    If a PBCCH exists and the mobile station is not in circuit switched active condition (e.g. a telephone call is in progress), the mobile station performs GPRS cell reselection and RAU (routeing area updates) when leaving one routeing area and entering another, or when timer T3312, the periodic routeing area update timer, expires.

    The mobile station monitors its PCCCH_GROUP, containing the paging group. The mobile station has calculated its paging group from its IMSI and the number of paging group channels available. It implements DRX parameters, including split paging.

    The mobile station can receive paging messages on the PCCCH. However, this is unlikely without a PDP context being established.

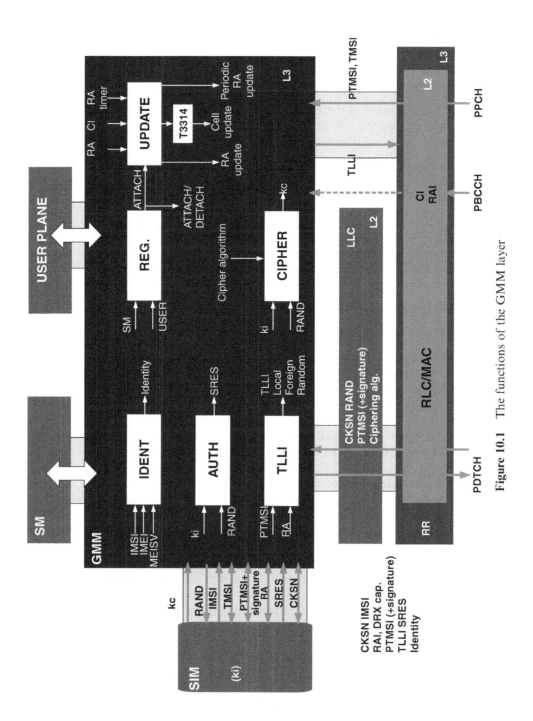

Figure 10.1   The functions of the GMM layer

A GMM context is said to exist. The mobile station can go to *PDP context Activation* from this GMM state. The mobile station can perform extended measurements in this state.

3. **GMM ready state** *(The mobile station may be in packet idle or packet transfer condition in this state)*.

   In this state:

   —radio resources have been allocated by the GPRS sub-network and an uplink or downlink TBF of user data may start. The ready timer is running;

   or

   —a TBF has been completed and the radio resources released, but the ready timer is running.

The mobile station remains in the ready state after a TBF is completed and the radio resources released. It remains in the ready state until the ready timer T3314 expires. This timer is always started when an LLC PDU is activated at the LLC layer.

Whilst T3314 is running, the mobile station performs cell reselection (or GPRS sub-network controlled cell reselection if NCO3 is in operation) and cell updates if it reselects a cell.

Figure 10.2 illustrates these aspects of GMM states.

The following sections of this chapter look at the GMM procedures in detail.

10.2 GMM procedure attach (GSM 23.060 section 6)

Figure 10.3 illustrates communication between the mobile station and GPRS sub-network GMM layers to enable *GPRS attach*. The mobile station initiates attach upon receiving the internal stimulus, GMM-REG-ATTACH-REQ or GMMSM-ESTABLISH-REQ from respectively the application (network) or session management (SM) layers. Either of these two primitives causes the GMM layer to generate the message *attach request* which is transported across the LLC, RR, and BSSGP layers with appropriate primitives (see Figure 5.8). The *attach request* message carries the information:

- **cksn, the ciphering sequence number**. This is a number that may be allocated with RAND by the GPRS sub-network. This identifies the current *ciphering key kc* used by the GPRS sub-network and the mobile station.

  When another RAND is sent by the GPRS sub-network (perhaps during a routing-area update) then the cksn is incremented. When the cksn sent by the mobile station (with *attach request* or *RA update request*) agrees with the cksn held by the SGSN, then the SGSN may omit authentication and ciphering commands and go straight to ciphered mode with the *attach accept* message.
- **attach type**. This can be GPRS, combined CS and GPRS or GPRS with CS IMSI attach already in place.
- **PTMSI or IMSI**. Either of these two identifiers may be used to identify the mobile station, PTMSI if a valid PTMSI is present in the GMM layer, IMSI if it is not.
- **PTMSI signature**. This is included if the GPRS sub-network has assigned it in a previous *attach accept* message.

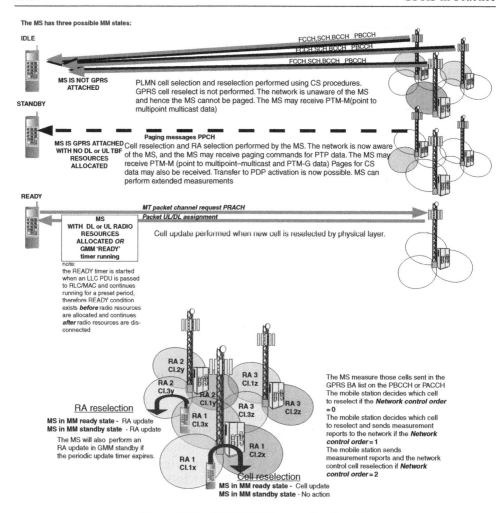

The MS has three possible MM states:

IDLE

FCCH,SCH,BCCH   PBCCH

FCCH,SCH,BCCH   PBCCH

FCCH,SCH,BCCH   PBCCH

MS IS NOT GPRS ATTACHED

PLMN cell selection and reselection performed using CS procedures. GPRS cell reselect is not performed. The network is unaware of the MS and hence the MS cannot be paged. The MS may receive PTM-M(point to multipoint multicast data)

STANDBY

Paging messages PPCH

MS IS GPRS ATTACHED WITH NO DL or UL TBF RESOURCES ALLOCATED

Cell reselection and RA selection performed by the MS. The network is now aware of the MS, and the MS may receive paging commands for PTP data. The MS may receive PTM-M (point to multipoint–multicast and PTM-G data) Pages for CS data may also be received. Transfer to PDP activation is now possible. MS can perform extended measurements

READY

MT packet channel request PRACH

Packet UL/DL assignment

MS WITH DL or UL RADIO RESOURCES ALLOCATED OR GMM 'READY' timer running

Cell update performed when new cell is reselected by physical layer.

note:
the READY timer is started when an LLC PDU is passed to RLC/MAC and continues running for a preset period, therefore READY condition exists *before* radio resources are allocated and continues *after* radio resources are disconnected

RA 2 Cl.2y

RA 2 Cl.3y

RA 3 Cl.1z

RA 2 Cl.1y

RA 3 Cl.3z

RA 3 Cl.2z

RA reselection

MS in MM ready state - RA update
MS in MM standby state - RA update

The MS will also perform an RA update in GMM standby if the periodic update timer expires.

RA 1 Cl.3x

RA 1 Cl.2x

RA 1 Cl.1x

Cell reselection
MS in MM ready state - Cell update
MS in MM standby state - No action

The MS measure those cells sent in the GPRS BA list on the PBCCH or PACCH
The mobile station decides which cell to reselect if the **Network control order** = 0
The mobile station decides which cell to reselect and sends measurement reports to the network if the **Network control order** = 1
The mobile station sends measurement reports and the network control cell reselection if **Network control order** = 2

Figure 10.2   GPRS MM states for the MS

- **DRX capability**. The mobile station indicates its capability for split paging on the CCCH (in case the CCCHs are used for paging – they will be if the PBCCH is closed).
- **Radio access capability**. The mobile station indicates its capability in the following areas: GSM900 P – primary, E – extended, R – railway, and GSM1800 working. (Later, this will include UMTS capability).
- **Pseudosynchronisation capability**. The mobile station indicates its capability, given the absolute time difference in TN0 transmission between the serving cell and a neighbour cell, to calculate the required timing advance for a handover to the neighbour cell in 'synchronous' mode.
- **A5 algorithms** available for CS mode – A5/1–7.
- **VGCS** – voice group calls capability.
- **VBS** – voice broadcast capability.

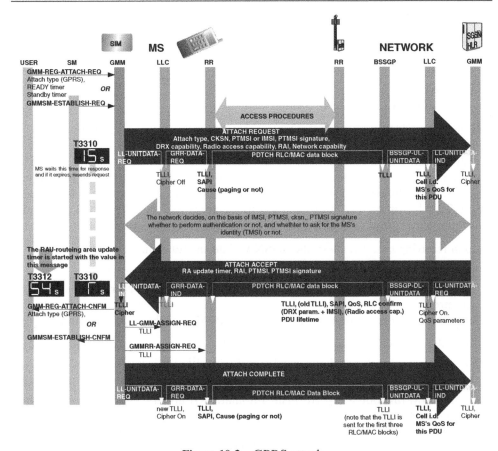

Figure 10.3   GPRS attach

- **RF power capability**.
- **Measurement capability**. SMS – switch measure switch value, the minimum time required for the mobile station to change frequency, perform a measurement on a neighbour cell and switch back to the serving frequency: range 1/8th of a burst to 2 bursts.

   Also included is the SM – switch measure value which is the minimum time needed for the mobile station to switch from one frequency and perform a measurement on another frequency in a neighbour cell.
- **Revision level** – GSM phase 1 or phase 2 capability.
- **Early sending of class mark capability** – yes or no.
- **Multislot capability**.
- **RAI, the routeing area identifier** is necessary because the PTMSI is only valid for a particular routeing area. This will be the old routeing area where the PTMSI was allocated. The GPRS sub-network will be able to identify a mobile station with a 'foreign' PTMSI/RAI pair, by going to the SGSN indicated by PTMSI/RAI.

- **Network capability**

  —**SS** (supplementary screening) indicator;
  —**Short message service** MT capability on CS channels;
  —**Short message service** MT capability on GPRS channels;
  —**UCS 2 support**;
  —**GEA/1** GPRS encryption algorithm 1 capability.

- **Ready timer T3314 value request**. This is a request for the setting of ready timer T3314, which is the maximum time that the mobile station is allowed to remain in GMM ready condition without an LLC PDU transfer with the GPRS sub-network. The sending of an LLC PDU to the GPRS sub-network restarts this timer, but if the timer expires, the mobile station is forced to *GMM standby*. The default value is 44 seconds. With T3314 running, the mobile station must report its new cell to the GPRS sub-network when a cell is reselected.

Upon receiving *attach request*, the SGSN will do the following:

- Check the PTMSI/RAI pair against that stored in the SGSN register.
- Check the PTMSI signature for agreement with the SGSN register's PTMSI signature.
- Check the cksn against that stored against the PTMSI of this mobile station.

If the above checks are in agreement, the GPRS sub-network GMM function then may generate an *attach accept* message (as in Figure 10.3) which contains:

- **T3312 periodic RAU timer** (default 54 minutes), which determines when the mobile station GMM function will perform an automatic RA update (irrespective of whether the RA has changed).
- **PTMSI reallocation** if appropriate (if this is not included in the *attach accept* message, then the mobile station retains its original PTMSI).
- **PTMSI signature** if this is used and a PTMSI is reallocated.

  If a new PTMSI is allocated, the GPRS sub-network GMM layer calculates the new TLLI and sends it to the GPRS sub-network LLC and BSSGP layers.

  If the SGSN does not recognise the PTMSI, it may send an *Identity* command to the mobile station asking for its IMSI. This is covered below.

  The GPRS sub-network will send to the mobile station an *authentication/ciphering request* before *attach accept*. This is also covered below.

When the mobile station receives the *attach accept* message, it does the following:

- If a new PTMSI is included, calculates the new TLLI;
- Informs the layer which requested attach that a GMM context is in place, and the mobile station is now GPRS attached;
- Assigns the newly calculated TLLI to the LLC and RLC layers;
- Sets T3312, the routeing area periodic update timer, to the value contained in the *attach accept* message;
- Generates the *attach complete* message and passes it to the mobile station LLC layer for onward transmission.

When the SGSN receives the *attach complete* message the GPRS attach is completed. The GPRS sub-network GMM layer now instructs its LLC and BSSGP layers to discard the mobile station's old TLLI. The old TLLI will continue to be used if, for any reason, the GPRS sub-network does not receive *attach complete* from the mobile station.

The mobile station remains in GMM ready condition until T3314, the ready timer, expires and then it will change to the GMM standby condition.

Before looking at the mechanics of TLLI calculation, *authentication/ciphering request* and *identity request*, we shall look at some of the inter-layer parameters transferred with the primitives, as shown in Figure 10.3.

10.2.1 Attach request primitives

The primitives used for transporting this message include:

- **LL-UNITDATA-REQ**. The GMM layer generates LL-UNITDATA-REQ and, in addition to carrying the GMM PDU *attach request*, the parameters TLLI and cipher off are passed with this primitive to the LLC layer.

  TLLI, the temporary logical link identifier, identifies the LLC logical link, but it is not included in any of the LLC headers; the TLLI is passed to the RLC layer for inclusion in all uplink blocks of a TBF for identification purposes if one-phase access is used, or in the *packet resource request*.

  If two-phase access is used, the TLLI is included in the initial single uplink block provided by the GPRS sub-network. The cipher off parameter is necessary because at this stage no ciphering key is agreed between the mobile station and GPRS sub-network.

  The primary reason why PTMSI is used is because it is sent unciphered across the UL GPRS sub-network and it can be intercepted by anyone with suitable equipment. However, when the PTMSI was allocated by the GPRS sub-network it was sent in ciphered form – this is its first appearance in unciphered form and it will be meaningless to an interceptor.

- **GRR-DATA-REQ**. The primitive used between the LLC and RR (RLC) layer. It carries the LLC PDU and also contains the TLLI for inclusion in the RLC/MAC message for identification (and also so that the RR layer can recognise downlink messages on the PCCCH which use TLLI as the identifier.) The SAPI is included to show the RR layer that the primitive is from the LLC layer.

  The 'cause' is included to tell the RR layer whether this is a response to a *packet paging request*. The RLC/MAC layer uses this information for the PRACH bursts in the access procedure.

- **PDTCH RLC/MAC data block**. The TLLI is extracted by the GPRS sub-network from the RLC/MAC data and is used by the GPRS sub-network to identify the data link from many others simultaneously in use.

  The 'old' TLLI will be used by the GPRS sub-network to address the mobile station until a new PTMSI is allocated to the mobile station, whereupon the GPRS sub-network and mobile station will calculate a new TLLI to be used.

- **BSSGP-UL-UNITDATA**. The primitive that is used on the uplink direction between the BSSGP to LLC. In this case it is carrying the primitive GRR-DATA-IND, which is carrying the LLC PDU, which encapsulates the *attach request* message.

  The BSSGP also transfers the cell identity with this parameter, and the TLLI.

- **LL-UNITDATA-IND**. Between the GPRS sub-network LLC layer and the GPRS sub-network GMM layer, this primitive carries the GMM PDU encapsulating *attach request*. The mobile station's 'old' TLLI is also transferred, together with an indication of whether ciphering was used for this PDU.
- **T3310**. This timer is started when *attach request* is sent, and if no *attach accept* is received before its expiry, the *attach request* message is resent.

10.2.2 Attach accept primitives

The primitives used for transporting this message include:

- **LL-UNITDATA-REQ**. Between the GPRS sub-network GMM layer and the LLC layer, this primitive carries the GMM PDU *attach accept* and also transfers the calculated 'new' TLLI that is used after the mobile station has signalled *attach complete*; for this communication the 'old' TLLI will be used.

    The LLC layer is ordered to cipher this message (the mobile station and GPRS sub-network now have an agreed ciphering key, kc).
- **BSSGP-DL-UNITDATA**. This is the downlink version of BSSGP-UL-UNITDATA above. It carries the LLC PDU and the parameters shown.

    The 'RLC confirm' parameter is instructing the RLC layer to map this primitive to the primitive GRR-DATA-REQ.

    The radio access capability is transferred to ensure that the RLC/MAC layer does not demand more from the mobile station than its capability.

    The 'PDU lifetime' tells the RLC/MAC layer for how long this PDU may be retained before it is discarded. (In cases of heavy traffic, there may be a delay before the GPRS sub-network can schedule transmission of this PDU.) The DRX parameters and the IMSI will only be transferred for paging messages as in that case the BSSGP layer must know these parameters to calculate which paging channel to place the paging message upon.

    PDTCH RLC/MAC data block no inter-layer parameters are exchanged on this channel.

    GRR-DATA-IND. Apart from the LLC PDU, the only other parameter transferred is the 'old' TLLI from the RLC/MAC layer. This enables a check on the status of the TLLI held by the RLC/MAC layer.

    LL-UNITDATA-IND. Between the mobile station LLC and GMM layers, this primitive delivers the GMM PDU containing the *attach accept* message. The only other parameters transferred are the 'old' TLLI and whether this PDU was ciphered (it was).

10.2.3 Attach complete

The only difference in parameter transfer between this PDU and the *attach request* PDU is the GMM PDU itself and the TLLI, which is now the 'new' TLLI. This is the TLLI which the RLC/MAC layer will use to recognise downlink PCCCH messages and identify itself during TBFs. If there has been no PTMSI reallocation then the new TLLI remains the same as the old TLLI.

10.3 TLLI construction (GSM 23.060)

The temporary logical link identifier (TLLI) uniquely identifies a mobile station within the GPRS sub-network allocating the PTMSI from which TLLI is derived. Used in conjunction with service access point identifier (SAPI), it uniquely, within that network, identifies a particular customer packet data transfer link.

- **Random TLLI**, a mobile station builds a random TLLI by setting bit 31 to 0, bits 30–27 to 1 and bits 26–0 are set randomly. The random TLLI is used when the mobile station has no valid PTMSI.
- **Local TLLI**. Based upon PTMSI and the RA in which the PTMSI was allocated, the mobile station or SGSN sets bits 31 and 30 to 1, bits 29 to 0 are set equal to bits 29 to 0 of the PTMSI.
  This PTMSI-associated local TLLI is valid only for the routeing area of the PTMSI allocation.
- **Foreign TLLI**. If a mobile station finds itself in a different RA to the RA in which the PTMSI was allocated, then bit 31 is set to 1, bit 30 is set to 0, bits 29 to 0 are set equal to the corresponding PTMSI bits.
  The foreign TLLI sent by a mobile station may inform a new SGSN that an old SGSN, identifiable from the TLLI and RAI, holds the subscriber details. The new SGSN may communicate with the old SGSN to extract the necessary information. The SGSN will pair the mobile station's TLLI to its IMSI and PTMSI and IP address.
- **Auxiliary TLLI**. This is generated by the SGSN in response to a random TLLI and is used as a TBF identity until a new PTMSI is allocated.

The TLLI is transmitted to identify the mobile station in all the RLC/MAC blocks of a one-phase access TBF and in the uplink single block allocation of a two-phase access TBF. The GPRS sub-network returns TLLI to the mobile station as a part of the contention resolution procedures.

Figure 10.4 illustrates the construction of local, foreign and random TLLIs.

10.4 Routing area update

Routeing area update (RAU) is completed by the mobile station GMM layer from either the GMM standby or GMM ready states when the LLC layer passes to it a PDU containing BCCH or PBCCH system information, indicating that the physical layer has reselected a cell in a different routeing area.

10.4.1 Normal routing area update

Figure 10.5 shows the procedure for a normal RAU, when the mobile station physical layer has reselected a cell in a different RA and passes the new RA to the GMM layer. The mobile station in this case is in the GMM ready condition, engaged in an uplink TBF.

The GMM puts the LLC layer on hold with the primitive LLGMM-SUSPEND-REQ. This stops all uplink transfers of LLC PDUs on the Um, air interface. This 'suspend' allows only GMM PDUs to be transferred across the air interface, inhibiting SN PDUs.

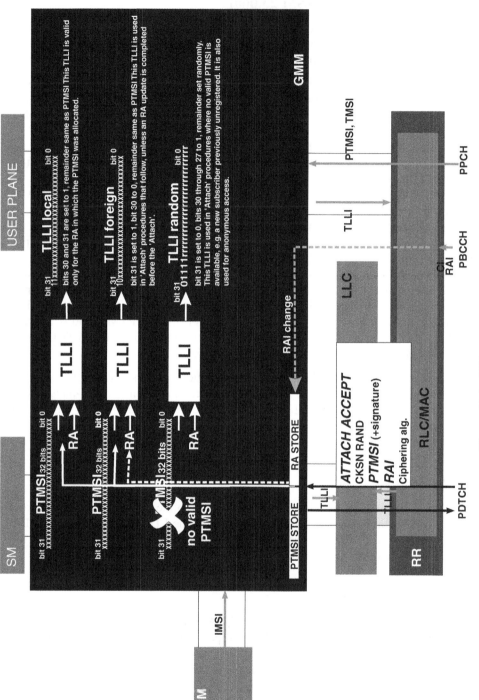

Figure 10.4 TLLI construction

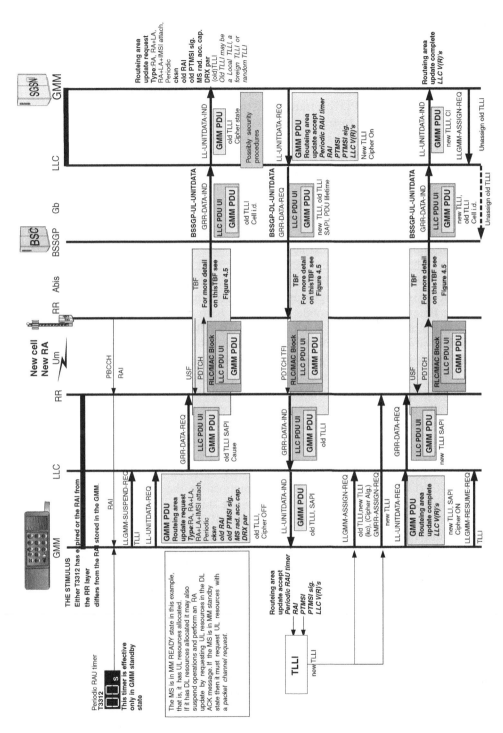

Figure 10.5 Routeing area update

The GMM layer passes the message *routeing area update request* to the LLC layer on the LL-UNITDATA-REQ primitive. The message contains the type of RAU – in this case RA alone, not RA plus location area, or RA plus IMSI attach; it indicates whether the RAU is for normal or periodic updating. The PTMSI/RAI pair and PTMSI signature are also sent in this message. The TLLI is also transferred with this primitive, and the ciphering is set to off.

The mobile station LLC PDU is transferred to its 'peer entity', the SGSN LLC layer, using the primitives shown to transit the lower layers. The cell identity of the mobile station's new cell is passed to the GPRS sub-network LLC layer from the BSSGP function, with the mobile station's TLLI.

On receiving the *routeing area update request*, the SGSN updates the mobile station location for this IMSI, allocates a new PTMSI and generates a new TLLI. (It may perform authentication/ciphering procedures at this point).

If the new cell (in the new routeing area) belongs to a different SGSN, then from the received foreign TLLI the new SGSN identifies and accesses the old SGSN to obtain the mobile station and PDP context details.

The new SGSN then passes the message *routeing area update accept* on the primitive LL-UNITDATA-REQ to the GPRS sub-network LLC layer. This message contains a new PTMSI, possibly a PTMSI signature, the RAI, settings for T3312 periodic routeing area update timer, and, if the mobile station has been transferring SN PDUs in the ABM condition, the LLC V(R) value for all NSAPI ABM transactions on this TLLI.

The GPRS sub-network LLC V(R) parameter tells the mobile station the next LLC PDU the GPRS sub-network is expecting to receive from the mobile station. If, for example, V(R) = 6, the GPRS sub-network has satisfactorily received LLC PDU 5.

If a mobile station is involved in a TBF and the physical layer reselects a cell then the TBF is lost and must be restarted in the newly selected cell. The exchange of the LLC register values allows the LLC links at both ends to resend missing LLC PDUs.

The new TLLI is transferred with this primitive, and the LLC layer is commanded to cipher the PDU.

The LLC layer passes this encapsulated GMM PDU to the BSSGP layer in the BSSGP-DL-UNITDATA primitive, which is mapped to the primitive GRR-DATA-REQ with the parameter RLC confirm. The mobile station's old TLLI and new TLLI are also transferred to the GPRS sub-network BSSGP layer with this primitive.

New and old TLLI are transferred to the BSSGP layer because:

1. If the mobile station does not respond to *routeing area update accept* with *routeing area update complete*, it may mean that the mobile station has not received *routeing area update accept* and will therefore use its old TLLI for further communication.
2. If the mobile station does not respond to *routeing area update accept* with *routeing area update complete*, it may mean that the mobile station *has* received *routeing area update accept* but its reply (*routeing area update complete*) is lost due to poor radio conditions and the mobile station will therefore use its new TLLI for further communication.

The receipt of *routeing area update complete* confirms that the mobile station will thereafter be using the new TLLI and the old TLLI can be deleted by the GPRS sub-network.

The *routeing area update accept* message is passed to the mobile station peer entity using the air interface and inter-layer primitives shown in Figure 10.5. On receiving this message the mobile station GMM layer derives the new TLLI, and assigns this TLLI to the LLC and RR layers with LLGMM-ASSIGN-REQ and GMMRR-ASSIGN-REQ respectively. It then sends the message *routeing area update complete* to the GPRS sub-network. This contains the mobile station's LLC V(R) status, so that the GPRS sub-network will know which of its SN PDUs' LLC frame numbers is next expected to be received by the mobile station.

The message is sent between layers using the primitives shown in the diagram. On receiving this message the GPRS sub-network GMM layer tells the LLC layer to use the new TLLI. The LLC layer must also pass this information to the BSSGP layer.

The mobile station GMM layer now sends the primitive LLGMM-RESUME-REQ with the new TLLI, instructing the LLC layer to resume its uplink TBF.

10.4.2 *Periodic routing area update*

Figure 10.5 shows the timer T3312, initially set to a value allocated by the GPRS sub-network and now expired. This timer is only operative when the mobile station is in the GMM standby state, when the mobile station is GPRS attached, has no radio resources allocated for a TBF and the ready timer T3314 has expired.

Timer T3312 is started when:

- T3314 ready state timer expires. The ready state timer expiry occurs when the mobile station is in GMM ready state but no TBF has taken place for the duration of this timer. In the ready condition, when an LLC PDU is transferred between the mobile station and GPRS sub-network, T3314 is restarted. When T3314 expires, the mobile station goes to GMM standby condition and the RAU timer T3312 is started.
- The mobile station has made a routeing area update from the GMM standby condition.

The procedure is then as shown in the diagram, but no suspension of TBF is necessary. The *routeing area update request* message indicates that this is a periodic update. The GPRS sub-network has a timer called the *mobile reachable timer*, which will expire if the mobile station does not perform a scheduled periodic update. The GPRS sub-network will not then attempt to send paging messages to the mobile station.

10.5 Cell update

A cell update takes place when a mobile station is in the GMM ready state and the RR layer reselects a new cell in the same RA as the old cell. If the new cell is in a different RA, then the mobile station performs an RAU.

The cell update is less demanding than the routeing area update, as the new cell in a cell update will always be in the same routeing area as the old cell, and the update will be handled by the same SGSN.

Figure 10.6 shows a mobile station in a TBF and the physical layer has reselected a new cell in the same RA. The TBF is abandoned and the RR layer informs the GMM layer of the new cell identity and RA. As the RA is the same, the GMM layer initiates a *cell update* by sending the primitive LL-GMM-TRIGGER-REQ. This primitive is used when the GMM layer wishes the LLC layer to send any LLC PDU.

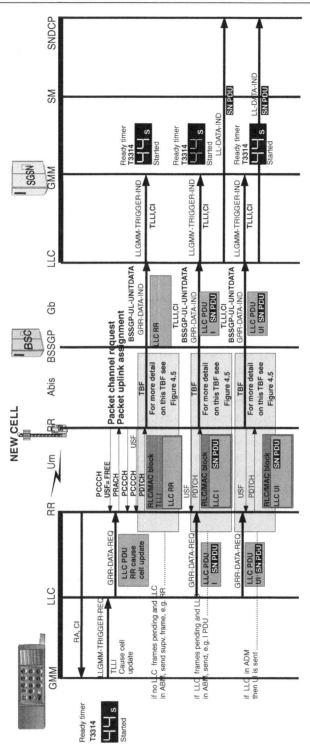

Figure 10.6 GMM procedures cell update

The TBF in this diagram is an uplink transfer; the same procedures will apply to a downlink transfer, the downlink TBF is abandoned and the GMM layer triggers the LLC layer to send an uplink LLC PDU.

The LLGMM-TRIGGER-REQ restarts the T3314 ready timer and causes the LLC layer to generate a GRR-DATA-REQ. This causes the RR layer to use the access procedures, and when uplink resources are granted to the mobile station, an LLC PDU is sent to the GPRS sub-network. The contents of the LLC PDU will depend on the SN PDUs waiting to be sent at the LLC layer:

- If there are no SN PDUs queuing at the LLC layer (there may be none if the TBF was previously a downlink TBF and there may be none at this time for an uplink TBF), the LLC layer will send an L2 RR – 'receiver ready' frame.
- If the LLC layer was transferring uplink PDUs in ABM (asynchronous balanced mode), and there is an SN PDU awaiting transmission, then an I – 'information' frame carrying that SN PDU will be sent.
- If the LLC layer was transferring uplink PDUs in ADM – asynchronous disconnect mode, and there is an SN PDU awaiting transmission, then a UI – 'unacknowledged information' frame is sent.

For identification, the LLC frames, when encapsulated by the RLC/MAC layer frames, will have the TLLI included in the TBF. This is extracted at the GPRS sub-network BSSGP layer and passed to the LLC layer with the cell identity.

The LLC layer sends the primitive LL-GMM-TRIGGER-IND, including the TLLI and CI, to the GPRS sub-network GMM layer. The GMM layer recognises the TLLI, updates the cell location and allows uplink SN PDUs to be passed by restarting the timer T3314 (the value of 44 seconds for T3314 shown in the diagram is the default value).

10.6 Paging procedures

In the mobile station GMM standby state the RR layer is responsible for managing the logical channels and extracting the information from the PBCCH, and in particular from the PCCCH-PPCH (packet common control channel – packet paging channel). The decoded information from this channel is passed to the GMM layer which decides, from the PTMSI or IMSI, whether it is being paged and the GMM layer then initiates a *paging response*.

10.7 Authentication and encryption procedures

The GPRS sub-network initiates the authentication and encryption procedures at any time, but always during the GPRS *attach* procedures.

Authentication is a GPRS sub-network function managed by the GMM layer with input from the HLR, and encryption is managed by the GMM layers in the mobile station and GPRS sub-network, with overall control from the GPRS sub-network GMM layer. Encryption is implemented in each case by the LLC layers in the mobile station and GPRS sub-network.

Figure 10.7 gives an overview of the GPRS authentication and ciphering procedures. These procedures are the same as in circuit switched operations. The GPRS sub-network obtains triplets, *RAND SRES* and the ciphering key, *kc*, from the HLR. As Figure 10.7 shows, the triplets are generated in the authentication centre (AUC), which holds the *ki*, customer identification key. The algorithms used by the AUC are the A3 and A8 algorithms illustrated.

The authentication centre passes a number of sets of triplets to the HLR, which downloads them to the SGSN. The SGSN is now in a position to authenticate the mobile station and order encryption.

The SGSN sends the message *authentication and ciphering request* to the mobile station. This carries one of the triplets RAND. When the mobile station receives RAND it generates SRES and the ciphering key, kc. These are produced by the SIM card as it holds algorithms A3, A8 and ki.

The mobile station sends the message *authentication ciphering response* to the GPRS sub-network with the mobile station generated triplet parameter SRES. The SGSN receives this message and compares the SRES received with the SRES stored in the SGSN. If they match, the mobile station is authenticated.

The SGSN now knows that the mobile station has generated the ciphering key kc and can order ciphering mode at any time by ciphering a message to the mobile station and indicating that ciphering is used by setting the *E* bit in the LLC header.

Figure 10.8 shows the GPRS authentication and encryption procedures in more detail. The top double-headed arrow on this diagram shows the step prior to *authentication and ciphering*, perhaps *attach request*, but a procedure which will have enabled the GMM layer to identify the mobile station through the PTMSI or IMSI or TLLI.

If the GMM layer has no remaining store of 'triplets' against the IMSI, it asks the HLR to provide these. The triplets are three numbers called:

- **SRES** – signed response;
- **RAND** – a random number; and
- **kc** – the ciphering key.

The diagram shows the GMM layer asking the HLR to provide these. The HLR requests the authentication centre (AUC) to generate triplets. The AUC firstly generates the random number RAND, and uses this as an input to the A3 algorithm, the other input being ki, subscriber identification key. The output from the A3 algorithm is SRES, the subscriber response to an authentication request.

The two inputs, RAND and ki, are also inputs to the A8 algorithm, and the output from this is the ciphering key, kc. This is the key applied to the ciphering algorithm.

A number of sets of triplets can be generated by the AUC against a mobile station's IMSI. The sets are delivered to the HLR, which delivers them to the GMM layer. The GMM layer indexes the sets against a *cksn, ciphering key sequence number*.

Each time *authentication and ciphering* is performed, the previous triplet set is discarded and the next set used. When all sets are used, the GMM layer requests another set.

The cksn allows the GPRS sub-network to short circuit authentication and go directly to ciphered mode. The cksn sent by the mobile station is compared with the

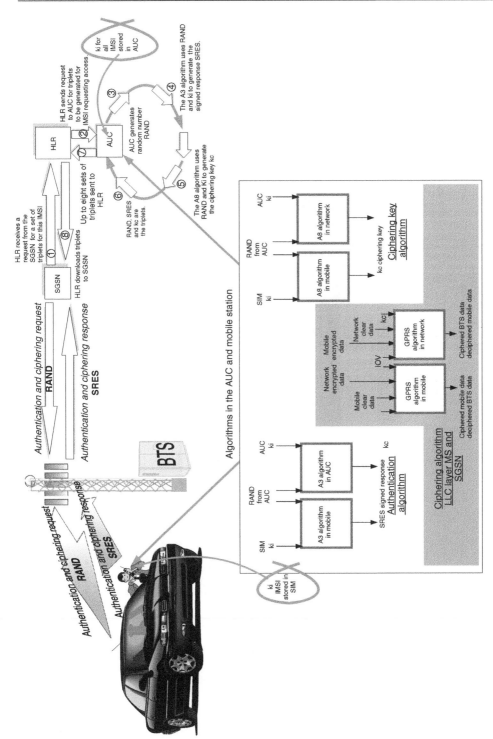

Figure 10.7 GMM procedures – authentication

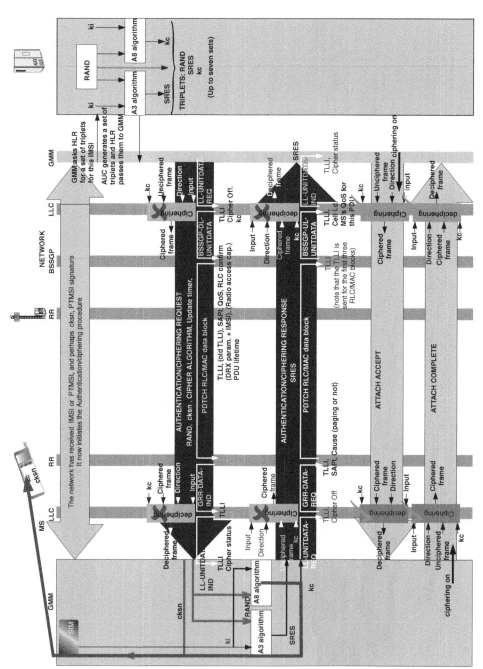

Figure 10.8 GMM procedures authentication and ciphering

cksn stored by the GMM layer; if they are the same, the GPRS sub-network GMM layer can immediately go into ciphered mode, knowing that the mobile station will use the same cipher key, kc.

Having obtained the triplets, the GPRS sub-network GMM layer is now able to send *authentication/ciphering request*. This contains RAND, cksn and the cipher algorithm to be used. Additionally, the ciphering key, kc, is delivered to the LLC layer with LL-UNITDATA-REQ.

The ciphering is implemented in the LLC layer as indicated in the diagram. Ciphering is not implemented for the message *authentication and ciphering request*. The mobile station receives the request at the GMM layer.

RAND is used as an input to the mobile station SIM card A3 and A8 algorithms along with ki, the mobile station identification key, which is stored on the SIM card. These two algorithms provide the ciphering key, kc, and SRES, the signed response.

The ciphering key and cksn are stored in the mobile station, and the mobile station sends the message *authentication and ciphering response* to the GPRS sub-network GMM layer. This contains the derived parameter SRES.

Simultaneously, the primitive between the mobile station's GMM and LLC layer delivers the derived ciphering key, kc, to the LLC layer. The message *authentication and ciphering response* is not ciphered.

On receiving this message, the GPRS sub-network has authenticated the mobile station if the SRES received is the same as the stored SRES in the GPRS sub-network GMM layer. The GPRS sub-network now also knows that the mobile station has ciphering key kc indexed to the cksn sent (in this case, the RAND is the first of a set, therefore the cksn is '0').

The authentication and ciphering procedures are now completed and the GPRS sub-network can go into ciphered mode at any time. In this case, it does so when completing the attach procedure by sending *attach accept*. With this message the LLC layer is instructed to implement ciphering, which it does using the algorithm illustrated on the LLC layer of the diagram.

The mobile station knows the contents of the LLC PDU are ciphered because the LLC header indicates ciphered mode. The mobile station LLC layer therefore operates the deciphering algorithm using the same ciphering key, kc, as was used by the GPRS sub-network.

The mobile station's reply in this case is *attach complete*, which is ciphered by the LLC layer.

These GMM messages are *not* encrypted by the LLC layer:

- Attach request;
- Attach reject;
- Authentication and ciphering request;
- Authentication and ciphering response;
- Authentication and ciphering reject;
- Identity request;
- Identity response;
- Routeing area update request;
- Routeing area update reject.

10.8 Identification

The identification procedure is initiated by the GPRS sub-network at any time during a TBF, but most commonly during GPRS sub-network access procedures. It can identify two mobile station parameters:

- IMSI – international mobile subscriber identity;
- IMEI – international mobile equipment identity.

IMSI checks may be necessary to the GPRS sub-network if, for example, the PTMSI sent by the mobile station during access procedures is not recognised by the GPRS sub-network. This might be the case where a mobile station is roaming between GPRS sub-networks.

It may also be convenient to the GPRS sub-network operator to use the identity command if a mobile station attempts to access the GPRS sub-network in a new SGSN service area; all SGSNs are connected together to exchange subscriber information, but it may be convenient to use the identity command instead of inter-SGSN data transfer.

Similarly, if a mobile station indicates a correct IMSI but the cksn or PTMSI signature does not correspond to those held by the SGSN, the GPRS sub-network operator might use the identity procedure.

IMEI checks may be necessary to identify suspected stolen handsets, or handsets which do not meet the GPRS sub-network operator's specifications for accessing the network.

10.9 Detach

GPRS detach may be initiated by the mobile station or by the GPRS sub-network. The mobile station initiates detach for the following reasons:

- The mobile station is switched off;
- The SIM card is removed;
- The GPRS TE is disabled.

The GPRS sub-network may *detach* an attached mobile station for many reasons.

Figure 10.9 illustrates the detach procedures for both mobile station originated detach and GPRS sub-network originated detach.

10.9.1 Mobile originated detach

The mobile station 'user' layer (or application layer – more correctly network layer) initiates the detach procedure. The access procedures must be completed before *detach request* can be sent across the air interface. This message indicates whether the detach is GPRS, IMSI or combined, or whether the mobile station has been switched off.

If the mobile station has been switched off, then the GPRS sub-network does not respond to the mobile station – it simply deletes the TLLI from the LLC layer. The mobile station also deletes the TLLI from its LLC layer before shutting down.

If the detach is for a reason other than power off, the GPRS sub-network responds with *detach accept*. Timer T3321 is started by the mobile station and this timer is

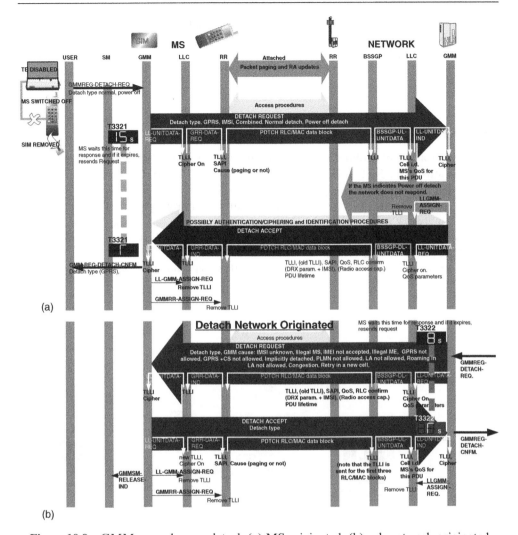

Figure 10.9 GMM procedures – detach (a) MS originated; (b) sub-network originated

stopped on reception of *detach accept*. If the timer expires, the mobile station resends the *detach request*. On receipt of *detach accept*, the mobile station informs the user layer that the detach is confirmed and deallocates the TLLI from the LLC and RR layers.

10.9.2 GPRS sub-network originated detach

Figure 10.9(b) illustrates GPRS sub-network originated detach. The GPRS sub-network can order a mobile station to detach for the reasons shown on the large arrow. *Implicit detach* means that the GPRS sub-network is sending this command some time after the *mobile reachable timer* in the GPRS sub-network has expired. The default setting for this timer is four minutes greater than the setting of T3312, the periodic RA update timer.

The timer T3322 GPRS sub-network detach timer is the GPRS sub-network equivalent of T3321. If this timer expires, the GPRS sub-network may retransmit the *detach request*.

10.10 GPRS roaming

The authentication of mobiles visiting a foreign GPRS sub-network is the same as for CS operations. The SGSN GMM layer of the visited PLMN asks for the mobile station's IMSI when the visiting mobile station attempts to attach. This is passed as a query to the HLR, which, after ascertaining that a roaming agreement exists with the GPRS sub-network indicated by the IMSI, establishes a data link to the HPLMN HLR. Subscription details and sets of triplets (kc, RAND and SRES) are passed from the home HLR to the visited HLR.

Figure 10.10 illustrates the procedures for GPRS roaming.

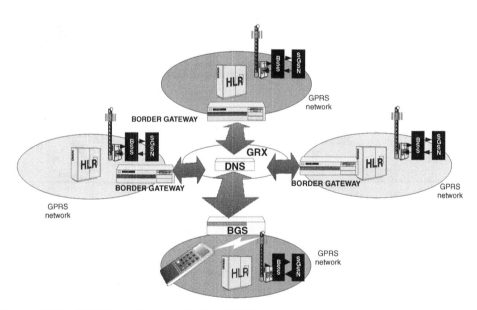

Figure 10.10 GPRS roaming, the GRX – GPRS roaming exchange. The GRX is used when a mobile station visits a foreign network and wishes to access an IP network which can only be accessed through the HPLMN. After registering with the VPLMN in the normal way, the mobile station establishes a PDP context specifying an IP address or APN which can only be accessed through the HPLMN. The VPLMN establishes contact with the HPLMN through the GRX, and a PDP context is then set up with the HPLMN. For Internet access, the mobile station sets up a PDP context with the VPLMN and the procedures are then as if the mobile station belongs to the VPLMN, with two exceptions: (i) further triplets must be obtained by the VPLMN from the HPLMN, (ii) the billing from the PLMN is transferred to the HPLMN

The GRX – GPRS roaming exchange – is a packet exchange based in one country, and GPRS network operators that wish to provide full roaming capabilities subscribe to this packet exchange. By specifying the APN of a network they wish to connect to, the GRX DNS will route packets to that network.

The GRX is used when a mobile station visits a foreign network and wishes to access an IP network (such as an Intranet), which can only be accessed through the HPLMN. After registering with the VPLMN in the normal way, the mobile station attempts to establish a PDP context specifying an IP address or APN, which can only be accessed through the HPLMN. The VPLMN establishes contact with the HPLMN through the GRX, and a PDP context is then set up with the HPLMN.

For Internet access, the mobile station sets up a PDP context with the VPLMN and the procedures are then as if the mobile station belongs to the VPLMN, with two exceptions:

1. Further triplets must be obtained by the VPLMN from the HPLMN;
2. The billing from the PLMN is transferred to the HPLMN.

10.11 Abbreviations used in this chapter

ABM	Asynchronous balanced mode
ADM	Asynchronous disconnect mode
APN	Access point name
AUC	Authentication centre
BCCH	Broadcast control channel
BSSGP	Base station sub-system GPRS protocol
CCCH	Common control channel
cksn	Ciphering key serial number
CS	Circuit switched
DL	Downlink
DRX	Discontinuous reception
E	Encryption bit
GEA	GPRS encryption algorithm
GMM	GPRS mobility management
GPRS	General packet radio service
GRX	GPRS roaming exchange
GSM	Global system for mobile communication
HLR	Home location register
HPLMN	Home public land mobile network
IMEI	International mobile equipment identity
IMEISV	International mobile equipment identity, software version
IMSI	International mobile subscriber identity
IP	Internet protocol
kc	Ciphering key
ki	Identification key
LLC	Logical link control
MM	Mobility management

NSAPI	Network service access point identity
PBCCH	Packet broadcast control channel
PCCCH	Packet common control channel
PDP	Packet data protocol
PDTCH	Packet data traffic channel
PDU	Protocol data unit
PPCH	Packet paging channel
PRACH	Packet random access channel
PTM-M	Point-to-multipoint multicast
PTMSI	Packet temporary mobile subscriber identity
RA	Routeing area
RAI	Routeing area indication
RAND	Authentication random number
RAU	Routeing area update
RF	Radio frequency
RLC	Radio link control
RR	Radio resources
RR	Receiver ready
SAPI	Service access point identity
SGSN	Serving GPRS support node
SIM	Subscriber identity module
SM	Session management
SM	Switch measure
SMS	Short message service
SMS	Switch measure switch
SN PDU	Sub-network protocol data unit
SRES	Signed response for authentication
SS	Supplementary screening
TBF	Temporary block flow
TE	Terminal equipment
TLLI	Temporary logical link identifier
TN	Timeslot number
UI	Unacknowledged information
UL	Uplink
UMTS	Universal mobile telecommunications system
VBS	Voice broadcast service
VGCS	Voice group calling service
VPLMN	Visited public land mobile network
V (R)	Next PDU expected to be received

11

SM Layer Procedures

(24.008 section 6)

This chapter is about the mobile station's session management's place in the scheme of things GPRS. It takes us from an overview to detailed operation of the SM layer, and its interactions with the network, GMM and SNDCP layers.

The by now familiar diagram of Figure 11.1 shows the session management (SM) layer sandwiched between the network layer and the GMM layer. The diagram shows the SM layer providing a service to the network layer by managing the establishment of a PDP context when asked to do so by the network layer sending *PDP context activation request*.

The SM layer relies upon the services of the GMM layer and will check that it is GPRS attached and when assured that it is, send down *PDP context activation request*, asking the GMM layer to deliver it to the SGSN session management layer.

The mobile station SM layer will receive from the GPRS sub-network *PDP context activation accept* (or reject), and the SM layer will signal to the mobile station network layer that its request is fulfilled. If the network layer does not like the allocated QoS profile, it can terminate the session.

Assuming that the network likes what it gets, it signals acceptance and the SM layer then tells the SNDCP layer about the PDP context. The SNDCP layer now accepts N PDUs from the application layer, and a TBF results!

Figure 11.2 shows the mobile station side of the session management layer in context. The functions of the session management layer are:

- To set up a session for transfer of SN PDUs and to deactivate the session when instructed.
- To modify the QoS and radio priority given to a PDP context when the GPRS sub-network instructs it to do so.
- To remove a PDP context when so instructed.

GPRS in Practice: A Companion to the Specifications   Peter McGuiggan
© 2004 John Wiley & Sons, Ltd   ISBN: 0-470-09507-5

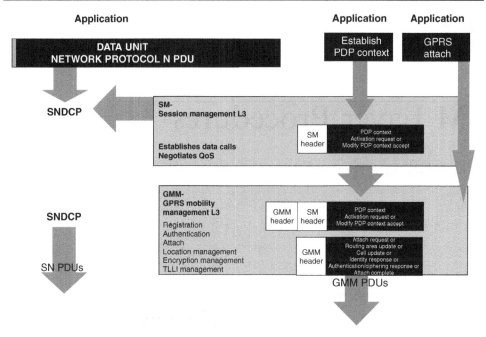

Figure 11.1   The session management (SM) layer

As Figure 11.2 shows, the SM layer is central to the establishment of SN PDUs between the mobile station and GPRS sub-network SNDCP layers. The SM layer is asked to establish a session through the SMREG-SAP with the primitive PDP-ACTIVATE-REQ. If the GMM layer has not established *GPRS attach*, then this primitive causes the primitive ESTABLISH-REQ to be sent on the GMMSM-SAP to the GMM layer; this will cause the GMM layer to GPRS attach.

Once a GMM context – *attach* – is in place (and any security procedures are completed), the SM layer can use the GMM context to establish a PDP session.

Let us look at Figure 11.2 in more detail, in particular the service access points, (SAPs) and the primitives they carry; for convenience we will start at the top left and work toward the right and down.

GMMREG-SAP

This is the service access point which allows the mobile station network layer to communicate with the GMM layer. The primitives which use the SAP and their functions are:

- **ATTACH-REQ**. This is the network layer directly instructing the GMM layer to GPRS attach.
- **ATTACH-CNFM**. This is the GMM's response to ATTACH-REQ, telling the network layer that GPRS attach is successful.
- **ATTACH-REJ**. The GMM layer tells the network layer that the GPRS attach request is rejected by the GPRS sub-network.

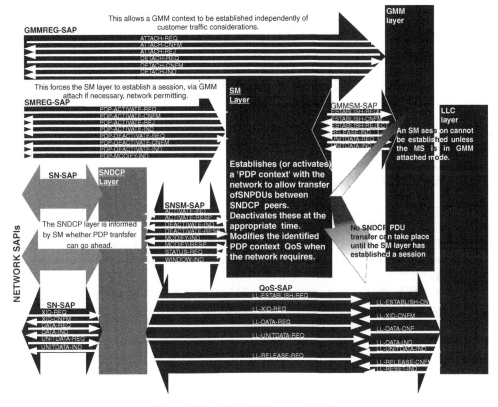

Figure 11.2 The SM layer primitives

- **DETACH-REQ**. The GPRS sub-network asks the mobile station GMM layer, which is GPRS attached, to detach. The reasons for this request are covered in Chapter 12. Detach destroys any established PDP context and the TLLI is deleted.
- **DETACH-CNFM**. The GPRS sub-network has confirmed to the mobile station GMM layer that it is now detached and the GMM layer tells the network layer.
- **DETACH-IND**. The GPRS sub-network has requested the mobile station to detach, and the GMM tells the network layer.

SMREG-SAP

The session management registration service access point allows the network layer to talk to the SM layer.

In the mobile station there are four SMREG-SAPs to access the four session management functions, which allow four QoS profiles to be activated simultaneously. Only one SAP is shown in Figure 11.2, but all four SAPs are identical to the one shown.

- **PDP-ACTIVATE-REQ**. The primitive used to instruct the SM layer to establish a PDP context. It includes all the parameters that the SM layer needs to do this.
- **PDP-ACTIVATE-CNFM**. The SM layer establishes a PDP context and informs the network layer. If the network layer doesn't like what it hears in this primitive (which includes the QoS profile imposed by the GPRS sub-network) it closes down the PDP context by sending PDP-DEACTIVATE-REQ.
- **PDP-ACTIVATE-REJ**. The SM layer has attempted to establish a PDP context, but it has resulted in failure.
- **PDP-ACTIVATE-IND**. The mobile station SM layer has received a request from the SGSN to establish a PDP context. It is now up to the mobile station network layer to initiate one through the PDP-ACTIVATE-REQ primitive.
- **PDP-DEACTIVATE-REQ**. The mobile station network layer asks the SM layer to discontinue an established PDP context.
- **PDP-DEACTIVATE-CNFM**. The SM layer indicates to the mobile station network layer that the PDP context has been removed as requested.
- **PDP-DEACTIVATE-IND**. The SGSN has asked the mobile station SM layer to deactivate an established PDP context and the SM informs the mobile station network layer.
- **PDP-MODIFY-IND**. The mobile station SM layer has been asked by the SGSN to modify the QoS profile for an established PDP context. The SM layer does this and informs the mobile station network layer.

GMMSM-SAP

This service access point allows communication between the SM layer and GMM layer. This is the point at which the GMM layer provides a service to the SM layer by delivering SM PDUs between the GPRS sub-network SM layer and the mobile station SM layer.

- **ESTABLISH-REQ**. The SM layer asks the GMM layer to GPRS attach. This primitive results from the SM layer receiving PDP-ACTIVATE-REQ from the mobile station network layer; the SM layer will try to establish a PDP context, but as it needs the services of the GMM layer to do this, must first ensure that the GMM layer is GPRS attached.
- **ESTABLISH-CNFM**. The GMM layer has GPRS attached and informs the SM layer that it can now deliver SM PDUs through the GMM layer.
- **ESTABLISH-REJECT**. The GMM layer has been unsuccessful in achieving GPRS attach. The SM layer must inform the mobile station network layer that a PDP context cannot be established.
- **RELEASE-IND**. The GMM layer has changed from GPRS attach to GPRS detached, and tells the SM layer that it must release the established PDP context.
- **UNITDATA-REQ**. This primitive carries mobile station SM PDUs for delivery to the SGSN, such as PDP context activation request. Unitdata indicates that SM PDUs are transferred by the LLC layer in unacknowledged mode.
- **UNITDATA-IND**. This carries SM PDUs delivered from the SGSN to the mobile station SM layer, such as *PDP context accept*.

SNSM-SAP

This SAP is for the SM layer to inform the sub-network dependent convergence protocol (SNDCP) layer of the establishment or deactivation or modification of a PDP context.

- **ACTIVATE-IND**. This tells the SNDCP layer that a PDP context is established. It tells the SNDCP the NSAP that is involved, the QoS profile involved and the radio priority. NSAP is the network SAP, the communication point of the user (strictly speaking the network) to the GPRS sub-network. This is the point through which N PDUs (network protocol data units) pass. N PDUs are the revenue earning customer data units. If a PDP context is established, the NSAPI information will tell the SNDCP which NSAPI port to activate. There are eleven NSAPs.
- **ACTIVATE-RESP**. The SNDCP layer indicates to the SM layer that it is ready to receive N PDUs from the stated N SAPI.
- **DEACTIVATE-IND**. The SM layer informs the SNDCP layer that the indicated NSAPI PDP context is now deactivated.
- **DEACTIVATE-RESP**. The SNDCP layer tells the SM layer that the PDP context for the indicated NSAPI is closed down.
- **MODIFY-IND**. The mobile station SM layer has received an instruction from the SGSN to modify the QoS profile of an established PDP context. The SM layer tells the SNDCP what this new profile is and the new radio priority.
- **MODIFY-RESP**. The SNDCP layer has accommodated the new QoS profile and radio priority for the indicated NSAPI and responds to the SM layer.

The SN SAP, the communication point between the user layer (or network layer) and the SNDCP layer, and QoS SAP, the communication point between the SNDCP and LLC layers, are covered in the next chapter.

11.1 PDP context activation by the mobile station

PDP context activation is a data call set-up for GPRS operations, akin to call set-up for circuit switched operations, with the difference that a circuit switched call set-up is removed when the call is completed, but a PDP context activation remains valid when the call, or session, which caused the context is completed.

A PDP context can be repeatedly reused, whereas a circuit switched call set-up is a one-off operation.

Figure 11.3 illustrates the procedures for PDP context activation by the mobile station. Upon receiving the primitive SMREG-PDP-ACTIVATE-REQ on the SMREG-SAP, the SM layer of the mobile station sends the message ACTIVATE-PDP-CONTEXT-REQ to the GPRS sub-network. This message contains the NSAP, the type of PDP, the requested PDP address, the requested access point name (APN), the requested LLC SAP and the requested QoS and PDP protocol configuration options. This diagram shows this message transferred across the various layers with the primitives for inter-layer communication.

Upon receiving this message, the GPRS sub-network SM layer speaks to the HLR and GGSN (gateway GPRS support node).

The HLR will indicate if the requested service is allowed for this subscriber. The HLR may also supply the destination address and the access GGSN (APN) to be used by the

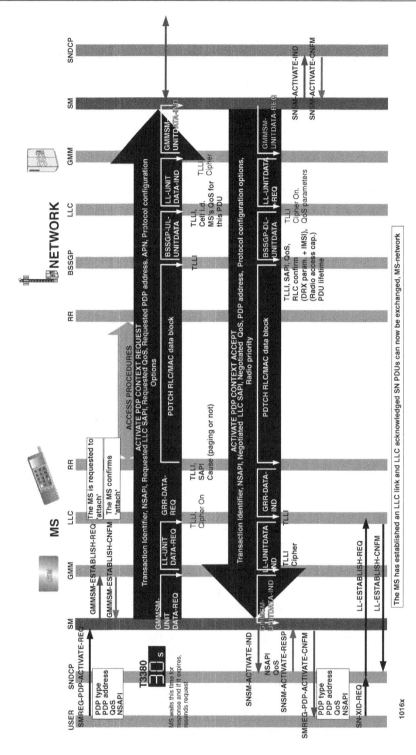

Figure 11.3 SM procedures PDP context activation

mobile station. It may also supply the subscriber's QoS profile. Indeed the ACTIVATE-PDP-CONTEXT-REQ sent by the mobile station may be quite empty (apart from the NSAP), and it is then left up to the HLR to supply the missing details!

When the SGSN has got all the data it requires for call set-up, it then passes the request to the GGSN. The GGSN is the interface to the outside world (The Internet or Intranet). A dynamic IP address is normally allocated by the GGSN. The GGSN sets up the call to the external network. For Internet communications this will involve authentication of the GGSN by the ISP it is accessing. When Internet communication is established, the GGSN informs the SGSN.

The SGSN obtains the working QoS and radio priority, and sends to the mobile station the message ACTIVATE-PDP-CONTEXT-ACCEPT. This is illustrated in Figure 11.3 with the primitives that carry it to the mobile station peer entity. It contains the working QoS, NSAPI, negotiated LLC SAP, negotiated QoS, protocol configuration options and the radio priority that the mobile station will use for the PDP transfer.

The mobile station SM layer receives this *accept* message and completes the procedure initiated by the SMREG-PDP-ACTIVATE-REQ by sending SMREG-PDP-ACTIVATE-CONFIRM to the 'user' (mobile station network) layer.

This message tells the application the QoS parameters that will be used for the PDP transfer. If this is not acceptable, the mobile station network layer initiates PDP-DEACTIVATION procedures.

The SM layer then informs the SNDCP layer of the context, giving the QoS profile and radio priority applicable, and it gets an acknowledgment response.

The PDP context is now nearly fully established; only the step of negotiating the SNDCP XID parameters remains. However, as this is the responsibility of the SNDCP layer, the SM layer task is completed, the PDP context is established.

Below is an explanation of the parameters exchanged with the two messages ACTIVATE-PDP-CONTEXT-REQ and ACTIVATE-PDP-CONTEXT-ACCEPT:

- **NSAPI** (network SAPI), this is the network service access point, which the mobile station network layer has selected. The GPRS sub-network will use this in conjunction with TLLI to address packets to the correct application in the mobile station.
- **APN** (access point name). The mobile station may use this to specify which external network it wishes to access. If the mobile station specifies this then it has been programmed into the SIM on subscription. The GPRS sub-network operator may have specific GGSNs working to specific external networks or even specific ISPs. If the mobile station does not include APN it may be provided by the HLR.
- **LLC SAPI** (this is covered in Chapter 9). There are six SAPs, one for GMM signalling, one for SMS and four for user data. The user data SAPs are labelled QoS 1, QoS 2, QoS 3 and QoS 4. In future these SAPs may be tied to a particular QoS profile (in which case the negotiation of LLC SAPI will have meaning), but in phase 1 they are not tied to QoS profiles. Quality of service parameters are covered in Chapter 6.
- **TI** (transaction identifier). This is used to identify a particular PDP transaction. This is included in the SM header. The TIs for downlink and uplink may be different. There are seven TIs that the mobile station and GPRS sub-network may choose from, 0–6.

- **PDP address**. This originating (for static IP address) and destination address may be an IPv4 or IPv6 address.
- **Radio priority**. The GPRS sub-network decides upon a radio priority 1–4, based upon the QoS granted and congestion parameters.
- **Protocol configuration options**. This specifies protocols such as header information to be used in communicating with an external network.

Figure 11.4 illustrates NSAPI and APN.

In phase 1 GPRS, there are eleven NSAPs that the mobile station may use, numbered as shown. Some of these SAPs will connect to user applications on a laptop PC, and others will be permanently connected to applications supported by the mobile station itself. The mobile station can simultaneously use multiple NSAPs (that is it can handle multiple separate PDP contexts).

Downlink packets are addressed to a particular NSAP, which discriminates between individual applications in a multiple application session.

For the SGSN, the combination of temporary logical link identity (TLLI) and NSAP completely specifies the address for delivery of packets to a mobile station.

The form of the access point name (APN) is shown in Figure 11.4 and consists of two parts, first, the 'network identifier', which is the address of the gateway GPRS support node which connects to a specific external network. This name is specified and supplied by the GPRS sub-network operator. The second part, the 'operator identifier' consists of the MNC – mobile network code, and MCC – mobile country code. The mobile station may specify the APN to use for a specific PDP context, or it may not, depending upon the SIM data.

If these fields are empty in the *PDP context activation request* (apart from the NSAP), the SGSN will extract them from the HLR. The HLR, or in some cases the SGSN, will have subscription records for the subscriber which will include the PDP types, details of the PDP addresses and APN of the subscription.

Figure 11.5 is a snapshot of what may happen when a PDP context is being established.

There are many permutations of the circumstances of PDP context activation. The mobile station sends *PDP context activation request* to the GPRS sub-network. This may include the following or be quite empty!

- PDP type;
- PDP address (of the mobile station and the destination);
- APN.

The SGSN/HLR checks the subscription record, and, if it is for a single subscription (e.g. an ISP), goes to the GGSN APN, which gives access to that subscription. The domain name server indicated by the single subscription is then interrogated and security authentication invoked. If this interrogation is successful, the GGSN signals back to the SGSN that the call is set up.

Alternatively, the SGSN checks the subscriber's subscription record, and, if it is for a multiple subscription (e.g. an ISP, an Intranet), the SGSN checks for a subscribed APN which fits the PDP type requested. Interrogation and authentication procedures are conducted to the network on the GGSN of the APN.

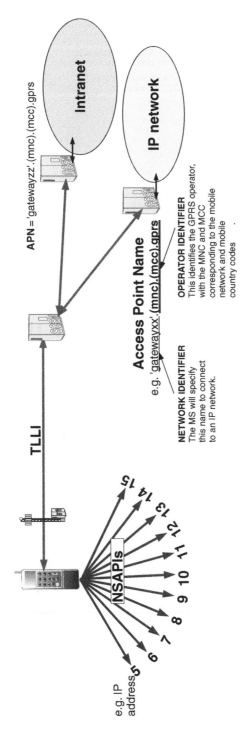

Figure 11.4 NSAP, TLLI and APN

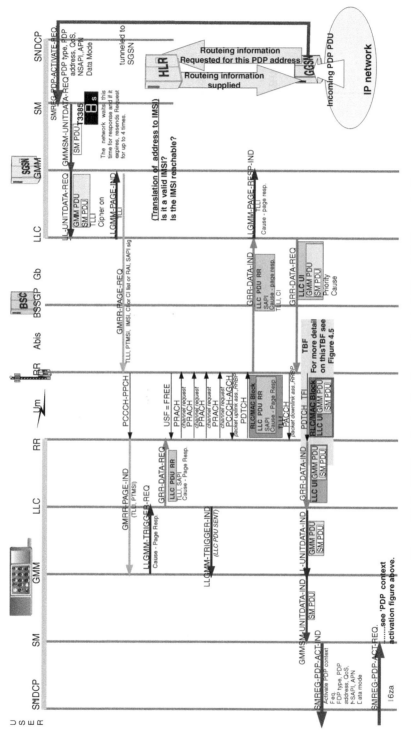

Figure 11.5 PDP context activation, network initiated (MS attached)

11.2   PDP context activation by the GPRS sub-network

Figure 11.5 shows the procedure for GPRS sub-network initiated PDP context activation. This procedure is valid only for Internet subscribers that are GPRS attached and have static IP addresses.

The gateway GPRS support node (GGSN) receives a PDP PDU from an IP network. It is the function of the GGSN in this case to find the SGSN serving the addressed subscriber. If the GGSN is already managing a PDP transfer for this address then it will know which SGSN is serving the addressed mobile station and can immediately pass the PDU to that SGSN. But of course in that case a PDP context will already exist.

In this case, the GGSN does not know the controlling SGSN and it must ask the HLR to provide the routeing details. In the diagram, the GGSN asks the HLR for routeing details, sending the IP address of the incoming PDU. The HLR translates the address into an IMSI and checks upon whether that IMSI is reachable, that is, whether the mobile station is GPRS attached. If it finds that the mobile station is GPRS detached, then it informs the GGSN and the GGSN will discontinue communication with the IP network.

If the HLR finds that the IMSI/mobile station is reachable, it sends to the GGSN the routeing information which includes the IMSI and the IP address of the serving SGSN. The GGSN initiates an SMREG-PDP-ACTIVATE-REQ primitive. The SM layer sends a unitdata primitive to the GMM layer (carrying the 'activate' request).

The GMM layer sends an LL-UNITDATA-REQ to the LLC layer. The LLC layer cannot send the PDU because it is suspended as the mobile station is in GMM standby. The LLC layer asks the GMM layer to page the mobile station. An alternative to paging the mobile station is applicable if the GPRS sub-network knows which cell the mobile station is currently in. It will know the cell if the ready timer is running. A *packet downlink assignment* can be addressed to the mobile station on the PCCCH, identifying the mobile station with TLLI.

The GMM layer asks the RR layer to page the mobile station, sending the mobile station's DRX parameters, PTMSI, RA to be paged and TMSI. The RR layer calculates the packet paging channel for each cell from the DRX parameters IMSI and PCCCH configuration, and sends the *packet paging request* on the appropriate PCCCH.

The mobile station reads its paging channel and passes this request to the GMM layer, which initiates access, indicating a *page response.*

The GPRS sub-network RR layer allocates an uplink PDCH. The mobile station sends an LLC frame on the allocated uplink which contains TLLI in the RLC/MAC block.

The GPRS sub-network LLC layer has identified the mobile station positively and can forward the *activate PDP* message. When this is received by the mobile station SM layer the user (network) layer receives an SMREG-PDP-ACTIVATE-IND primitive. The mobile station user layer generates the primitive SMREG-PDP-ACTIVATE-REQ message and the procedure is now as in Figure 11.3.

11.3   PDP context modification

A PDP transfer between the GPRS sub-network and mobile station uses a set of parameters which include the QoS profile and radio priority which were negotiated when the PDP context was established. The GPRS sub-network, for various reasons, may need to modify the QoS profile and radio priority. The reasons may include high demand on the GPRS sub-network resources in the busy hours, or the inverse of this as the busy hours pass (allowing the QoS offered to a mobile station to be improved).

The GPRS sub-network initiates the modification by sending SMREG-PDP-MODIFY-REQ to the GPRS sub-network session management layer. This primitive includes the required QoS and the mobile station NSAP in use. This primitive causes the SM layer to generate GMMSM-UNITDATA-REQ which encapsulates the PDP modify request and conveys it to the mobile station's SM layer. (At the same time, timer T3386 (8 second default) is started, if no response is received before expiry, the message is resent up to four times.)

The mobile station SM layer incorporates the received new QoS parameters, informing the mobile station user (network) layer. Its SM layer responds by generating the primitive GMMSM-UNITDATA-REQ carrying the message *modify PDP context accept*. The mobile station SNDCP layer is informed of the new QoS profile and radio priority.

If the mobile station user (network) layer does not find the new QoS values acceptable, it will send the primitive SMREG-PDP-DEACTIVATE to the SM layer and the PDP session is discontinued.

Figure 11.6 illustrates these processes and, in addition, shows the LLC link being discontinued to allow another LLC link to be established using different parameters. This is not possible if the particular NSAP that has had its PDP context modified is sharing the LLC link with another NSAP PDP context.

11.4 PDP context deactivation by the mobile station

PDP context deactivation deletes a PDP context previously established. For a mobile station initiated PDP context deactivation, the mobile station network (application, user) layer decides to delete the PDP context. The process is illustrated in Figure 11.7.

When an internal clock expires, or another mechanism within the mobile station 'user' (network) layer decides upon deletion, it generates the primitive SMREG-PDP-DEACTIVATE-REQ to the SM layer. Simultaneously, timer T3390 is started.

The session management layer generates the message *deactivate PDP context request*, which is delivered as a unitdata message as shown to the GPRS sub-network SM layer.

The cause in this message may be *regular PDP context deactivation*, but could also be *insufficient resources* if the mobile station has been asked to activate a PDP session which it cannot do because the particular resources required are in use; *QoS not accepted* might be the cause if the mobile station has been given a QoS with which it disagrees, which might occur at the initial establishment or in response to a *PDP context modification* request.

Upon receiving the *deactivate PDP context request*, the GPRS sub-network SGSN informs the SNDCP layer, and PDP transfers to that NSAP are discontinued. The GPRS sub-network SM layer responds with *deactivate PDP context accept*. Receiving this, the mobile station SM layer informs the mobile station SNDCP layer which disconnects that NSAP. If this is the only PDP transfer taking place on the LLC, the logical link connection can be closed.

As shown in Figure 11.7, the SNDCP layer in the mobile station sends the primitive LL-RELEASE-REQ to the LLC layer. The LLC layer closes the logical link with primitive GRR-DATA-REQ to the RR layer. This primitive carries the layer 2 frame

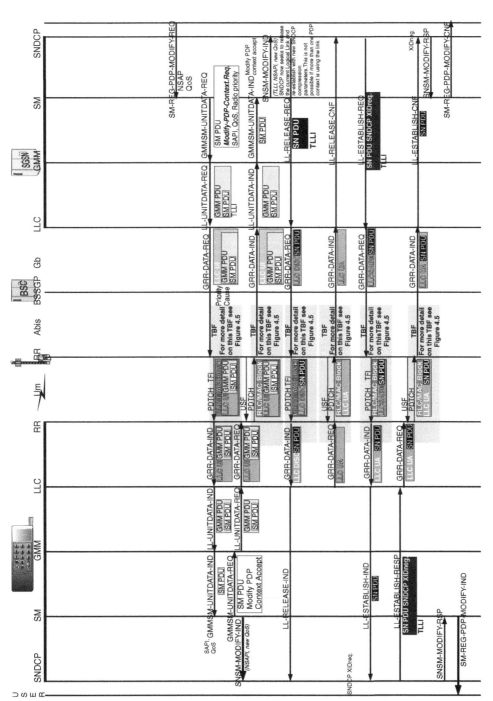

Figure 11.6 PDP context modification

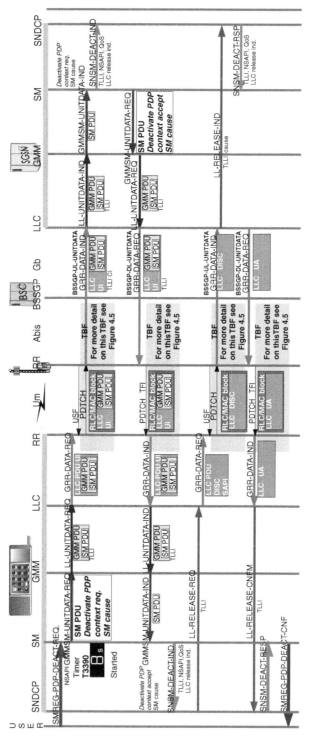

Figure 11.7   PDP context deactivation, MS initiated

LLC DISC. This is received by the GPRS sub-network. LLC layer responds with a layer 2 LLC UA frame and disconnects the LLC connection. The GPRS sub-network LLC layer indicates to the GPRS sub-network SNDCP layer that this TLLI is disconnected. The mobile station LLC layer receives UA, disconnects the LLC link and confirms the release to the SNDCP. The SNDCP layers of both the GPRS sub-network and mobile station inform their SM layers that the disconnection is registered. The SM layer confirms the deactivation to the mobile station.

11.5 PDP context deactivation by the GPRS sub-network

The GPRS sub-network can deactivate a PDP context to a mobile station at any time. The circumstances in which this may happen are GPRS sub-network operator dependent. If it is a PDP context that has not been used for some time (hours, days, a week) then the GPRS sub-network operator will close it down as it is using GPRS sub-network resources (in particular a dynamic IP address) without revenue.

The subscription details of the PDP context may have a bearing on the decision, with an individual having an expensive subscription being less likely to be deactivated than one with a cheap subscription.

It may be necessary to let the customer know precisely how long they can have an inactive PDP context before it is closed down.

Figure 11.8 illustrates a PDP context deactivation initiated by the GPRS sub-network. It is similar to the PDP context deactivation initiated by the mobile station.

11.6 Negotiated QoS profiles and radio priority

This negotiation is rather intransigent. The supplicant (mobile station) asks for what it wants, the respondent (GPRS sub-network) tells the supplicant what it is going to get!

The mobile station requests a QoS profile, which the GPRS sub-network might allocate if it is within its capability, taking into consideration GPRS sub-network congestion for both circuit switched and GPRS circuits. The GPRS sub-network may refuse service if it is very busy.

The QoS profile granted by the GPRS sub-network is returned in the ACTIVATE-PDP-CONTEXT-ACCEPT message. If the mobile station does not find the offered QoS profile acceptable, it may deactivate the session.

11.6.1 QoS

The mobile station requests a set of QoS parameters with the ACTIVATE-PDP-CONTEXT-REQUEST message. The range of these parameters is shown in Chapter 8, and is repeated here.

11.6.1.1 Service precedence

The mobile station may ask for high priority precedence 1, normal priority precedence 2, low priority precedence 3, or simply the subscribed priority, which will be one of these

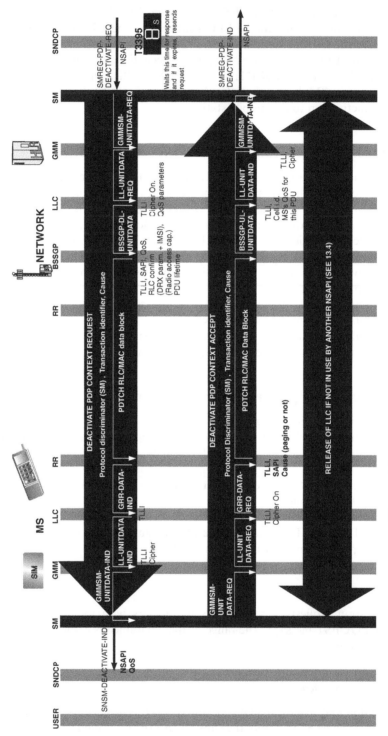

Figure 11.8 PDP context Deactivation by the network

three. If the GPRS sub-network experiences difficulties in communication, for example degrading interference on some physical channels, it will attempt to keep the communication at the best possible level for the highest precedence PDP transfers, giving less effort to maintain the QoS for lower precedence.

11.6.1.2 Reliability class

These parameters are shown in Figure 8.1, and to achieve operation within these limits, the mobile station requests:

- Reliability class 1, which invokes LLC acknowledged mode, LLC data protection, RLC acknowledged mode, GTP acknowledged mode.
- Reliability class 2, which invokes the same as class 1 for LLC and RLC, but unacknowledged mode for GTP.
- Reliability class 3, which invokes LLC unacknowledged mode, LLC data protection, RLC acknowledged mode and GTP unacknowledged mode.
- Reliability class 4, which invokes LLC unacknowledged mode, LLC data protection, RLC unacknowledged mode and GTP unacknowledged mode.
- Reliability class 5, which invokes unacknowledged mode on LLC, RLC and GTP and unprotected data on the LLC link.

11.6.1.3 Delay class

The mobile station asks for subscribed delay class, delay class 1, delay class 2, delay class 3 or delay class 4. These delays are defined in Figure 8.1.

11.6.1.4 Throughput class

The mobile station asks for a peak throughput class subscribed of 1–9, and a mean throughput class subscribed of 1–19. These are listed in Figure 8.2.

11.6.2 Radio priority

The radio priority class is in the range 1–4, where 1 is the highest priority. This is given by the GPRS sub-network to the mobile station in the ACTIVATE-PDP-CONTEXT-ACCEPT message. If the layer 3 messages are higher layer signalling messages (from the GMM layer), then the radio priority used by the RLC/MAC layer is the highest priority – priority 1. This applies to messages such as *GPRS attach*, security procedures and *PDP context activation*.

When N PDUs are to be transferred, the SM layer has previously conveyed the radio priority for them to the SNDCP layer, which sends it to the GRR layer to be used in GPRS sub-network access procedures. The mobile station RLC/MAC layer uses this parameter as described in Section 7.9 to determine when the *packet channel request* should be sent. The mobile station may also indicate the radio priority (if the PRACH is using the 11-bit format) in the *packet channel request*.

The GPRS sub-network may use this radio priority indication in allocating radio resources for a TBF.

The network may further use the radio priority for congestion control by limiting access to a cell to the higher priorities. This information is broadcast on system information 13, and is detailed in Appendix 2 AS 13.7

11.7 Abbreviations used in this chapter

APN	Access point name
DISC	Disconnect
DRX	Discontinuous reception
GGSN	Gateway GPRS support node
GMM	GPRS mobility management
GPRS	General packet radio service
GRR	GPRS radio resources
GTP	GPRS tunnelling protocol
HLR	Home location register
IMSI	International mobile subscriber identity
IP	Internet protocol
ISP	Internet service provider
LLC	Logical link control
MAC	Medium access control
MCC	Mobile country code
MNC	Mobile network code
N PDU	Network protocol data unit
NSAP	Network service access point
NSAPI	Network service access point identity
PCCCH	Packet common control channel
PDCH	Packet data channel
PDP	Packet data protocol
PDU	Protocol data unit
PRACH	Packet random access channel
PTMSI	Packet temporary mobile subscriber identity
QoS	Quality of service
RA	Routeing area
RLC	Radio link control
RR	Radio resources
SAP	Service access point
SGSN	Serving GPRS support node
SIM	Subscriber identity module
SM	Session management
SMS	Short message service
SNDCP	Sub-network dependent convergence protocol
SN PDU	Sub-network protocol data unit

TBF   Temporary block flow
TI   Transaction identifier
TLLI   Temporary logical link identifier
UA   Unnumbered acknowledge
XID   Exchange identities

12

SNDCP Procedures

(GSM 44.065)

The sub-network dependent convergence protocol layer is where the GPRS sub-network meets the 'outside world'. In the case of the mobile station, the outside world is the 'network' layer (also referred to in this book as the 'user' and 'application' layer). The major 'network' is the IP network and the 'packets' entering the SNDCP layer from the IP network are typically TCP/IP data units, called N PDUs (network protocol data units).

The SNDCP title can be analysed as:

- **SND** (sub network dependent). In our case the sub-network is the GPRS network. There will be a similar layer for UMTS, where the sub-network will be the UMTS radio access network.
- **CP** (convergence protocol). The convergence is the adaptation of the N PDUs to GPRS sub-network PDUs, so the SNDCP layer's purpose is to process incoming (typically) TCP/IP data packets to make them suitable to the GPRS sub-network. The resultant SN PDU is the data packet which will be transferred across the air interface in a TBF. (In the opposite direction the SNDCP layer receives SN PDUs from the LLC layer and converts them back into 'network' data packets – N PDUs.)

This chapter describes how the SNDCP layer processes the incoming N PDUs before transferring them to the LLC layer for delivery over the air interface.

Figure 12.1 shows the mobile station SNDCP layer within the context of the other layers. The function of the SNDCP is:

- To multiplex N PDUs from more than one NSAP onto the LLC connection in use. This requires the SNDCP layer to be fully aware of the QoS profile for each user on the 'network' layer. As there can be multiple users with different negotiated QoS profiles, the SNDCP must ensure that each N PDU from each user (which access the SNDCP layer through the 'network' service access point) is processed by the SNDCP layer and passed to the LLC layer at a rate which keeps the communication rate equal to the rate of the QoS profile. However, as the SNDCP layer has no direct control of

GPRS in Practice: A Companion to the Specifications Peter McGuiggan
© 2004 John Wiley & Sons, Ltd   ISBN: 0-470-09507-5

the radio resources, it cannot guarantee that the QoS profile rate will be met. It will ensure that the delivery of an application's N PDUs are delivered as SN PDUs to the LLC layer at a rate which, if the radio systems permit, will result in the QoS profile being met.

- To map SN-DATA (acknowledged) primitives and SN-UNITDATA (unacknow-ledged) primitives on to the LL-DATA (acknowledged) and LL-UNITDATA (unacknowledged) primitives.
- To initiate the establishment, re-establishment and release of acknowledged LLC operations.
- To buffer N PDUs for acknowledged services (N PDUs are retained until they are acknowledged).
- To compress N PDU headers for transmission and decompress received SN PDUs before forwarding them to the network layer.
- To segment N PDUs into SN PDUs (maximum 1520 octets) for transmission to the LLC layer and to reassemble received SN PDUs into N PDUs for forwarding to the network layer.
- To negotiate XID parameters between SNDCP layers in the mobile station and network.

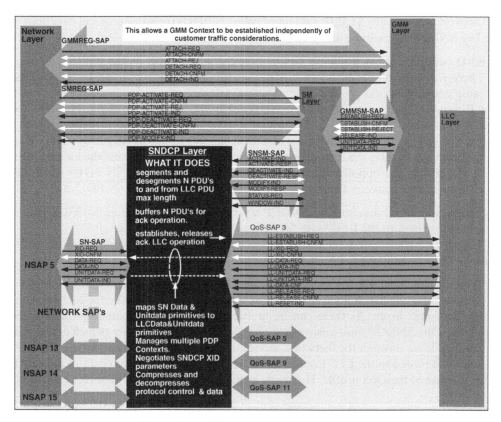

Figure 12.1 The SNDCP layer in context

12.1   SNDCP operation overview

Figure 12.2 gives an indication of the sequence of operation of the SNDCP layer. When the user (network) layer wishes to transfer a network PDU, it first asks the SM layer to establish a PDP context with the primitive SMREG-PDP-ACTIVATE-REQUEST. As covered in Chapter 11, this primitive contains the NSAP, the type of PDP, the PDP IP address, the requested LLC SAP, PDP protocol configuration options, APN and the requested QoS parameters. (Some, or most, of this information may be missing and may be extracted by the SGSN from the HLR. The NSAP must, of course, always be included).

The SM establishes a session, receiving from the network an activation acceptance which includes the network layer dynamic IP address, QoS parameters and the radio priority to be used for this PDP transfer.

The SM layer informs the user (network) and SNDCP layers that the PDP context is established, passing to both the QoS parameters received from the network. The user layer may reject the 'negotiated' QoS parameters by sending the primitive SMREG-PDP-DEACTIVATE-REQUEST to the SM layer, in which case the session is discontinued.

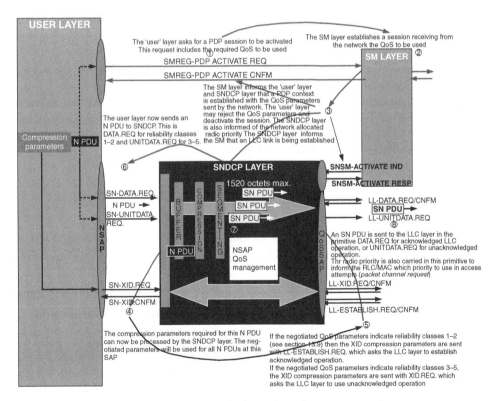

Figure 12.2   The SNDCP layer in action – an overview

The radio priority allocated by the network is also passed to the SNDCP. The SNDCP responds to this primitive with SNSM-ACTIVATE-RESPONSE.

The SNSM-ACTIVATE-IND from the SM to SNDCP layer allows the SN-XID. REQ from the user (network) layer to be processed by the SNDCP. This primitive contains the compression parameters the network layer wishes to use for N PDU transfer across the sub-network.

The SNDCP receives SN-XID.REQ and sends it to the LLC layer:

- *LL-XID.REQ if the negotiated QoS reliability is in class 3–5.* This primitive, which carries the SN-XID parameters, asks the LLC layer to send the information to the network in unacknowledged (connectionless) mode.
- *LL-ESTABLISH.REQ if the QoS reliability class is 1–2.* This primitive asks the LLC layer to establish an acknowledged mode data link for transferring the SN XID message. (The mobile station LLC layer does this by sending a SABM frame which piggy-backs the SN XID message); the sub-network LLC layer responds to SABM with UA and the interchange of these two layer 2 frames establishes the connection mode data link.) Confirmation of the XID parameters to be used is received from the network. The received SN XID parameters are used by the SNDCP layer to compress the headers and data of incoming N PDUs. The user (network) layer is informed of these parameters.

The network layer is now in a position to transfer N PDUs to the sub-network (the GPRS PLMN). It does this according to the QoS reliability:

- SN-DATA.REQ if the reliability is classes 1–2;
- *SN-UNITDATA.REQ if the reliability is classes 3–5.*

These two primitives respectively ask the SNDCP layer to send the N PDUs in LLC acknowledged mode or LLC unacknowledged mode.

The N PDU carried by each SN-DATA.REQ or SN-UNITDATA.REQ is buffered in the SNDCP layer, compressed according to the negotiated SN XID parameters and then segmented into lengths for processing by the LLC as a number of single PDUs. The segment length is determined by the n201 parameter of the LLC layer. These PDUs, which are segments of the N PDU, are called sub-network PDUs, SN PDUs.

Each SN PDU within an SN-DATA.REQ primitive or an SN-UNITDATA.REQ primitive will cause the primitive LL-DATA.REQ or LL-UNITDATA.REQ to be generated by the SNDCP layer:

- *LL-DATA.REQ* causes the LLC layer to send the SN PDU in acknowledged mode;
- *LL-UNITDATA.REQ* causes the LLC layer to send the PDU in unacknowledged mode.

12.2 Buffering, segmentation and acknowledged mode transmission of network PDUs

Figure 12.3 illustrates the action of the SNDCP layer for acknowledged mode. Acknowledged mode transmission of N PDUs is used if the QoS reliability for the NSAP is class 1–2. The user layer delivers N PDUs in an SN-DATA.REQ primitive; this indicates that the N PDU must be sent in acknowledged mode.

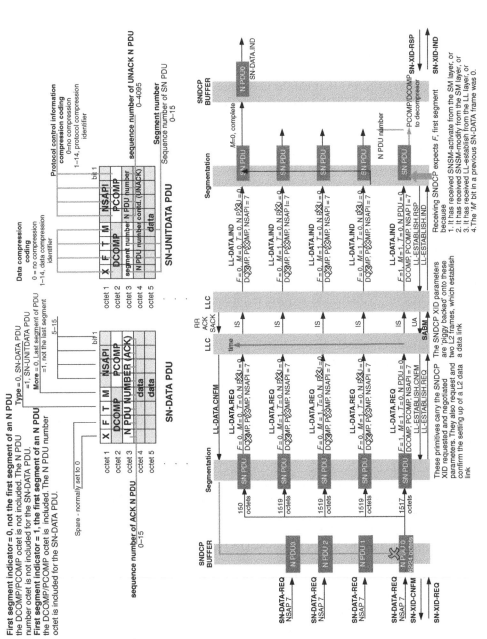

Figure 12.3 Segmentation of N PDUs into SN PDUs, acknowledged transmission and reassembly

The lower part of Figure 12.3 shows N PDUs delivered to the SNDCP layer in the primitives SN-DATA.REQ on NSAP number 7. The N PDUs are placed in a buffer store until their transmission is acknowledged by the network. We shall examine the segmentation of the first network data protocol unit, N PDU0.

Now N PDU0 contains 6224 octets. This is for illustrative purposes only. The maximum size of a TCP/IP data packet on the Ethernet is 1520 octets, and this is probably the maximum size used for TCP/IP or UDP packets on a GPRS NSAP. The maximum number of octets that can be accommodated by an LLC frame is 1520 (see Chapter 10). Therefore, N PDU0 must be segmented into blocks of this length. (This block size is determined by LLC parameter n201). These blocks are called sub-network protocol data units, SN PDUs. But there are headers to be added to the SN PDU blocks, the SN PDU headers. These headers are shown in the upper part of Figure 12.3, and will be considered before proceeding further.

12.2.1   SN PDU headers

There are two types of header, one for acknowledged mode transmission of the SN PDU and the other for unacknowledged transmission. The 'acknowledged' and 'unacknowledged' modes refer to the data link mode of the LLC layer. The header content is similar in both the acknowledged and unacknowledged modes, with the exception of the operation of the F, first bit, N PDU and segment numbering. These differences are brought out in the sections below.

12.2.1.1   Acknowledged SN-DATA PDU description

This is shown to the top left of Figure 12.3. The first SN-DATA PDU of a segmented N PDU has three header octets, and the following SN-DATA PDUs of the same segmented N PDU have just one header octet. The information contained in these octets is:

- **First octet**

  X, spare, 1 bit.
  F, first segment (of the segmented N PDU, also the first SN PDU of the segmented N PDU). If this is set to 0, then this SN PDU is not the first SN PDU. Only the first octet of the header is included. If this is set to 1, then this is the first SN PDU. Three octets are included in the header.
  T, type, 0 indicates the SN PDU is SN-DATA (acknowledged) or 1 indicates SN-UNITDATA (unacknowledged).
  M, more, 0 indicates this is the last SN PDU of the segmented N PDU, 1 indicates that more SN PDUs belonging to the segmented N PDU are to come.
  NSAPI indicates the network service access point in use by the N PDU. Service access points 5 to 15 are currently valid.

- **Second octet**

  The second and third octets are only transmitted with the first segment (SN PDU), not with subsequent SN PDUs.

DCOMP, an identifier of the data compression coding in use.

PCOMP, an identifier of the protocol control information compression coding in use.

- **Third octet**

  The second and third octets are only transmitted with the first segment (SN PDU), not with subsequent SN PDUs.

  N PDU number, the number of the network PDU of which this SN PDU is a segment. For acknowledged operations this falls in the range 0–15.

12.3   Acknowledged mode description of operation

Before acknowledged mode transmission of SN PDUs commences, two preliminary steps are required:

- The layer 2 LLC must establish acknowledged mode of operation. This is also known as establishing a data link connection.
- XID (exchange identity) parameters must be set up between the two communicating SNDCP layers. These will determine the compression to be used on customer data PDUs.

As indicated at the bottom of Figure 12.3, these two steps are performed simultaneously as follows:

- The mobile station network layer sends SN-XID-REQ to the SNDCP layer. This request indicates that the network layer requires data and header compression as contained in the parameters of the primitive.
- Because the SNDCP layer knows the reliability class is 1 or 2 for this PDP context, this request causes the SNDCP layer to send LL-ESTABLISH-REQ to the LLC layer. This request asks the LLC layer to establish a data link connection and also carries the XID parameters from the network layer.
- Upon receiving this request the LLC layer establishes a data link connection by sending SABM – set asynchronous balanced mode. SABM carries the layer 3 XID-REQ parameters received from the mobile station SNDCP layer.
- The SGSN LLC layer receives SABM and sends LL-ESTABLISH-IND to its SNDCP layer, indicating that a data link connection is being established. This primitive carries the requested layer 3 XID parameters received from the mobile station.
- The SGSN SNDCP layer informs the network layer of the XID requested with SN-XID-IND.
- The network layer responds with SN-XID-RSP. This contains the XID parameters that will be used.
- The SGSN SNDCP layer now responds to LL-ESTABLISH-IND with LL-ESTABLISH-RSP, carrying the layer 3 XID parameters.
- This causes the LLC layer to generate the layer 2 frame UA, which carries the XID parameters.
- The mobile station LLC layer receives UA and a data link connection is established. The exchange of SABM and UA frames sets up the layer 2 timers and counters and

other parameters as described in Chapter 10, allowing acknowledged transfer of layer 2 IS – information plus supervisory frames. The frames SABM and UA may also carry layer 2 XID parameters. These establish agreed layer 2 parameters (discussed in Chapter 10) for the data link.

- The mobile station LLC layer now informs its SNDCP layer that a layer 2 data link connection is established with the primitive LL-ESTABLISH-CNFM. This conveys the received XID parameters.
- The SNDCP layer takes the received XID parameters to apply to subsequent N PDUs and informs its network layer of the XID parameters in use with SN-XID-CNFM.

With a layer 2 connection in place and agreed layer 3 XID parameters, the mobile station network layer can now start to send N PDUs which contain customer data. It does this for acknowledged mode with the primitive SN-DATA-REQ. The first N PDU to be sent is N PDU0. This is seen entering the SNDCP layer at NSAP7.

N PDU0 in Figure 12.3 contains 6224 octets (ignoring, in this example, any compression) and it is segmented into five SN PDUs:

- The first SN PDU takes 1517 octets from this N PDU and adds three header octets, giving the maximum 1520 octets that the LLC layer can accept.
- Subsequent SN PDUs to the first SN PDU use only one header octet, so 1519 octets can be fitted into each of these.
- The final SN PDU takes 150 octets.

Each SN PDU is sent individually to the LLC layer in the primitive LL-DATA.REQ, indicating that acknowledged LLC transmission is required.

The first SN PDU has the header settings:

- *F*, first = 1, indicating that this is the first segment of an N PDU;
- *M*, more = 1, indicating that there are more SN PDUs to follow this one;
- *T*, type = 0, indicating that this is an SN DATA type PDU which is acknowledged;
- *NSAPI* = 7, indicating that the network is using network service access point number seven;
- *DCOMP* and *PCOMP* in the second octet; and
- *N PDU* number in the third octet.

The subsequent SN PDUs have the header settings:

- *M*, more = 1, except for the last SN PDU where it is set to '0', indicating that this is the last segment of N PDU0
- *F*, first = 0, for all the SN PDUs which follow the first SN PDU, indicating that they are not the first SN PDU;
- *T*, type = 0;

The SN PDUs are transported on the primitive LL-DATA.REQ to the mobile station LLC layer where they are 'wrapped' in a layer 2 data link IS frame for transmission between the LLC peers. Each primitive LL-DATA.REQ carries a reference number. At some point the GPRS network LLC layer will acknowledge reception of LLC frames from the mobile station. This is shown in Figure 12.3 as one of the frames labelled *RR – receiver ready, ACK – acknowledge, SACK – selective acknowledge.*

One of these frame types acknowledges the reception of LLC IS frames. Receiving one of these causes the mobile station LLC layer to send an LL-DATA.CNFM primitive to the SNDCP layer carrying the reference of the LL-DATA.REQ primitive. If this reference number belongs to the final segment SN PDU, then the SNDCP knows that the SGSN LLC layer has received all the segments of N PDU0 and deletes it from the buffer store.

On the GPRS sub-network side, the SN PDUs are delivered from the LLC layer to the SNDCP layer in the primitive LL-DATA.IND. The SNDCP layer is waiting to receive an SN PDU with the F bit set to 1, the first SN PDU, as indicated in Figure 12.3.

The PCOMP and DCOMP indicators are extracted from the first SN PDU and sent to the decompression function. As the following SN PDUs are received, the final one is indicated by $M = 0$ and the SN PDUs are reassembled, numbered with the N PDU number extracted from the first SN PDU, and passed with the primitive SN-DATA.IND to the network layer as the complete N PDU0.

12.4 Buffering, segmentation and unacknowledged mode transmission of network PDUs

Unacknowledged mode transmission of N PDUs is used if the QoS reliability for the PDP context is Class 3–5. The mobile station network layer delivers N PDUs for transmission in unacknowledged mode in the SN-UNITDATA.REQ primitive.

12.4.1 The unacknowledged SN-DATA PDU

This is shown to the top right of Figure 12.4. The first SN-UNITDATA PDU of a segmented N PDU has four header octets, and following SN-UNITDATA PDUs of the same segmented N PDU have just three header octets. The information contained in these octets is:

- **First octet**

  X, spare, 1 bit.
  F, first segment (of the segmented N PDU, also the first SN PDU of the segmented N PDU).
  If this is set to 0, then this SN PDU is not the first, and only the first octet and segment number and N PDU number of the header is included. If this is set to 1, then this is the first SN PDU and all four octets are included in the header.
  T, type, 0 indicates the SN PDU is SN-DATA (acknowledged) or 1 indicates SN-UNITDATA (unacknowledged).
  M, more, 0 indicates this is the last SN PDU of the segmented N PDU, 1 indicates that more SN PDUs belonging to the segmented N PDU are to come.
  NSAPI indicates the network service access point in use by the N PDU, 5 to 15 being currently valid.

- **Second octet**

  The second octet is transmitted only in the first segment SN PDU, not in subsequent SN PDUs.

DCOMP, an identifier of the data compression coding in use.

PCOMP, an identifier of the protocol control information compression coding in use.

• **Third octet**

The third octet is transmitted only in the first segment SN PDU, not in subsequent SN PDUs.

Segment number, in unacknowledged mode, each SN PDU is numbered. The receiving end keeps count of the received SN PDU number (but does not attempt to correct the situation if numbered SN PDUs are missing). It will attempt to reorder the SN PDUs if they are received out of order. The 'segment number' numbers the SN PDUs which comprise an N PDU in the range 0–16.

N PDU number, for the unacknowledged mode of operation, the network PDUs are numbered in the range 0–4095. This is continued into the fourth octet.

12.5 Unacknowledged mode description of operation

Before unacknowledged mode layer 2 transmission of SN PDUs takes place, there is only one step to complete, the exchange of layer 3 XID parameters between the mobile station and network. The mobile station network layer initiates this by sending SN-XID-REQ containing the required XID parameters to the SNDCP layer.

The SNDCP layer knows that the PDP context for this NSAP is reliability class 3–5 and therefore reacts to this primitive by sending LL-XID-REQ carrying the requested XID parameters to the mobile station LLC layer. The mobile station LLC layer reacts to this by sending a layer 2 XID frame to the GPRS sub-network LLC layer. This layer 2 frame carries the requested layer XID information. It may also carry layer 2 XID parameters to negotiate the settings for the LLC link.

The GPRS sub-network LLC layer reacts to the reception of XID by sending the layer 2 XID parameters to the SNDCP layer in the LL-XID-IND primitive, which passes them to the network layer in the primitive SN-XID-IND. The network layer responds by sending SN-XID-RESP, carrying the compression parameters that will be used, to its SNDCP layer. The GPRS sub-network SNDCP layer extracts these parameters for compression and decompression of subsequent N PDUs/SN PDUs and passes them to the LLC layer in the primitive LL-XID-RESP. This primitive causes the LLC layer to respond to the mobile station layer 2 XID frame with a further XID frame carrying the layer 3 XID parameters to be used. (This frame may also include layer 2 XID parameters).

The mobile station LLC layer receives the XID frame and passes the contents to its SNDCP layer in the primitive LL-XID-CNFM. The mobile station SNDCP layer stores the contents for compression and decompression and informs its network layer of the successful completion of XID with SN-XID-CNFM.

The mobile station network layer can now send customer data N PDUs for transmission to their address.

Figure 12.4 illustrates unacknowledged transfer of N PDUs.

The mobile station network layer delivers N PDU0 in the primitive SN-UNITDATA. REQ. which indicates to the SNDCP layer that the N PDU should be transferred in LLC unacknowledged mode in accordance with the negotiated QoS reliability classes (3–5) for this NSAP.

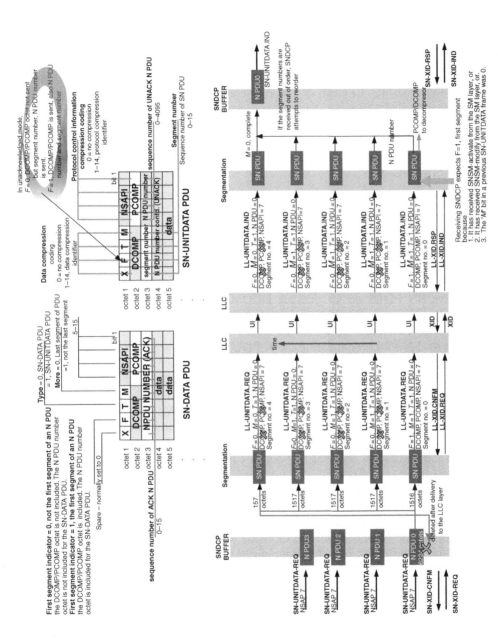

Figure 12.4   Segmentation of N PDUs into SN PDUs, unacknowledged transmission and reassembly

The SNDCP layer segments them as shown. The segments in this case differ from the acknowledged mode segments because there are more header octets in the unacknow-ledged SN PDU frame, four in the first SN PDU, three in subsequent SN PDUs.

In unacknowledged mode, each segment is numbered. This is not the case for acknowl-edged mode. The reason for this is that, in acknowledged mode, the LLC layer guarantees delivery of the SN PDUs it receives from the SNDCP layer; the LLC layer effectively numbers each SN PDU and demands acknowledgment of them. In unacknowledged mode, the sending LLC layer again numbers SN PDUs, but the numbers are disre-garded by the receiving LLC layer and delivered to the receiving SNDCP layer as received, regardless of whether any have been lost. The only way the receiving SNDCP layer knows whether any are missing is if the sending LLC layer numbers them.

The SNDCP layer delivers the SN PDUs to the LLC layer in the primitive LL-UNI-TDATA.REQ, which tells the LLC layer that they should be sent in unacknowledged mode.

The first SN PDU has:

- Segment number 0
- N PDU0, with
    — F, first = 1, indicating the first segment (SN PDU);
    — $M = 1$, indicating there are more SN PDUs belonging to this N PDU after the first;
    — $T = 1$, indicating that this is an SN-UNITDATA PDU;
    — DCOMP/PCOMP is included.

Subsequent SN PDUs do not include DCOMP/PCOMP. The segment number is incremented each time an SN PDU belonging to this N PDU is sent to the LLC layer. The final SN PDU is indicated by $M = 0$.

When all the SN PDUs comprising the N PDU have been sent to the LLC layer, the N PDU is deleted from the buffer store.

The LLC peer transfer uses unacknowledged information (UI) frames. The GPRS sub-network LLC layer delivers each received SN PDU to the SNDCP layer in the primitive LL-UNITDATA.IND. The SNDCP removes DCOMP/PCOMP from the first SN PDU and sends them to the decompression function. The N PDU number is extracted and is used in the reassembled SN PDUs for numbering of the N PDU sent to the network layer.

The SNDCP tracks the segment number in each SN PDU, and rearranges the SN PDUs if they arrive out of order.

The GPRS sub-network SNDCP layer delivers the received reassembled SN PDUs as a numbered N PDU to the network layer in the primitive SN-UNITDATA.IND.

As there is no guarantee of delivery to the end user of N PDUs, then the application which generated the N PDUs and the receiving application must have error checking and correction facilities for reliable communication.

12.6 Management of multiple PDP contexts

A mobile station can have multiple simultaneous PDP contexts operating. If the PDP contexts for the transfers have the same QoS profile, then SN PDUs may be routed to the same access point on the LLC layer. In that case the SNDCP layer must manage the delivery of SN PDUs from separate network service access points so that the PDUs of each transfer are delivered to the LLC layer at a rate which will meet the data throughput requirement for each NSAPI. This is illustrated in Figure 12.5.

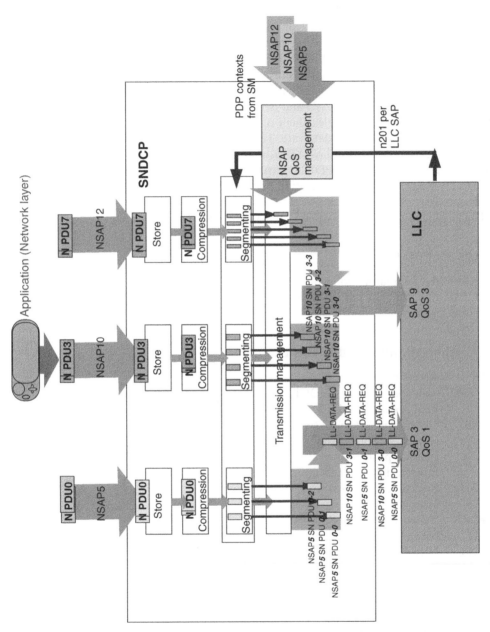

Figure 12.5 Multiplexing and flow control management in the mobile station SNDCP layer

Here, the mobile station is shown with three active PDP contexts in operation on network service access points 5, 10 and 12. NSAP5 is sending the first PDU of its session, N PDU0. NSAP10 is sending the fourth PDU of its session, N PDU3. NSAP12 is sending the eighth PDU of its session, N PDU7.

NSAP5 and NSAP10 have the same QoS profile for their PDP contexts and are sharing the same LLC service access point, SAP3, but NSAP12 has a different QoS profile and is routed to LLC SAP9.

The N PDUs are processed by the storage, compression and segmentation functions. The resulting SN PDUs are transferred individually to the LLC layer. However, as the SN PDUs originating from NSAP5 and NSAP10 share the same LLC access point, then the SNDCP layer must multiplex them onto this access point, ensuring that they are delivered to maintain the QoS data rate. This is illustrated in Figure 12.5, where the SN PDUs belonging to NSAP5 and NSAP10 are shown entering LLC SAP3, multiplexed by the SNDCP layer. Each SN PDU is carried to the LLC layer on its own primitive LL-DATA-REQ, in this case indicating that LLC SAP3 is operating in acknowledged mode.

The labelling on the sub-network PDUs is explained as follows:

NSAP5 SN PDU-0-0

- **NSAP5**, the originating point of the SN PDU.
- **SN PDU-0-0**, the first zero indicates that this SN PDU is a segment of N PDU0; the second zero indicates that this SN PDU is the first segment of N PDU0.

12.7 Abbreviations used in this chapter

APN	Access point name
DCOMP	data compression SNDCP
F	First segment SNDCP
GPRS	General packet radio service
HLR	Home location register
IP	Internet protocol
IS	Information/supervisory
LLC	Logical link control
M	More
N PDU	Network protocol data unit
NSAP	Network service access point
NSAPI	Network service access point identity
PCOMP	Protocol compression SNDCP
PDP	Packet data protocol
PDU	Protocol data unit
PLMN	Public land mobile network
QoS	Quality of service
RR	Receiver ready
SABM	Set asynchronous balanced mode
SACK	Selective acknowledge

SAP	Service access point
SGSN	Serving GPRS support node
SM	Session management
SNDCP	Sub-network dependent convergence protocol
SN PDU	Sub-network protocol data unit
T	Type of SNDCP
TBF	Temporary block flow
TCP/IP	Transmission control protocol/Internet protocol
UA	Unnumbered acknowledge
UI	Unacknowledged information
UMTS	Universal mobile telecommunications system
XID	Exchange identities

Appendix 1: GMSK and EDGE

Although derived from a somewhat contrived phrase – *enhanced data rates for GSM evolution*, the acronym EDGE is perhaps the most apt within the GSM lexicon. EDGE takes GSM radio to the limit, or edge, of its practicable data rates (59.2 kb/s for each physical channel of the radio channel). To go beyond this edge requires other radio systems, which is the cue for UMTS to make its stage appearance.

GMSK – Gaussian minimum shift keying – is the original circuit switched GSM modulation system, allowing the GSM radio channel to be modulated at a data rate of 271 kb/s whilst keeping the radio channel within a 200 kHz bandwidth. This is the modulation system used for circuit switched and GPRS operations. The maximum possible customer data rate is about 22.8 kb/s on one physical channel without channel encoding.

EDGE when used with enhanced GPRS services will be called EGPRS.

A1.1 MSK

Figure A1.1 illustrates the principles of minimum shift keying (MSK). The top diagram shows the phasor representation of MSK. The phasor moves by 90 degrees in a clockwise direction for each modulating 'one' bit, and 90 degrees in an anti-clockwise direction for each 'zero' modulating bit. The maximum instantaneous shift in the phase of the modulated waveform is 90 degrees and this limitation on the shift keeps the bandwidth of the modulated waveform within reasonable limits. The mechanism of how this is achieved is illustrated in the waveform diagrams of Figure A1.1.

GPRS in Practice: A Companion to the Specifications Peter McGuiggan
© 2004 John Wiley & Sons, Ltd   ISBN: 0-470-09507-5

CS1	9.05 kb/s	GMSK
CS2	13.40 kb/s	GMSK
CS3	15.60 kb/s	GMSK
CS4	21.40 kb/s	GMSK
EGPRS MCS1	8.80 kb/s	GMSK
EGPRS MCS2	11.20 kb/s	GMSK
EGPRS MCS3	14.8 kb/s	GMSK
EGPRS MCS4	17.60 kb/s	GMSK

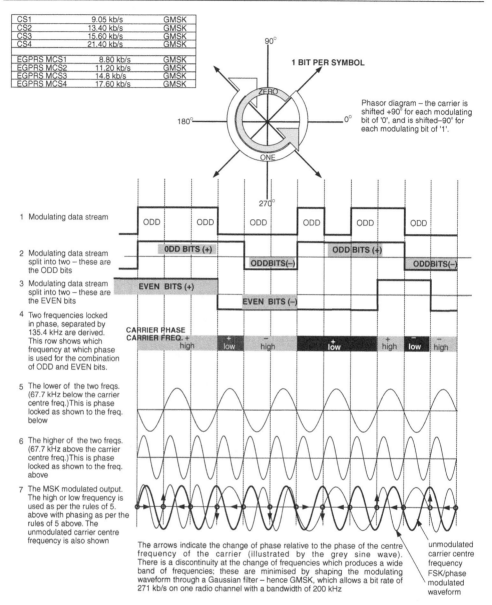

Figure A1.1 Minimum shift keying

- **Waveform 1** shows the modulating data stream. This comprises, for each radio burst, 114 bits of a 13 kb/s GSM voice channel signal after channel encoding. To this is added a 26-bit training sequence code, six 'tail' bits and two flag (FACCH) bits, giving an aggregate rate of 33.8 kb/s. This rate is clocked out at 271 kb/s, the data rate of waveform 1.
- **Waveforms 2 and 3** show the encoded voice signal split into two half-rate digital signals, waveform 2 is from the *odd* bits of waveform 1 and waveform 3 is from the *even* bits of waveform 1. Waveforms 2 and 3 are shown as bipolar waveforms.

- **4** shows the rules for the modulation of the carrier by the waveforms 2 and 3.

  —When *odd* bits and *even* bits are both of positive polarity, the carrier phase is *positive* with the carrier the *higher* of two possible frequencies (this is the nominal ARFCN centre frequency plus 67.7 kHz).

  —When *odd* bits are *positive* and *even* bits *negative*, then the carrier phase is *positive* with the carrier the *lower* of two possible frequencies (nominal ARFCN centre frequency minus 67.7 kHz).

  —The modulating rules for all combinations of odd and even bits are shown.

- **Waveform 5** shows the lower frequency (ARFCN − 67.7 kHz) of the two possible carrier frequencies.
- **Waveform 6** shows the higher frequency (ARFCN + 67.7 kHz) of the two possible carrier frequencies. Note that the lower frequency is inverted compared to the higher frequency. Also note that the 'centre frequency' does not actually exist! The carrier shifts between the upper and lower frequencies.
- **Waveform 7** shows the transmitter modulated output after applying the rules above. The waveform is frequency and phase shifted. The (hypothetical) unmodulated carrier centre frequency is superimposed to show the relative phase shift. Phasor arrows are shown indicating the relative phase shift.

A1.2   GMSK

Figure A1.1 shows that there is a discontinuity between frequency shifts. This increases the occupied bandwidth and is minimised by passing the modulating pulse train through a Gaussian filter which rounds the pulses, making transitions between carrier phases less sharp, reducing spurious emission.

A1.3   EDGE – EGPRS

EGPRS, in addition to using GMSK, transits to 8-phase PSK if the condition of the air interface is appropriate (high C/I).

In 8-PSK each of the eight phases carries three bits of information. Each phase is called a symbol and the symbol rate is fixed at 271 ksymbols per second. As the symbol separation is only 45 degrees, increased demand is placed upon the receivers in the mobile station and TRX. Dependent upon the channel encoding used (and the C/I ratio), 8-PSK EGPRS will provide data rates from 22.4 kb/s to 59.2 kb/s per physical channel. The full set of data rates for EGPRS, using a mixture of GMSK and 8-PSK are given in Table 1.1. The data rate used is adaptable to the prevailing C/I.

Figure A1.2 illustrates 8-PSK. The eight phases are shown in the phasor diagram at the top, and the three bits per phase or symbol are indicated. It is evident that the main modulating bit stream must be split into three bit streams, each at a bit rate of one-third of the main stream. These three lower rate streams are combined to move the phase of the carrier to the angles shown. Using the EGPRS modulation and coding system 9 – MCS-9, then a customer may have a data rate of 59.2 kb/s on one physical channel.

EGPRS may provide a useful transition between GPRS data speeds and UMTS because of the network infrastructure.

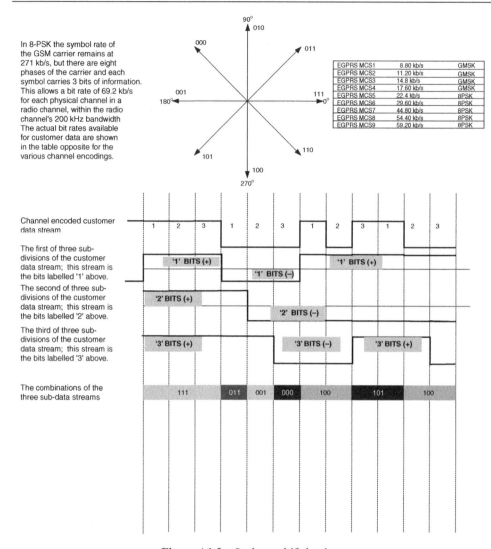

In 8-PSK the symbol rate of the GSM carrier remains at 271 kb/s, but there are eight phases of the carrier and each symbol carries 3 bits of information. This allows a bit rate of 69.2 kb/s for each physical channel in a radio channel, within the radio channel's 200 kHz bandwidth. The actual bit rates available for customer data are shown in the table opposite for the various channel encodings.

EGPRS MCS1	8.80 kb/s	GMSK
EGPRS MCS2	11.20 kb/s	GMSK
EGPRS MCS3	14.8 kb/s	GMSK
EGPRS MCS4	17.60 kb/s	GMSK
EGPRS MCS5	22.4 kb/s	8PSK
EGPRS MCS6	29.60 kb/s	8PSK
EGPRS MCS7	44.80 kb/s	8PSK
EGPRS MCS8	54.40 kb/s	8PSK
EGPRS MCS9	59.20 kb/s	8PSK

Figure A1.2   8-phase shift keying

Current GSM infrastructure is based upon the 13 kb/s voice channel. In particular, this determines the data rate across the A-bis interface (and largely across the A interface too.) With a 3 kb/s signalling channel required between the transcoder rate adapter unit (TRAU) and TRX, each physical channel on the air interface requires a 16 kb/s data rate on the A-bis interface. Four air interface physical channels can therefore be accommodated within one standard E1 64 kb/s channel. This is currently the basis of network infrastructure capacity planning. As the data rates for air interface physical channels rise to the higher GPRS rates and EGPRS rates, it will be necessary to add extra channel capacity in the network infrastructure. This is not a trivial consideration as it may cost the network operators millions of pounds.

Now UMTS, with its very high customer data rates will demand a high network infrastructure capacity from the equivalent of the A-bis and A interface, and of course this capacity must be in place before the UMTS services can be offered. It is very convenient if the network operators put the high capacity network infrastructure in place well before UMTS services are ready and use it simply by changing the GPRS TRXs to EGPRS TRXs (assuming the handset manufacturers have the EGPRS product ready).

The UMTS system can then be rolled out, first sharing the infrastructure capacity (and giving lower air interface data rates than the maximum), then as GPRS and EGPRS are phased out, taking over completely the infrastructure capacity and offering customers the full services of UMTS.

However, the higher data rate GPRS and EGPRS will probably not be universally on offer to customers. This is because the higher data rates, because they give less redundancy to the information, demand better C/I (carrier to interference ratios) than the normal 9 dB of circuit switched operations.

The higher C/I can be achieved by:

- Raising the BTS transmit power for GPRS and EGPRS physical channels. This has the great disadvantage of increasing the network interference, so it is an option to be approached with extreme caution.
- Putting the GPRS and EGPRS services upon physical channels which are experiencing only a small amount of interference. If such channels can be found, this is a possible solution. However, the current practice, in GSM900 systems certainly, is to implement frequency hopping. This has the effect of spreading any interference a cell is experiencing amongst all the customers using that cell and the possibility of finding a 'quiet' physical channel for GPRS/EGPRS use is therefore remote.

If neither of the above options is available, then coverage will be limited as the effect of requiring a higher C/I ratio is to reduce the cell's working radius. GPRS takes advantage of this by varying the customer data rates dependent upon the C/I ratio. But this does mean that the higher rates are only available near to the cell antenna.

If full coverage at the higher data rates is required, the only real option is to increase the number of cells, and this is a very expensive option, which may not be transferable to UMTS working.

Network operators may, therefore, offer the higher rate GPRS and EGPRS services only in limited areas.

A1.4 Abbreviations used in this appendix

ARFCN   Absolute radio frequency channel number
BTS   Base transceiver station
C/I   Carrier to interference ratio
E1   European 2 Mb/s transmission standard
EDGE   Enhanced data rate for GSM evolution
EGPRS   EDGE applied to GPRS
FACCH   Fast associated control channel

GMSK   Gaussian minimum shift keying
GPRS   General packet radio service
GSM   Global system for mobile communication
MCS   Modulation and coding system EDGE
MSK   Minimum shift keying
PSK   Phase shift keying
TRAU   Transcoder rate adapter unit
TRX   Transceiver
UMTS   Universal mobile telecommunications system

Appendix 2: System Information and Packet System Information

A2.1 Key

This information is compiled from GSM 04.60, 44.060, ETSI TS 144 060v4.5.0 MS-BSS Interface RLC/MAC Protocol.

AS = appendix system information

The GPRS information is broadcast on the BCCH as *system information*.

> **Example:** AS3 gives the contents of system information messages type 3.

AP = Appendix packet system information

GPRS information is broadcast on the PBCCH as *packet system information*.

> **Example:** AP5 gives the contents of packet system information messages type 5.

Figure A2.1 shows the contents lists of the system information messages. This appendix contains the information parameters sent on the BCCH or PBCCH or PACCH.

GPRS in Practice: A Companion to the Specifications Peter McGuiggan
© 2004 John Wiley & Sons, Ltd ISBN: 0-470-09507-5

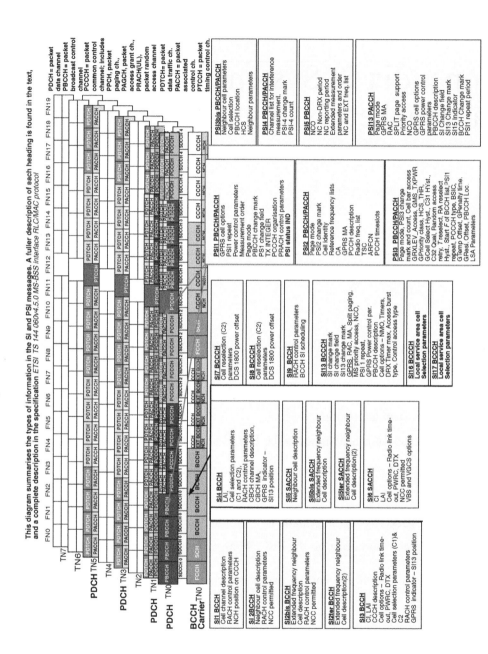

This diagram summarises the types of information in the SI and PSI messages. A fuller description of each heading is found in the text, and a complete description in the specification ETSI TS 144 060v4.5.0 MS-BSS Interface RLC/MAC protocol

PDCH = packet data channel
PBCCH = packet broadcast control channel
PCCCH = packet common control channel; includes PPCH, packet paging ch.,
PAGCH, packet access grant ch.,
PRACH(UL), packet random access channel
PDTCH = packet data traffic ch.
PACCH = packet associated control ch.
PTCCH = packet timing control ch.

SI1 BCCH
Cell channel description
RACH control parameters
NCH position on CCCH

SI2BCCH
Neighbour cell description
RACH control parameters
NCC permitted

SI2bis BCCH
Extended frequency neighbour
Cell description
RACH control parameters
NCC permitted

SI2ter BCCH
Extended frequency neighbour
Cell description(2)

SI3 BCCH
CI, LAI
CCCH description
Cell options – Radio link time-out, PWRC, DTX
C2
Cell selection parameters (C1)&
GPRS indicator – SI13 position

SI4 BCCH
LAI,
Cell selection parameters (C1 and C2),
RACH control parameters,
CBCH channel description,
CBCH MA,
GPRS indicator –
SI13 position

SI5 SACCH
Neighbour cell description

SI5bis SACCH
Extended frequency neighbour
Cell description

SI5ter SACCH
Extended frequency neighbour
Cell description(2)

SI6 SACCH
CI
LAI
Cell options – NMO, Timers,
DRX Timer max, Access burst
type, Control access type
NCC permitted
VBS and VGCS options

SI7 BCCH
Cell reselection (C2)
parameters
DCS 1800 power offset

SI8 BCCH
Cell reselection (C2)
parameters
DCS 1800 power offset

SI9 BCCH
RACH control parameters
BCCH SI scheduling

SI13 BCCH
SI change mark
SI change field
SI13 change mark
GPRS, RAC, MA, Split paging,
MS priority access, NCO,
PSI 1 repeat,
GPRS Power control par.
PBCCH description
Cell options – NMO, Timers,
DRX Timer max, Access burst
type, Control access type

SI16 BCCH
Local service area cell
Selection parameters

SI17 BCCH
Local service area cell
Selection parameters

PSI1 PBCCH/PACCH
GPRS cell options
PSI1 repeat
Power control parameters
Measurement order
Page mode
PBCCH change mark
PSI1 change field
TX INTEGER
PCCCH organisation
PRACH control parameters
PSI status IND

PSI2 PBCCH/PACCH
Page mode
PSI2 change mark
Cell identity
Reference frequency lists
CA
GPRS MA
PCCH description
Radio freq. list
TSC
ARFCN
PCCH timeslots

PSI3 PBCCH/PACCH
Page mode, PSI3 change
mark and count, Cell bar access
GRXLEV_Access, GMS_TXPWR
G Priority class, HCS_THR,
GCell Select Hyst, C31 H1st.,
C32 Qual, Random access
retry, T reselect, RA reselect,
Hyst., Start F of BCCH list,>SI1
repeat, PCCCH type, BSIC,
G Temp Offset, GPenalty time,
GResl. Offset, PBCCH Loc.
LSA Parameters

PSI3bis PBCCH/PACCH
Neighbour cell parameters
Cell selection
PBCCH location
HCS
Neighbour parameters

PSI4 PBCCH/PACCH
Channel list for interference
measurement.
PSI-4 change mark
PSI-4 count

PSI5 PBCCH
NCO
NC Non-DRX period
NC reporting period
Extended measurement
parameters and order
NC and EXT freq. list

PSI13 PACCH
Page mode
GPRS MA
RAC
SPLIT page support
Priority access
NCO
GPRS cell options
GPRS power control
parameters
PBCCH description
SI13 change field
SI13 change mark
SI15 Indicator
BCCH Change mark
PSI1 repeat period

Figure A2.1 Phase 2 + system information

AS.3 SI3 and SI4 messages on BCCH

Either of these messages indicates to the MS whether the cell supports GPRS services. They do this by including two information elements:

- **AS.3.1** RA colour code, a three-bit code indicating the colour code of the GPRS routeing area. If an MS changes cells (cell reselection) and this colour code changes, then it knows it has entered a different routeing area (RA) and will perform RA update.
- **AS.3.2** SI13 position. This one-bit code indicates whether SI13, which contains essential GPRS PSI is located on the BCCH (= 0), or on the extended BCCH (= 1). For GPRS services the MS will now read the SI13 messages.

AS.13 SI13 messages on BCCH or extended BCCH

- **AS.13.1** BCCH change mark. This three-bit code indicates whether any changes have been made to BCCH SI messages.
- **AS.13.2** SI change field. The value of this field indicates which particular SI messages have changed – see specification GSM 04.08/10.5.2.37b.
- **AS.13.3** SI13 change mark. This identifies changes to GPRS mobile allocation (MA) contained in SI13 or PSI13 messages. MA is the range over which PDCHs – the physical channels allocated to GPRS, are frequency hopping.
- **AS.13.4** GPRS MA. When used in SI13 or PSI13 messages, this refers to the CA – cell allocation (the total number of frequencies in use in the cell) as defined in system information messages SI1 and PSI2.
- **AS.13.5** RAC – if a PBCCH is not present in a cell. This eight-bit field gives the routeing area code for the cell.
- **AS.13.6** SPGC_CCCH_SUP – if a PBCCH is not present in a cell. This one-bit field indicates whether the CCCH of the BCCH carrier supports split paging. Split paging is covered in detail in Chapter 7.

$$0 = \text{SPLIT\_PG\_CYCLE not supported}$$
$$1 = \text{supported.}$$

- **AS.13.7** Priority_Access_Threshold – if a PBCCH is not present in a cell. This three-bit field indicates the MS radio priority allowed to access the cell. There are four MS radio priorities, with 1 the highest to 4 the lowest. These priorities are determined by the SGSN using QoS subscription, and network congestion. The network operator may use these priorities to control congestion. The values transmitted on SI13 may be:

> 000–010 packet access not allowed in this cell
> 011 access allowed to priority1
> 100 access allowed to priorities1–2
> 101 access allowed to priorities1–3
> 110 access allowed to priorities1–4
> 111 access allowed (spare).

- **AS.13.8** NCO (Network_Control_Order) – if a PBCCH is not present in the cell. Network control orders inform the MS of the types of measurement and cell reselection to be performed:

   00 = NCO0 = MS controlled cell reselection when the mobile station is in GMM ready condition, no measurement reporting.

   01 = NCO1 = MS controlled cell reselection when the mobile station is in GMM ready condition, MS sends measurement reports.

   10 = NCO2 = Network controlled cell reselection when the mobile station is in GMM packet transfer, MS sends measurement reports.

   11 = NCO3 = Reserved (= NCO0).

- **AS.13.9** GPRS cell options – if a PBCCH is not present in the cell. These include:

—**The network mode of operation (NMO)**:

   00 = NMO1
   01 = NMO2
   10 = NMO3

—The network modes of operation are illustrated in Figures 7.17–7.18.

—**T3166_T3168**, The timer settings for these two timers are in the range 0–7, in units of setting +1 units of 500 ms.

—**T3166** is a timer that starts when the MS sends its first RLC/MAC block and stops when it receives a *packet uplink ack*. Message. If it expires without receiving this message it must discontinue the TBF. Default 5 s.

—**T3168** starts upon the MS sending a *packet resource request* or *channel request description* in *packet downlink ack*. It stops on receiving a *packet uplink assignment* from the network. If it expires before this, the mobile station retransmits the messages.

—**T3192**, timeout value for this MS timer, range 0–7 setting +1 in units of 500 ms. It starts when the final RLC block is received in a DL TBF, with no retransmissions requested. When it expires, the MS releases the radio resources and monitors the paging channel.

—**DRX_TIMER_MAX**, range 0–7. The parameter value is interpreted as a value of $[2^{(setting-1)}]$ in units of 1 second. The parameter value 0 indicates non-DRX mode is not supported in this cell.

   This parameter tells the MS the maximum time that it is allowed to remain in the non-DRX mode after transferring from packet transfer mode to packet idle mode. The MS can negotiate a time for non-DRX mode that is less than this value. The purpose of the non-DRX period is that, after the completion of a TBF session where the MS returns to packet idle mode, the MS monitors all downlink CCCHs during this period and the network may, if it receives downlink network PDUs for that MS, initiate a packet downlink assignment on any of the CCCHs.

—**ACCESS_BURST_TYPE**, the number of bits in the short access bursts to be used by the MS for *packet channel requests*.

$$0 = 8 \text{ bits}$$
$$1 = 11 \text{ bits}$$

—**CONTROL_ACK_TYPE**, the type of *packet control acknowledgment* the MS should use.

$$0 = \text{four access bursts}$$
$$1 = \text{one RLC/MAC block}$$

—**BS_CV_MAX**, range 0–15, the maximum countdown value for data blocks on the uplink to be used by the mobile station. The example below shows how this is effected.

$TBC = 19$ is the total block count for this uplink TBF and only one TN is used

$BS\_CV\_MAX = 15$

BSN is the block sequence number.

TN1 BSN	TN1 CV
0	15
1	15
2	15
3	15
4	14
5	13
6	12
7	11
8	10
9	9
10	8
11	7
12	6
13	5
14	4
15	3
16	2
17	1
18	0

—**PAN_MAX**, range 0–15, this sets the maximum value of counter N3102 in the MS.

—**PAN_INC**, range 0–15, whenever an MS receives a *packet acknowledge* that allows V(S) to be advanced, N3102 is incremented by the value of PAN_INC, but PAN_MAX is never exceeded.

—**PAN_DEC**, range 0–15, there is a timer T3182 in the MS which starts when the final uplink data block is transmitted during a TBF. If an *uplink packet acknowledge* is received before its expiry, all is well; however, if this timer expires without the acknowledgment being received, then N3102 is decremented by the value of PAN_DEC. If N3102 reaches 0 then the MS will perform an abnormal release, and reselect another cell. If T3182 expires with N3102 > 0, then the MS releases the resources and attempts to re-establish communication on the same cell.

- **AS.13.10** SI15_IND – if a PBCCH is not present in the cell, indicates the presence of SI15 messages on the BCCH:

$$0 = \text{not present}$$
$$1 = \text{present}$$

- **AS.13.11** PSI1_REPEAT_PERIOD – if a PBCCH is present in the cell, indicates the repeat period for PSI1 messages,

$$0000 = 1 \text{ multiframe}$$
$$\text{to } 1111 = 16 \text{ multiframes}$$

- **AS.13.12** GPRS power control parameters – if a PBCCH is present in the cell,

—**ALPHA**, α, is a parameter for control of the MS output power, its range is 0–10, and units 0.1 second.

—**T_AVG_W** is a parameter for MS output power control with range 0–25. This parameter sets the signal strength filter period for power control in packet idle mode

$$2^{(k/2)/6} \text{ multiframes}, \ k = 0, 1, \ldots, 25$$

—**T_AVG_T** is a parameter for MS output power control with range 0–25. This parameter sets the signal strength filter period for power control in packet transfer mode

$$2^{(k/2)/6} \text{ multiframes}, \ k = 0, 1, \ldots, 25$$

—**PC_MEAS_CHAN** indicates the type of channel to be used by the MS for downlink measurements for power control purposes.

$$0 = \text{BCCH}$$
$$1 = \text{PDCH}$$

—**N_AVG_I** is a parameter for MS output power control with range 0–15. This parameter sets the interference signal strength filter constant for power control

$$2^{(k/2)}, \ k = 0, 1, \ldots, 15$$

- **AS.13.13** PBCCH description – if a PBCCH is present in the cell,

  —**TSC** field, range 0–7. The training sequence code used by the PBCCH and PCCCH.
  —**TN** field, range 0–7. The timeslot number (physical channel PDCH) containing PBCCH and associated PCCCHs.
  —**ARFCN** field is the representation of the frequency used by PBCCH in non-hopping mode, its range is 0–1023.
  —**MAIO** field is the mobile allocation index offset of the PBCCH if in the hopping mode, its range is 0–63.

Having obtained these parameters from the BCCH or extended BCCH, the MS now switches to the indicated position for PBCCH and obtains the packet system information messages as indicated in AP.1 onwards.

AP.1 PSI1 messages on the PBCCH

- **AP.1.1** GPRS cell options – see AS.13.9.
- **AP.1.2** Global power control parameters. These include:

  —α, see AS.13.12.
  —**T_AVG_W**, see AS.13.12.
  —**T_AVG_T**, see AS.13.12.
  —**Pb**, the level by which the PBCCH is below the BCCH, range $0, -2, -4, \ldots, -30\,\mathrm{dB}$.
  —**PC_MEAS_CHAN**, see AS.13.12.
  —**INT_MEAS_CHANNEL_LIST_AVAIL**, indicates if PSI4 is broadcast, if so it will contain a list of channels to be measured for interference.

- **AP.1.3** PSI1 repeat period, range 1 to 16 in multiframes.
- **AP.1.4** PSI_COUNT_LR. The number of PSI messages carried on PBCCH at low repetition rates, range 0–63.
- **AP.1.5** PSI_COUNT_HR. The number of PSI messages carried on PBCCH at high repetition rates, range 0–16.
- **AP.1.6** Measurement order. If set to 0 the MS is in control of cell selection and reselection and will not send measurement reports. This is equal to NCO0. It also indicates PSI5 messages are not broadcast. If set to 1, the MS shall send measurement reports for cell selection and reselection. PSI5 must be read for more information on cell reselection.
- **AP.1.7** Page mode – normal, extended, reorganisation or same as before.
- **AP.1.8** PBCCH_CHANGE_MARK. A three-bit counter which changes each time information messages have been changed in PSI2 and above.
- **AP.1.9** PSI_CHANGE FIELD. A four-bit indicator which shows which PSI message was changed to change the PBCCH_CHANGE_MARK counter. If more than one PSI has been changed, this indicates unspecified messages.

$$0 = \text{unspecified messages}$$
$$1 = \text{unknown}$$
$$2 = \text{PSI2}$$
$$3 = \text{PSI3/PSI3bis}$$
$$4 = \text{PSI4}$$
$$5 = \text{PSI5}$$

- **AP.1.10** TX_INTEGER. A parameter used by the MS when using the PRACH.
- **AP.1.11** PCCCH organisation parameters. These include:

—**BS_PCC_REL**,

> = 0 indicates that the PBCCH/PCCCH will not be released
>
> = 1 indicates that these are about to be released and MSs on PCCCH
> should go to CCCH, and follow the instructions on the BCCH.

—**BS_PBCCH_BLKS**, this indicates the number of blocks allocated to PBCCH in a 52-frame multiframe,

> 0 = block 0 only
>
> 1 = block 0 and block 6
>
> 2 = block 0 and block 6 and block 3
>
> 3 = block 0 and block 6 and block 3 and block 9.

—**BS_PAG_BLKS_RES**, indicates the PCCCH blocks reserved for PAGCH, PDTCH and PACCH in the 52-frame multiframe structure, and for PAGCH in the 51-frame multiframe structure. If this field is not included, the MS interprets this as 0 blocks reserved.

> 0 = 0 blocks reserved for PAGCH, PDTCH and PACCH
>
> 1 = block reserved for PAGCH, PDTCH and PACCH
>
> 2 = blocks reserved for PAGCH, PDTCH and PACCH
>
> $\vdots$
>
> 12 = 12 blocks reserved for PAGCH, PDTCH and PACCH

—**BS_PRACH_BLKS**, the number of blocks reserved for PRACH on any PDCH carrying PCCCH and PBCCH (52-frame multiframe structure only). If this field is not included, the MS interprets it as 0 blocks reserved for PRACH

> 0 = no blocks reserved for PRACH
>
> 1 = block 0 reserved for PRACH
>
> 2 = block 0 and block 6 reserved for PRACH
>
> 3 = block 0 and block 6 and block 3 reserved for PRACH
>
> 4 = block 0 and block 6 and block 3 and block 9 reserved for PRACH
>
> 5 = block 0 and block 6 and block 3 and block 9 and block 1
> reserved for PRACH
>
> 6 = block 0 and block 6 and block 3 and block 9 and block 1
> and block 7 reserved for PRACH

    7 = block 0 and block 6 and block 3 and block 9 and block 1
      and block 7 and block 4 reserved for PRACH

    8 = block 0 and block 6 and block 3 and block 9 and block 1
      and block 7 and block 4 and block 10 reserved for PRACH

    9 = block 0 and block 6 and block 3 and block 9 and block 1
      and block 7 and block 4 and block 10 and block 2
      reserved for PRACH

    10 = block 0 and block 6 and block 3 and block 9 and block 1
      and block 7 and block 4 and block 10 and block 2
      and block 8 reserved for PRACH

    11 = block 0 and block 6 and block 3 and block 9 and block 1
      and block 7 and block 4 and block 10 and block 2
      and block 8 and block 5 reserved for PRACH

    12 = block 0 and block 6 and block 3 and block 9 and block 1
      and block 7 and block 4 and block 10 and block 2
      and block 8 and block 5 and block 11 reserved for PRACH.

- **AP.1.12** PRACH parameters

—**TX_INT**, a number which determines the 'spread' of transmissions – this determines the probability of secondary 'clashes' of MSs whose first PRACH attempt has failed because more than one MS PRACH burst has arrived simultaneously at the BTS.

Parameter value	Slots used to spread transmissions
0000	2
0001	3
0010	4
0011	5
0100	6
0101	7
0110	8
0111	9
1000	10
1001	12
1010	14
1011	16
1100	20
1101	25
1110	32
1111	50

—S, this parameter gives the minimum number of slots between repetitions of *packet channel request* messages on the PRACH.

Parameter value	S value
0000	12
0001	15
0010	20
0011	30
0100	41
0101	55
0110	76
0111	109
1000	163
1001	217

—**MAX_RETRANS**, the number of *packet channel request* reattempts, after the first, that an MS is allowed to make, values 1, 2, 4, 7.

—**PERSISTENCE_LEVEL**, this field maps a value in a range of 1–16 to a radio priority class in the range 1–4. Expressed as $P(i) = y$, where (i) is the radio priority class and p the persistence level. The mapping of $P(i)$ to y is broadcast by the cell. Thus, the couple 10,2 maps priority class 2 to persistence level 10. This is then used to regulate access attempts (and ease cell congestion) according to the radio priority class of an MS. After the first attempt, this parameter is examined to see if further attempts are allowed.

—**ACC_CONTR_CLASS**, MSs are allocated an access class code. This parameter tells an MS whether its class is allowed to make PRACH access attempts in this cell. This can be used as a method of cell congestion control. The parameters are sent as a bit-map with EC indicating whether emergency calls are allowed in this cell.

Bit position	16	15	14	13	12	11	10	9	8	7	6	5	4	3	2	1
Access class	15	14	13	12	11	10	9	8	7	6	5	4	3	2	1	0
1 = barred			1											1		

In this example access classes 2 and 13 are not allowed to use the cell. Access classes 0–9 are the normal subscriber classes, 11–15 are reserved for emergency, security and network services.

- **AP.1.13** PSI_STATUS_IND. The MS may report the status of received packet system information if this parameter is set to 1. If it is set to 0 then this feature is not offered by this cell. If the MS does not have a complete set of PSI, it can indicate this to the network upon establishment of TBF and the network will send on the PACCH the PSI messages required to make up the deficit.

AP.2 PSI2 messages on the PBCCH

- **AP.2.1** PAGE_MODE, see AP.1.7.
- **AP.2.2** PSI2 change mark indicates changes in PSI2 messages.

- **AP.2.3** PSI2 index and PSI2 count indicate the index and count for PSI2 messages.
- **AP.2.4** Cell i.d. Consists of location area identity – MCC (three digit) + NCC (two or three digits) + LAI (0–16383) + routeing area code (eight bits) + cell identity (16 bits).
- **AP.2.5** Non-GPRS cell options skip.
- **AP.2.6** Reference frequency lists. These frequency lists are indexed by RFL_NUMBERS. An MS is told which RFL_NUMBER to use to identify a particular list of frequencies. The RFL contents may be a bit-map of frequencies or variable bit-map. The table below shows a typical bit-map 0 format with the RFL containing ARFCNs 4, 7, 116 and 122.

				ARFCN 124	ARFCN 123	ARFCN 122	ARFCN 121	OCTET 2
						1		
ARFCN 120	ARFCN 119	ARFCN 118	ARFCN 117	ARFCN 116	ARFCN 115	ARFCN 114	ARFCN 113	OCTET 3
				1				
ARFCN 8	ARFCN 7	ARFCN 6	ARFCN 5	ARFCN 4	ARFCN 3	ARFCN 2	ARFCN 1	OCTET 17
	1			1				

- **AP.2.7** Cell allocation. This is the set of frequencies used within the cell and is referred to by an RFL number which identifies a particular RFL content.
- **AP.2.8** GPRS mobile allocation (MA). If frequency hopping is used, this identifies the hopping parameters. These include:

  —**HSN**, hopping sequence number 0–63, 0 = cyclic, 61–63 = PR codes.
  —**RFL number list**, an RFL number identifying a list which specifies the range of frequencies the MS must hop. If hopping is in use and this information element is not included, the MS will use the cell allocation.
  —**MA_BITMAP**, a bit-map of the hopping frequencies.

- **AP.2.9** PCCCH description. This describes the timeslots carrying PCCCHs.

  —The parameter **KC** (range 1–16) gives the total number of timeslots carrying PCCCHs.
  —**PCCCH_Group** one group occupies one timeslot. (An MS IMSI effectively selects a PCCCH_Group). These are numbered from 0 to (KC − 1), with the lowest group occupying the lowest numbered timeslot, the next lowest group the next lowest timeslot etc. This continues in the same way on the next carrier carrying PCCCHs up to a possible total of 16 groups.

- **AP.2.10** Training sequence code (TSC). One of eight training sequence codes to be used on PCCCH.
- **AP.2.11** Mobile allocation index offset (MAIO). The index offset to be used by an MS when PDCHs are frequency hopping.
- **AP.2.12** ARFCN. The frequency of the PCCCH if no frequency hopping is in use.
- **AP.2.13** PCCCH_TIMESLOT. This gives the TN allocation for PCCCHs.

TIMESLOT:	7	6	5	4	3	2	1	0
PCCCH allocation:							1	1

This allocation corresponds to TN0 and TN1 having PCCCH allocated.

AP.3 PSI3 messages on the PBCCH

This conveys information on the BA_GPRS–BCCH allocation for GPRS neighbour cells and cell selection parameters.

- **AP.3.1** PAGE_MODE, see AP.1.7.
- **AP.3.2** PSI3_change_mark indicates changes in PSI3 or 3-bis messages. Two-bit code.
- **AP.3.3** PSI1_3-bis_count. The last, highest indexed message in 3-bis.
- **AP.3.4** Cell_Bar_Access

$$0 = \text{normal access for cell reselection,}$$
$$1 = \text{barred access for cell reselection.}$$

- **AP.3.5** GPRS_RXLEV_ACCESS_MIN. The GPRS_RXLEV_ACCESS_MIN is the minimum RXLEV an MS must receive in order for an MS to access a cell.

Parameter setting	−dBm	Parameter setting	−dBm	Parameter setting	−dBm	Parameter setting	−dBm
0	110	16	94	32	78	48	62
1	109	17	93	33	77	49	61
2	108	18	92	34	76	50	60
3	107	19	91	35	75	51	59
4	106	20	90	36	74	52	58
5	105	21	89	37	73	53	57
6	104	22	88	38	75	54	56
7	103	23	87	39	71	55	55
8	102	24	86	40	70	56	54
9	101	25	85	41	69	57	53
10	100	26	84	42	68	58	52
11	99	27	83	43	67	59	51
12	98	28	82	44	66	60	50
13	97	29	81	45	65	61	49
14	96	30	80	46	64	62	48
15	95	31	79	47	63	63	47

- **AP.3.6** GPRS_MS_TXPWR_MAX_CCH. The GPRS_MS_TXPWR_MAX_CCH is the maximum power that the MS may transmit on PRACH bursts.

Parameter setting	dBm	Parameter setting	dBm
0	43	10	23
1	41	11	21
2	39	12	19
3	37	13	17
4	35	14	15
5	33	15	13
6	31	16	11
7	29	17	9
8	27	18	7
9	25	19	5

- **AP.3.7** GPRS_PRIORITY_CLASS. This parameter gives the hierarchical cell priority for the cell. Its range is 0–7, 7 being the highest priority. In certain circumstances, a cell with a higher GPRS_PRIORITY_CLASS will be selected in preference to a cell with a lower priority.
- **AP.3.8a** GPRS_HCS_THR. This GPRS_HCS_THR parameter sets the RXLEV threshold for the cell related to the cell priority. Its range is −110 dBm to −48 dBm in 2 dB steps.
- **AP.3.8** GPRS_CELL_RESELECT_HYSTERESIS. This is applied to cells in the same RA when the MS is in the GMM ready condition. Its range is 0 db to 14 B in 2 dB steps.
- **AP.3.9** C31_HYST. This indicates, when the setting = 1, that GPRS_CELL_RESELECT_HYSTERESIS will be applied to the C31 criterion.
- **AP.3.10** C32_QUAL. This indicates, when the setting = 1, that GPRS_RESELECT_OFFSET will only be applied to the cell with the greatest RLA – receive level average.
- **AP.3.11** RANDOM_ACCESS_RETRY. If the setting =1, the MS is allowed to try to access another cell if available.
- **AP.3.12** T_RESEL is the time for which the MS is not allowed to attempt to reselect this cell after an abnormal release from this cell, if another cell is available for selection. The default is 5 seconds, which will apply if this parameter is not broadcast. The range is 5, 10, 15, 20, 30, 60, 120, 300 seconds.
- **AP.3.13** Neighbour cell parameters

  —**AP.3.13.1** RA_RESELECT_HYSTERESIS. Applying to the MS in both GMM standby and ready states, this RA_RESELECT_HYSTERESIS applies when selecting a neighbour cell in a new RA. In the absence of this parameter, GPRS_CELL_SELECT_HYSTERESIS applies. The range is 0 to 14 dB in 2 dB steps.
  —**AP.3.13.2** START_FREQUENCY defines the ARFCN for the BCCH frequency of the first neighbour cell in the list (10 bits).
  —**AP.3.13.3** NR_OF_REMAINING_CELLS. The number of neighbour cells remaining in the list. For each of these, FREQ_DIFF and CELL_SELECTION_PARAMS will be shown. The range is 1–16.

—**AP.3.13.4** FREQ_DIFF. Specifies the neighbour cell difference in frequency from the previous listed neighbour cell frequency using the FREQ_DIFF_LENGTH field.

—**AP.3.13.5** FREQ_DIFF_LENGTH. Specifies the neighbour cell frequency difference from the previously listed neighbour cell frequency. The range is 1 to 8 bits.

—**AP.3.13.6** PSI1_REPEAT_PERIOD. Indicates the neighbour cell PSI1 repeat period. The range is 1, 2, 3,...,16 multiframes.

—**AP.3.13.7** PCCCH type. Indicates the type of PCCCH for the neighbour cell.

$$0 = 52 \text{ multiframe}$$
$$1 = 51 \text{ multiframe}$$

—**AP.3.13.8** Neighbour cell selection parameters

○ **BSIC (n)**
○ **Same RA as serving cell**,

$$0 = \text{different RA, } 1 = \text{same RA}$$

○ **GPRS_RXLEV_ACCESS_MIN (n)**
○ **GPRS_MS_TXPWR_MAX_CCH (n)**
○ **GPRS_TEMPORARY_OFFSET (n)**, the negative offset applied to C32 for the duration of **GPRS_PENALTY_TIME (n)**, the time for which **GPRS_TEMPORARY_OFFSET (n)** is applied for C32 calculations. 10 – 320 seconds in 10 second steps.
○ **GPRS_RESELECT_OFFSET (n)**,

$-52\,$dB, $-48\,$dB, $-44\,$db, $-40\,$dB, $-36\,$dB, $-32\,$dB, $-28\,$dB, $-24\,$dB, $-20\,$dB, $-16\,$dB, $-12\,$dB, $-10\,$dB, $-8\,$dB, $-6\,$dB, $-4\,$dB, $-2\,$dB, $0\,$dB, $\cdots +16\,$dB, $+20\,$dB, $+24\,$dB $\cdots +48\,$dB.

$0\,$dB is the default used if this parameter is not transmitted.

○ **HCS parameters(n)**
○ **SI13 _PBCCH _Location(n)**,

$$0 = \text{BCCH normal, } 1 = \text{BCCH extended}$$

○ **PBCCH_LOCATION(n)**, the location of PBCCH on the BCCH carrier,

$$00 = TN1$$
$$01 = TN2$$
$$10 = TN3$$
$$11 = TN4$$

If either of the above two parameters are missing, then the information is obtained from the neighbour cell. PSI1 repeat period(n), 1–16 multiframes.

- **AP.3.14** LSA parameters. This adds a local service area (LSA) identity to

—the BA_GPRS list;
—the frequency list in a cell change order; or
—a packet measurement order.

These lists are used by subscribers belonging to local area subscriptions.

AP.3bis PSI3-bis messages on the PBCCH

These messages contain the same sets of information as PSI3. They are used to supplement PSI13.

AP.4 PSI4 messages on the PBCCH and PACCH

This is optional and is only sent if INT_MEAS_CHANNEL_LIST_AVAIL is set on PSI1 – see AP.1.2.

This PSI contains a list of channels within the cell to be used by the MS for interference measurements in packet idle mode

- **AP.4.1** PSI4_change_mark. This is changed each time information in PSI4 is changed. The MS should reread the messages if this has changed since the messages were last read. The range is 0–3.
- **AP.4.2** PSI4_count. This gives the last (highest indexed) PSI4 message. The range is 0–7.
- **AP.4.3** PSI4_Index. This discriminates individual PSI4 messages. The range is 0–7.
- **AP.4.4** ARFCN. The binary coded ARFCN of the PDCHs to be measured when hopping is not used (10 bits).
- **AP.4.5** MA_Number. This refers to a PSI2 list for mobile allocation and HSN if the PDCHs to be measured are hopping. The range is 0–15.
- **AP.4.6** MAIO. The mobile allocation index offset to be used with MA to measure the PDCHs when hopping is in use. The range is 0–63.
- **AP.4.7** TIMESLOT_ALLOCATION. The TNs to be measured. For format see AP.2.13.

AP.5 PSI5 messages on the PBCCH

These give information for measurement reporting and network controlled cell reselection.

- **AP.5.1** PSI5 change_mark. This is changed each time information in PSI5 is changed. The MS should reread the messages if this has changed since the messages were last read. The range is 0–3.
- **AP.5.2** PSI5_count. This gives the last (highest indexed) PSI5 message. The range is 0–7.
- **AP.5.3** PSI5_Index. This discriminates individual PSI5 messages. The range is 0–7.
- **AP.5.4** Network_Control_Order

0 = NCO0, the MS shall perform autonomous cell reselection

1 = NCO1, the MS performs cell reselection and sends
    measurement reports

2 = NCO2, the MS sends measurement reports and
    the network controls cell reselection

3 = Reserved.

- **AP.5.5** NC_NON_DRX_PERIOD. This gives the maximum time that an MS shall stay in non-DRX mode after an NC measurement report has been sent.

$$0 = \text{normal DRX}$$
$$1 = 0.24\,\text{s}$$
$$2 = 0.48\,\text{s}$$
$$3 = 0.72\,\text{s}$$
$$4 = 0.96\,\text{s}$$
$$5 = 1.2\,\text{s}$$
$$6 = 1.44\,\text{s}$$
$$7 = 1.92\,\text{s}$$

- **AP.5.6** NC_Reporting_Period_I (and T). The time periods for cell reselection measurement reporting for packet idle modes and packet transfer modes.

  Range $0-7 = 0.48, 0.96, 1.92, 3.84, 7.68, 15.36, 30.72, 61.44$ seconds.

- **AP.5.7** Ext_Measurement_Order

  0 = EM0, the MS shall not perform extended measurements

  1 = EM1, the MS shall send extended measurement reports
      to the network. If set to EM1, the MS must read the remaining
      extended measurement parameters.

- **AP.5.8** Ext_Reporting_Type

  0 = Type 1 measurement report. Carriers will be reported if they
      are among the strongest six whether or not BSIC was decoded.
      The report contains signal strength (SS) and BSIC if available.

  1 = Type 2 measurement report. Carriers will be reported if they
      are among the strongest six if BSIC was decoded with allowed
      NCC. The report contains (SS) and BSIC if available.

  2 = Type 3 measurement report. Carriers will be reported without
      BSIC decoding. The report contains SS.

- **AP.5.9** Ext_Reporting_Period. The time period between extended measurement reports.

  Range $0-7 = 1$ min, 2 mins, 4 mins, 8 mins, 16 mins, 32 mins, 64 mins, 128 mins.

- **AP.5.10** Int_Frequency, the interference frequency to be measured. This points to a frequency in the Ext_Freq_List. The range is $0-31$.
- **AP.5.11** NC_Frequency_List and Ext_Frequency_List

  —**AP.5.11.1** Start_Frequency. Ten bits, the first frequency in the list.
  —**AP.5.11.2** Freq_Diff_Length. The number of bits to be used for the frequency difference field in the current frequency group. The range is $1-8$.
  —**AP.5.11.3** Frequency_Difference. The difference in frequency from the previous frequency.

  $$\text{ARFCN}(n) = \text{ARFCN}(n-1) + W$$

where W is the binary number of Frequency_Difference.

AP.13 PSI13 messages on the PACCH

Provides the MS with cell specific information. This is the same information as in the SI3 messages, and Section AS.13 should be referred to for the messages on PSI13.

A2.2 Abbreviations used in this appendix

ARFCN	Absolute radio frequency channel number
BCCH	Broadcast control channel
BSIC	Base station identity code
BSN	Block sequence number
BTS	Base transceiver station
CA	Cell allocation
CCCH	Common control channel
CV	Countdown value
DL	Downlink
DRX	Discontinuous reception
EC	Emergency call
EMO	Extended measurement order
GMM	GPRS mobility management
GPRS	General packet radio service
HCS	Hierarchical cell structure
HSN	Hopping sequence number
IMSI	International mobile subscriber identity
LAI	Location area identifier
LSA	Local service area
MA	Mobile allocation

MAC	Medium access control
MAIO	Mobile allocation index offset
MCC	Mobile country code
MS	Mobile station
NC	Network controlled
NCC	Network colour code
NCO	Network control order
NMO	Network mode of operation
PACCH	Packet associated control channel
PAGCH	Packet access grant channel
PBCCH	Packet broadcast control channel
PCCCH	Packet common control channel
PDCH	Packet data channel
PDTCH	Packet data traffic channel
PRACH	Packet random access channel
PSI	Packet system information
RA	Routeing area
RAC	Routeing area code
RFL	Radio frequency list
RLA	Receive level average
RLC	Radio link control
RXLEV	Receive level
S	Spreading for PRACH
SS	Signal strength
TBF	Temporary block flow
TN	Timeslot number
TSC	Training sequence code
V(S)	Next PDU to be sent

Appendix 3: Inter-Layer Primitives

A3.1 The interface user to mobility management – the GMMREG-SAP and primitives

Figure A3.1 shows the primitives used in communication between the mobile station user (network) and the (G)MM layer. This uses the service access point GMMREG-SAP (GPRS mobility management registration service access point).

Registration is always the first service to be used by a mobile station. It tells the network that the mobile station is now contactable. When the procedure is completed, the network knows where the mobile station is located, and the mobile station has been authenticated and given a ciphering key. The mobile station enters the GMM ready mode when this procedure is completed.

Figure A3.1 lists the primitives used on the mobile station side and shows some of the parameters sent with these primitives.

A3.2 The interface user to SM – SMREG-SAP and primitives

Before the mobile station SNDCP layer can transfer N PDUs or receive SN PDUs, a PDP context activation must be performed. This simply means that the mobile station and network must establish a link for transfer of packet data. The session management layer takes care of this and the negotiation of QoS.

The session management registration service access point (SMREG-SAP) is used to set up a packet transfer session.

GPRS in Practice: A Companion to the Specifications   Peter McGuigann
© 2004 John Wiley & Sons, Ltd   ISBN: 0-470-09507-5

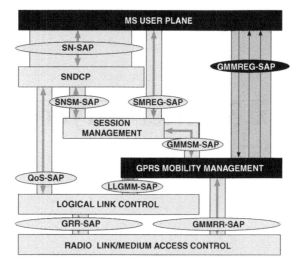

Primitives between session management GMM Layer and MS on the MS side at GMMREG-SAP MS side		
Primitive	Parameter	Use
GMMREG-ATTACH.REQ	Attach type, READY timer, STANDBY timer	MS-GMM, initiates GPRS and/or IMSI attach. GMM is asked to send ATTACH REQUEST to the network. Registration is pending
GMMREG-ATTACH.CNFM	PLMN's MT capability, attach type	GMM-MS. The REQ. is confirmed. The attach is confirmed by the network. The LLC and RR sub-layers are given the TLLI to be used by GMM.
GMMREG-ATTACH.REJ	Cause	GMMM-MS. REQ. rejected. The attach has failed. Network has rejected the request
GMMREG-DETACH.REQ	Detach type, power off or normal detach	MS-GMM requesting GPRS and/or IMSI detach. The GMM layer is asked to send DETACH. If it is power off detach, procedure is terminated in the MS after this message is sent.
GMMREG-DETACH.CNFM	Detach type	GMM-MS, network confirms the detach. Completes normal detach. Any PDP context is deactivated
GMMREG-DETACH.IND	Detach type	GMM-MS, either a network detach has been performed or the detach is performed locally by the MS due to STANDBY timer expiry or failed routeing area update

Figure A3.1   GPRS user – GMM (GMMREG-SAP) primitives

Figure A3.2 shows the primitives used by the mobile station at the SMREG-SAP action.

A3.3   The interface user to SNDCP – the SN-SAP and primitives

The SNDCP is responsible for converting the external network formats (or N PDUs) into the sub-network (GPRS) formats (SN PDUs). This service is accessed through the SN-SAP – sub-network service access point.

The SNDCP is also responsible for compression of the N PDUs to make for more efficient data transmission within the GPRS network.

The SNDCP layer also negotiates the compression parameters to be used in communication between a mobile station and the network – this is done using the XID (exchange identity) primitives.

Figure A3.3 shows the primitives used between the user and SNDCP across the SN-SAP.

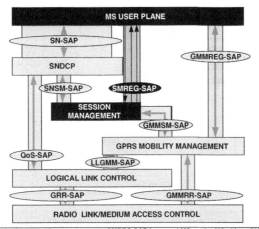

Primitives between session management SMREG SAP layer and MS on the MS side at SMREG-SAP		
Primitive	Parameter	Use
SMREG-PDP-ACTIVATE.REQ	PDP type QoS, NSAPI APN, Data Mode	MS-SM, initiates a PDP context activation. The SM is requested to send ACTIVATE PDP CONTEXT REQ. to the network, which it does via the GMM layer
SMREG-PDP-ACTIVATE.CNF	PDP type, PDP address, QoS, NSAPI, PDP	SM-MS. The REQ. is confirmed. ACTIVATE PDP CONTEXT ACCEPT is received from the network and the SM has asked the SNDCP to establish LLC
SMREG-PDP-ACTIVATE.REJ	Cause	SM-MS. REQ. rejected. Could be network rejection with return of ACTIVATE PDP CONTEXT FAILURE or not possible to establish LLC links.
SMREG-PDP-ACTIVATE.IND	PDP typ, ,QoS, NSAPI, APN, Data Mode	SM-MS -the network requests PDP CONTEXT ACTIVATION. The MS responds with SMREG-PDP-ACTIVATE.REQ or SMREG-PDP-ACTIVATE.REJ
SMREG-PDP-DEACTIVATE.REQ	NSAPI	MS-SM, initiates a PDP context deactivation. The SM is requested to send DEACTIVATE PDP CONTEXT REQ. to the network, which it does via the GMM layer
SMREG-PDP-DEACTIVATE.CNF	NSAPI	SM-MS. The REQ. is confirmed. ACTIVATE PDP CONTEXT ACCEPT is received from the network and the SM has asked the SNDCP to locally release LLC links. The PDP context is deactivated.
SMREG-PDP-DEACTIVATE.IND	NSAPI	SM-MS -the network requests PDP CONTEXT DEACTIVATION. The MS responds with SMREG-PDP-DEACTIVATE.ACCEPT
SMREG-PDP-MODIFY.IND	NSAPI, QoS	SM-MS. MODIFY PDP CONTEXT REQUEST is received from the network. This has been ack. with MODIFY PDP CONTEXT ACCEPT. LLC links are adjusted.
SMREG-AA-PDP-ACTIVATE-REQ	Server address, QoS, NSAPI, Data mode	MS-SM. MS sets up PDP context activation. SM asked to send ACTIVATE AA PDP REQUEST to network. AA PDP context pending.
SMREG-AA-PDP-ACTIVATE-CNF	QoS	SM-MS. The REQ. is confirmed. ACTIVATE AA PDP CONTEXT ACCEPT is received from the network and the SM has asked the SNDCP to establish LLC
SMREG-AA-PDP-ACTIVATE-REJ	Cause	SM-MS. REQ. rejected. Could be network rejection with return of AA ACTIVATE PDP CONTEXT FAILURE or not possible to establish LLC
SMREG-AA-PDP-DEACTIVATE.REQ	NSAPI	MS-SM, the MS initiates anonymous PDP context deactivation.
SMREG-AA-PDP-DEACTIVATE.IND	NSAPI	SM-MS, the MS anon. PDP context deactivation has been performed.
SMREG-PDP-ACTIVATE.REQ	PDP type QoS, NSAPI APN, Data mode	Network -SM, network initiates a PDP context activation. The SM is requested to send ACTIVATE PDP CONTEXT REQ. to the MS, which it does via the GMM layer. As the MS executes context activation no SMREG-ACTV-CNFM is required
SMREG-PDP-ACTIVATE.REJ	Cause	SM-Network. REQ. rejected. Could be MS rejection with return of ACTIVATE PDP CONTEXT FAILURE or lower layer failure or timer expiry
SMREG-PDP-DEACTIVATE.REQ	NSAPI	Network-SM, initiates a PDP context deactivation. The SM is requested to send DEACTIVATE PDP CONTEXT REQ. to the network, which it does via the GMM layer
SMREG-PDP-DEACTIVATE.CNF	NSAPI	SM-Network. The REQ. is confirmed. DEACTIVATE PDP CONTEXT ACCEPT is received from the MS and the SM has asked the SNDCP to locally release LLC links. The PDP context is deactivated.
SMREG-PDP-MODIFY.REQ	NSAPI, QoS	Network-SM. SM is asked to send MODIFY PDP CONTEXT REQUEST
SMREG-PDP-MODIFY.CNFM	NSAPI, QoS	SM-Network. MS has sent MODIFY PDP CONTEXT ACCEPT. The MS has modified the PDP context and SM asked SNDCP to adjust LLC links. Modification is complete
SMREG-PDP-MODIFY.REJ	NSAPI, QoS	SM-Network. Failure due to lower layer or timer expiry

Figure A3.2   GPRS user – session management (SMREG-SAP) primitives

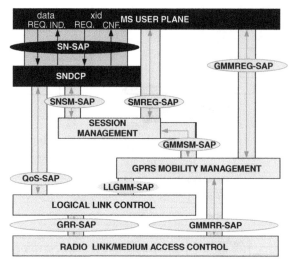

Primitives between SNDCP layer and SNDCP user layer		
Primitive	Parameter	Use
SN-DATA.REQ.	N PDU, NSAPI	SNDCP user requests acknowledged transmission of N PDU. Confirmation is received by the LLC layer
SN-DATA.IND	N PDU, NSAPI	Delivery by the SNDCP of N PDU to the other end user.
SN-UNITDATA.REQ.	N PDU, NSAPI	SNDCP user requests unacknowledged transmission of N PDU
SN-UNITDATA.IND.	N PDU, NSAPI	Delivery by the SNDCP of N PDU to the other end user.
SN-XID.REQ	Requested XID parameter	SNDCP user requests delivery of requested exchange identity parameters to far end peer.
SN-XID.IND	Requested XID parameter	Delivery of requested exchange identity parameters to far end peer.
SN-XID.RESP.	Negotiated XID parameter	Response used to deliver list of negotiated exchange identity parameters to far end peer.
SN-XID.CONFIRM	Negotiated XID parameter	Confirmation used to deliver list of negotiated exchange identity parameters to far end peer.

Figure A3.3   GPRS SNDCP – user (SN-SAP) primitives

A3.4   The interface SNDCP to SM – the SNSM-SAP and primitives

The layer 3 sub-network dependent convergence protocol is responsible for adapting external network protocols to the GPRS protocols, where the GPRS network (in this case) is the sub-network. This is the layer which initiates packet data transfer from mobile station to SGSN and vice versa.

Before the SNDCP can transmit and receive SN PDUs, a PDU transfer session must be established. The SNDCP layer communicates with the SM layer to use a session which will have been initiated through the SMREG-SAP.

Figure A3.4 lists the primitives which are used for communication between the SNDCP layer and the SM layers.

A3.5   The interface SNDCP to LLC – the QoS-SAP and primitives

When a TBF session has been established through the user plane asking the SM layer to establish a TBF session, the SM layer setting up a session and informing the SNDCP

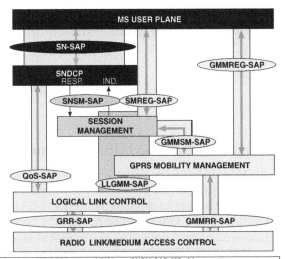

Note 1
Upon reception of the SNSM-MODIFY.IND. from the SM; the SNDCP shall if necessary:
establish ack peer–peer LLC operation for the indicated SAPI.
release the ack. peer–peer LLC operation for the originally assigned SAPI.

If ack. peer–peer operation is used after receipt of SNSM-MODIFY.IND., then the NSAPI involved shall enter the recovery state

If the peer–peer LLC operation is changed from ack. to unack. then all buffered N PDUs shall be deleted and the assignment of N PDU number for unack. mode shall start from 0, and

If the peer–peer LLC operation mode is changed from unack. to ack. then the assignment of N PDU number for ack. mode shall start from 0.

If the newly assigned SAPI differs from the original SAPI: LL-DATA.IND., LL-DATASENT.IND., LL-DATA.CNFM., LL-UNITDATA.IND. received on the old SAPI are ignored. LL-DATA.REQ., LL-UNITDATA.REQ shall be sent on the new SAPI, and; if ack. peer–peer LLC operation is used before and after the receipt of SNSM-MODIFY.IND, then all buffered N PDUs (whose reception has not been ack. or not transmitted) shall be transmitted starting from the oldest N PDU.

Primitives between SNDCP layer and SM layer-SNSM-SAP, MS side

Primitive	Parameter	Use
SNSM-ACTIVATE.IND.	TLLI, NSAPI	SM to SNDCP that an NSAPI is activated for
	QoS, SAPI	data transfer, and the QoS profile, and in the
	Radio priority	MS, the radio priority used by RLC/MAC.
		Upon receipt of this, the SNDCP shall if
		necessary, establish peer to peer LLC.
SNSM-ACTIVATE.RESP.	TLLI, NSAPI	SNDCP to SM that the indicated NSAPI is
		now in use and acknowledged LLC peer to
		peer operation, if necessary, is established.
SNSM-DEACTIVATE.IND.	TLLI, NSAPI	SM to SNDCP that an NSAPI is deactivated
	QoS, NSAPI,	and cannot be used by SNDCP. Stored N PDUs
	LLC Release	are deleted. The SNDCP shall, if necessary
	Indicator	release the ack. LLC peer–peer operation.
SNSM-DEACTIVATE.RESP.	TLLI, NSAPI	SNDCP to SM that the indicated NSAPI is
		no longer in use and acknowledged LLC peer to
		peer operation, if necessary, is released
SNSM-MODIFY..IND.	TLLI, NSAPI	SM to SNDCP to change QoS profile,
	QoS, SAPI	indicating SAPI to be used.
	Radio priority	In the case of an inter-SGSN routeing area
		update, in the new SGSN it is also used by
		the SM to inform the SNDCP that an NSAPI
		be created with the renegotiated QoS profile,
		the assigned SAPI, and in the MS the radio
		priority to be used by RLC/MAC.
SNSM-MODIFY..RESP.	TLLI, NSAPI	SNDCP to SM that the indicated NSAPI and
		QoS profile is now in use and acknowledged
		LLC peer to peer operation, if necessary, is
		released or established
SNSM-STATUS.REQ.	TLLI, SAPI	SNDCP to SM that the SNDCP cannot continue
	Cause.	operation due to LLC layer errors (as indicated
		with LLC release.ind.) or at SNDCP layer.
SNSM-WINDOW.IND.	TLLI, SAPI	Used in SGSN for inter-SGSN RA updates.
	V(R)	SM to SNDCP that the LLC PDU sequence
		number is correctly received by the MS. This
		is shown by theV(R) value received from the
		MS. SNDCP uses this, deleting all stored
		NPDUs with confirmed reception.
SNSM-SEQUENCE.IND	N PDU No	SM-SNDCP
SNSM-SEQUENCE.RSP	MS N PDU No	SNDCP-SM
SNSM-STOP.IND		NETWORK ONLY

(see note 1 — refers to SNSM-MODIFY..IND. rows)

Figure A3.4 GPRS sub network dependent convergence protocol – session management (SNSM-SAP) primitives

layer when a TBF session is ready, the SNDCP layer passes the SN PDUs down to the LLC (logical link control) layer, which assembles them into packets, adds control functions and establishes (if the requirement is indicated to the SNDCP layer by the negotiated QoS parameters) a highly reliable layer 2 data link to its peer entity (the LLC layer in the network or mobile station).

Figure A3.5 lists the primitives used for communication between the SNDCP and LLC layers through the QoS-SAP.

A3.6   The interface SM to GMM – GMMSM-SAP and primitives

The mobility management (MM) layer can be requested to take a mobile station from the MM idle state to the MM standby state, that is, from unattached to attached state either by the user plane requesting this (implicitly) through the SMREG-SAP or the user plane requesting this (explicitly) through the GMMREG-SAP. GMM attach procedures are then followed. When attach is completed (indicated by the GMM layer receiving *attach accept*), the SM layer can then use the GMM connection to set up the TBF session.

Figure A3.6 shows the primitives used for communication between the SM and GMM layers through the GMMSM-SAP. The session management messages passed between the two peer SM layers to set up a TBF session are unacknowledged messages.

Note that when the SM layer has set up a PDP session, it indicates this to the SNDCP layer with an SNSM-ACTIVATE.IND primitive message. The SNDCP then uses the QoS-SAP to negotiate the compression parameters to be used for the session. When this is completed, the SNDCP informs the SM layer and the SM layer informs the user layer that the session is activated. The SN-SAP can now be used to transfer SN PDUs (customer traffic).

A3.7   The interface GMM to LLC – the LLGMM-SAP and primitives

The purpose of the LLC layer is:

- To provide a highly reliable layer 2 data link between the mobile station and SGSN;
- To provide both acknowledged and unacknowledged message transfer;
- To provide ciphering for customer data;
- To provide user identity confidentiality;
- To provide logical link identification DLCI (data link connection identity) comprising TLLI and SAPI;
- To provide multiple logical connections;
- To provide sequential control of frames across the logical link;
- To detect and correct errors occurring on the link or declare the link unusable;
- To provide flow control of frames across the link.

Across the LLGMM-SAP, the GMM layer uses the LLC layer to transfer messages relating to GMM attach and for messages relating to session management for setting up a TBF for customer data.

The messages that are conveyed between peer entities using the LLGMM-SAP are unacknowledged messages (UNITDATA). The GMM layer is also responsible for assigning a TLLI to the LLC layer through this SAP. Figure A3.7 shows the primitives which flow across this SAP.

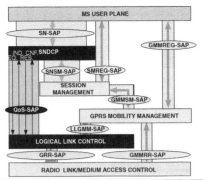

Primitives between SNDCP layer and LLC layer		
Primitive	**Parameter**	**The use is**
LL-RESET.IND	TLLI,	In the SGSN, LLC to SNDCP that the reset XID parameter is transmitted; in the MS LLC to SNDCP that the reset XID parameter is received. The SNDCP resets SNDCP XID par. to defaults and resets assignment of N PDU to start from 0 for every PDP context
LL-EST. REQ.	TLLI, XID requested	SNDCP to LLC to establish or re-establish acknowledged LLC layer peer–peer operation for a SAPI. XID requested delivers the requested SNDCP XID parameters to LLC. An LLC SABM frame is sent.
Network LL-EST. IND.	TLLI, XID requested n201-I,n201-U *n201-i and n201-U respectively define the maximum number of octets in the information fields of SN-DATA and SN-UNITDATA PDUs for a specific SAPI*	LLC to SNDCP, to inform SDNCP about establishment or re-establishment of ack.peer–peer operation for a SAPI in the LLC. XID requested delivers the SNDCP XID pars. In case of re-establishment, all NSAPIs for the affected SAPI enter the recovery state and all stored N PDUs not yet ack. or transmitted are transmitted when the link is re-established starting with oldest N PDU. All compressions used in ack. peer–peer LLC operations for this SAPI are reset. An LLC SABM frame is received.
Network LL-EST. RESP.	TLLI, XID negotiated	SNDCP to LLC, responds to LL-EST.IND from LLC; XID negotiated delivers these parameters to LLC. A UA frame is sent by the LLC, ABM mode is established.
LL-EST. CONFIRM	TLLI, XID negotiated n201-I,n201-U *If the 'local' parameter is included the LL enters ADM-asynchronous disconnect mode, timer T200 reset and L3 informed by LL.RELEASE.CNFM*	LLC to SNDCP, confirms initiation of ack. peer–peer operation for an LLC SAPI. XID negotiated delivers the SNDCP XID pars. to SNDCP. In case of re-establishment, all NSAPIs for the affected SAPI enter the recovery state and all stored N PDUs not yet ack. or transmitted are transmitted when the link is re-established starting with oldest NPDU. All compressions used in ack. peer–peer LLC operations for this SAPI are reset. A UA frame is received by the LLC and ABM mode is established
LL-RELEASE. REQ.	TLLI, Local	SNDCP to LLC, to release ack. peer–peer operation for an LLC SAPI. Local parameter indicates whether the termination is local. A LLC DISC frame will be sent to change to LLC ADM mode.
LL-RELEASE. CONFIRM	TLLI	LLC to SNDCP, confirms termination of ack. Peer–peer operation for an LLC SAPI. On receipt of this, compressed N PDUs awaiting forwarding to their SAPI are deleted at SNDCP. All compressions using ack. peer–peer operation on this SAPI are reset. The LLC link has been disconnected. LLC in ADM mode
LL-RELEASE. IND.	TLLI, Cause	LLC to SNDCP, that peer–peer ack. operation for a SAPI in LLC is terminated. Cause indicates why. On receipt, compressed N PDUs awaiting forwarding to their SAPI are deleted
LL-XID.REQ.	TLLI, XID requested	SNDCP to LLC, to deliver requested SNDCP XID parameters to LLC. An LLC XID frame will be sent.

(a)

Figure A3.5   Continued

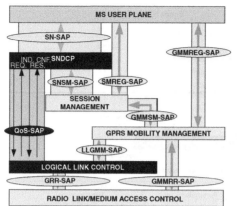

Primitives between SNDCP layer and LLC layer		
Primitive	**Parameter**	**The use is**
LL-XID.IND Network Network	TLLI, XID requested n201-I,n201-U	LLC to SNDCP, to deliver the requested SNDCP XID parameters to SNDCP. An LLC XID frame has been received.
LL-XID.RESP.	TLLI, XID negotiated	SNDCP to LLC, to deliver negotiated SNDCP XID parameters to LLC. An LLC XID frame is sent in response to a received XID frame
LL-XID.CONFIRM	TLLI, XID negotiated N201-I,N201-U	LLC to SNDCP, to deliver negotiated SNDCP XID parameters to SNDCP. An LLC XID frame has been received as response to a sent XID frame.
LL-DATA.REQ.	TLLI, SN PDU QoS param. Radio priority	SNDCP to LLC, to ask for transmission of an SN PDU. SNDCP associates a reference param. for each LL-DATA.REQ. SGSN QoS include the classes, precedence, delay, and peak throughput. MS QoS include peak throughput, radio priority – see 04.08. An LLC I frame is sent.
LL-DATA.IND.	TLLI, SN PDU	LLC to SNDCP, to deliver received SN PDU to SNDCP. An LLC I frame has been received.
LL-DATA.CONFIRM	TLLI, Reference	LLC to SNDCP, to confirm transmission of SN PDU. Includes a reference parameter so SNDCP can identify the LL-DATA.REQ. All stored N PDUs with completed reception are deleted. Successful reception of an LLC I frame has been acknowledged.
LL-DATASENT.IND	TLLI, Reference V(S)	Used in ack. mode in the SGSN, LLC to SNDCP, to inform SNDCP of the LLC send sequence number assigned to an LLC PDU. SNDCP associates the LLC send sequence number with the sent N PDU. The sent LLC frame was sent with V(S) indicated.
LL-UNITDATA.REQ.	TLLI, SN PDU QoS param. Radio priority Cipher	SNDCP to LLC, for unack. transmission of an SN PDU. Unconfirmed transmission is used by LLC without the necessity for ack. LLC peer–peer operation. QoS include in the SGSN precedence, delay, reliability, peak throughput; in MS peak throughput and reliability. The reliability class indicates whether the LLC frame is transmitted in protected or unprotected mode and whether RLC/MAC ack. or unack. mode is used. Radio priority is only in the MS and indicates the RLC/MAC radio priority. An LLC UI frame will be sent to the peer entity.
LL-UNITDATA.IND	TLLI, SN PDU	LLC to SNDCP, to deliver received SN PDU to the SNDCP. Acknowledged peer–peer LLC operation is unnecessary for unack. transmission of SN PDU. An LLC UI frame has been received from the peer entity.

(b)

Figure A3.5   GPRS SNDCP-LLC (QoS-SAP) primitives

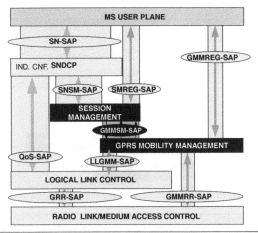

Primitives between SM layer and GMM layer – GMMSM-SAP, MS side		
Primitive	Parameter	The use is
GMMSM-ESTABLISH.REQ	-	SM to GMM, to send an ATTACH REQUEST message to network to set up a GMM connection. Only used if MS is not attached. GPRS attach is then caused by a non-anonymous PDP context activation.
GMMSM-ESTABLISH.CNF	-	GMM to SM, the establish request is confirmed and the MS is attached. SM can now proceed with PDP context activation.
GMMSM-ESTABLISH.REJ.	Cause	GMM to SM, the network has rejected attach. The MS has received the ATTACH REJECT message
GMMSM-RELEASE.IND.	-	GMM to SM, to indicate the MS is now detached (eg, timer expiry).
GMMSM-UNITDATA.REQ.	SM PDU	SM to GMM, to send an SM PDU to LLC for forwarding to peer entity in unack. mode.
GMMSM-UNITDATA.IND.	SM PDU	GMM to SM, forwards an unack. received SM PDU from LLC .

GPRS session management – GMM management (GMMSM-SAP)primitives, network side

Primitives between SM layer and GMM layer–GMMSM-SAP, network side		
Primitive	Parameter	The use is
GMMSM-RELEASE.IND.	-	GMM to SM, to indicate the MS is now detached (eg, timer expiry).
GMMSM-UNITDATA.REQ.	SM PDU	SM to GMM, to send an SM PDU to LLC for forwarding to peer entity in unack. mode.
GMMSM-UNITDATA.IND.	SM PDU	GMM to SM, forwards an unack. received SM PDU from LLC .

Figure A3.6   GPRS session management – GMM management (GMMSM-SAP) primitives, MS side

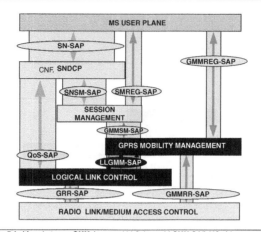

Primitives between GMM layer and LLC layer–LLGMM-SAP, MS side		
Primitive	**Parameter**	**The use is**
LLGMM-ASSIGN.REQ	Old TLLI New TLLI kc, RAND Ciph. Alg.	GMM to LLC, to assign a new TLLI/kc/Ciph. Alg. to LLC layer
LLGMM-TRIGGER.REQ	Cause	GMM to LLC, to request to send an LLC PDU to the network. Cause indicates whether the primitive triggers a page response.
LLGMM-TRIGGER.IND		LLC to GMM , to indicate that an LLC frame has been transmitted to the network.
LLGMM-SUSPEND.REQ	TLLI	GMM to LLC, to request suspension of all LLC links sending PDUs in ABM mode. GMM messages are not affected.
LLGMM-RESUME.REQ	TLLI	GMM to LLC, normal LLC frames may be sent again.
LLGMM-WINDOW.REQ	TLLI and old SGSNs V(R) per SAPI	GMM to LLC, requesting LLCs V(R)s
LLGMM-WINDOW.CNF.	TLLI and actual MSs V(R) per SAPI	LLC to GMM , give to the GMM the actual V(R)s for each LLC in ABM mode.
LL-UNITDATA.REQ.	TLLI and GMM PDU, protect, cipher	GMM to LLC, requesting to send a GMM message in unack. mode to far end peer.
LL-UNITDATA.IND..	TLLI and GMM PDU, cipher	LLC to GMM , a GMM message in unack. mode is delivered to GMM.
LLGMM-STATUS.IND.	TLLI and cause	LLC to GMM , LLC failure indication to GMM. Failure may be due to LLC or RLC/MAC
Primitives between GMM layer and LLC layer–LLGMM-SAP, network side		
LLGMM-ASSIGN.REQ	Old TLLI New TLLI kc, Ciph. Alg.	GMM to LLC, to assign a new TLLI/kc/Ciph. Alg. to LLC layer. An old TLLI can be unassigned.
LLGMM-TRIGGER.IND	TLLI	LLC to GMM , to indicate that an LLC frame has been received from the MS.
LLGMM-SUSPEND.REQ	TLLI, page	GMM to LLC, to request suspension of all LLC links sending PDUs in ABM mode. The page parameter indicates paging will be necessary to send data, or cause indicates no data to be sent until RESUME.REQ is received.
LLGMM-RESUME.REQ	TLLI	GMM to LLC, normal LLC frames may be sent again.
LLGMM-WINDOW.REQ	TLLI	GMM to LLC, requesting LLCs N(R)s
LLGMM-WINDOW.CNF.	TLLI and actual LLCs N(R) per SAPI	LLC to GMM , give to the GMM the actual N(R)s for each LLC in ABM mode.
LL-UNITDATA.REQ.	TLLI and GMM PDU, protect, cipher	GMM to LLC, requesting to send a GMM message in unack. mode to far end peer.
LL-UNITDATA.IND..	TLLI and GMM PDU, cipher	LLC to GMM , a GMM message in unack. mode is delivered to GMM.
LLGMM-STATUS.IND.	TLLI and cause	LLC to GMM , LLC failure indication to GMM.
LLGMM-PAGE.IND.	TLLI	LLC to GMM , a paging message is required for an MS..
LLGMM-PAGE-RESP.IND.	TLLI	LLC to GMM , a paging response is received from the MS.

Figure A3.7   GPRS GMM management – LLC (LLGMM-SAP) primitives, MS side

A3.8   The interface GMM to RR (RLC/MAC) – the GMMRR-SAP and primitives

The purpose of the RLC (radio link control) component of the RLC/MAC layer is:

- To interface the LLC primitives, allowing transfer of LLC protocol data units between the MAC and LLC;
- To segment LLC PDUs into RLC data blocks, and reassemble received RLC data blocks into LLC PDUs for transfer to the LLC layer;
- To segment and desegment RLC/MAC control messages into and out of RLC/MAC control blocks;
- To use, where appropriate, BEC – backward error correction (ARQ) for retransmission of RLC/MAC data blocks received with errors.

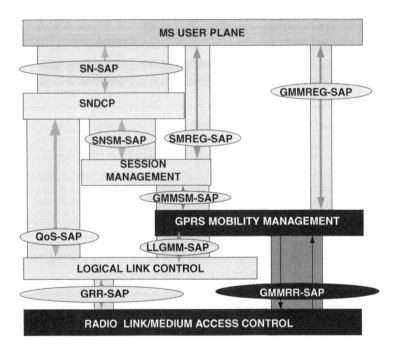

Primitives between GMM layer and RLC/MAC layer–GMMRR-SAP, MS side		
Primitive	Parameter	The use is
GMMRR-ASSIGN.REQ	New TLLI	GMM to RLC/MAC, to assign a new TLLI to RR (RLC/MAC) layer
GMMRR-PAGE.IND	TLLI	RLC/MAC to GMM, an RR paging message has been received by the RR layer.

Primitives between GMM layer and RLC/MAC layer–GMMRR-SAP, network side		
GMMRR-PAGE.REQ	TLLI, IMSI CI or CI list or RAI, priority	GMM to RLC/MAC, to request a paging message be sent.

Figure A3.8   GPRS GMM management – RLC/MAC (GMMRR-SAP) primitives

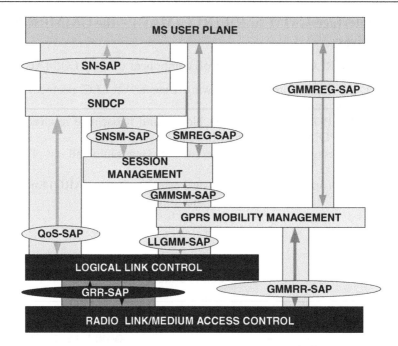

Primitives between LLC layer and RLC/MAC layer–GRR-SAP, MS side		
Primitive	**Parameter**	**The use is**
GRR-DATA.REQ	LLC PDU Priority Cause	LLC to RLC/MAC, to request ack. data transmission with a defined priority. Cause indicates if the GRR-DATA.REQ. is triggered as an implicit page response.
GRR-DATA.IND.	LLC PDU	RLC/MAC to LLC, to transfer received data from the RR to LLC layer
GRR-UNITDATA.REQ.	LLC PDU Priority	LLC to RLC/MAC, to request unack. data transmission with a defined priority.
GRR-UNITDATA.IND.	LLC PDU	RLC/MAC to LLC, to transfer unack received data from the RR to LLC layer
GRR-STATUS.IND.	Cause	RLC/MAC to LLC, to transfer RLC/MAC failures to the LL layer
Primitives between LLC layer and RLC/MAC layer–GRR-SAP, network side		
GRR-DATA.REQ	LLC PDU TLLI,CI, DRX MSCLM, QoS Priority	LLC to RLC/MAC, to request ack. data transmission with a defined priority.
GRR-DATA.IND.	LLC PDU TLLI,CI	RLC/MAC to LLC, to transfer ack. received data from the RR to LLC layer
GRR-UNITDATA.REQ.	LLC PDU TLLI,CI, DRX MSCLM, QoS Priority	LLC to RLC/MAC, to request unack. data transmission with a defined priority.
GRR-UNITDATA.IND.	LLC PDU TLLI,CI,	RLC/MAC to LLC, to transfer unack. received data from the RR to LLC layer
GRR-STATUS.IND.	TLLI,Cause,	RLC/MAC to LLC, to indicate to LLC that an error has occurred on the radio interface. The cause is indicated

Figure A3.9   GPRS logical link control – RLC/MAC (GRR-SAP) primitives

The purpose of the MAC component of the RLC/MAC layer is:

- To allow shared use of the radio medium (physical channels) between multiple mobile stations;
- To allow TBF of data packets carrying signalling or customer data between mobile stations and the network;
- To allow contention resolution and arbitration for mobile stations originating calls;
- To allow queuing and scheduling for mobile stations terminating calls;
- To receive PBCCH and PCCCH, allowing the mobile station to autonomously select and reselect cells.

The GMM layer uses the GMMRR-SAP to initiate (network side) and receive (mobile station side) paging messages. The GMM layer also uses this SAP to allocate the TLLI to the RLC/MAC layer, as, although this identifies the logical link used by the LLC, the RLC/MAC data blocks carry this identifier when necessary.

Figure A3.8 shows the primitives used by these two layers across the GMMRR-SAP.

A3.9   The interface LLC to RR (RLC/MAC) – the GRR-SAP and primitives

Customer data and control/signalling PDUs flow between the LLC and RR layers through this SAP. Figure A3.9 shows the primitives in use across this SAP.

A3.10   Abbreviations used in this appendix

ARQ	Automatic request for retransmission
BEC	Backward error correction
DLCI	Data link connection identity
GMM	GPRS mobility management
GPRS	General packet radio service
LLC	Logical link control
MAC	Medium access control
MM	Mobility management
N PDU	Network protocol data unit
PBCCH	Packet broadcast control channel
PCCCH	Packet common control channel
PDP	Packet data protocol
PDU	Protocol data unit
QoS	Quality of service
RLC	Radio link control
RR	Radio resources
SAP	Service access point
SAPI	Service access point identity
SGSN	Servicing GPRS support node

SM	Session management
SNDCP	Sub-network dependent convergence protocol
SN PDU	Sub-network protocol data unit
TBF	Temporary block flow
TLLI	Temporary logical link identifier
XID	Exchange identities

Appendix 4: Mobile Station Uplink Power Control

The MS controls its uplink power on any PDCH using the equation:

$$P_{ch} = \min\ (\Gamma_0 - \Gamma_{ch} - \alpha(C + 48),\ \text{PMAX}) \qquad (A4.1)$$

where, P_{ch} is the power to be used on the uplink PDCH; Γ_0 is 39 dBm for GSM900 and 36 dBm for GSM1800; Γ_{ch} is given with the *packet UL/DL assignment* and has range 0, 2, 4, ..., 64 dB; α is a PBCCH parameter with range 0, 0.1, 0.2, ..., 1.0; PMAX is GPRS_MS_TXPWR_CCH, with range 5–33 dBm and C is the normalised MS Rx level that the mobile station measures and processes.

This equation applies to all logical channels except the PRACH, for which the MS uses PMAX = GPRS_MS_TXPWR_CCH.

The value of C is derived for two MS states, packet idle mode and packet transfer mode, and is derived as follows.

A4.1 Packet idle mode

The MS measures the received level of the PCCCH when it leaves 'sleep' mode.

$$C_{\text{block } n} = SS_{\text{block } n} + Pb$$

where $SS_{\text{block } n}$ is the signal strength of an RLC/MAC block and Pb is the power reduction of the PCCCH compared to the BCCH.

This calculated C value is then processed by a running average filter

$$C_n = (1 - a)C_{n-1} + aC_{\text{block } n} \qquad C_0 = 0$$

GPRS in Practice: A Companion to the Specifications   Peter McGuigann
© 2004 John Wiley & Sons, Ltd   ISBN: 0-470-09507-5

where a is $1/\min$ $(n, \max(5, T_{AVG\_W}/T_{DRX}))$; T_{DRX} is the MS DRX period = 64 $(SPLIT\_PG\_CYCLE)^{-1}$ or BS_PA_MFRS. If DRX is not used, $SPLIT\_PG\_CYCLE = 768$ for PCCCH or 256 for CCCH; T_{AVG_W} is $2^{k/2}/6$, $k = 0$, $1, 2, \ldots, 64$. (See Table A4.1 for SPLIT_PG_CYCLE values).

Table A4.1 SPLIT_PG_CYCLE values

Parameter	Value
0	No DRX
1–64	1–64
65	71
66	72
67	74
68	75
69	77
70	79
71	80
72	83
73	86
74	88
75	90
76	92
77	96
78	101
79	103
80	107
81	112
82	116
83	118
84	128
85	141
86	144
87	150
88	160
89	171
90	178
91	192
92	214
93	224
94	235
95	256
96	288
97	320
98	352
99–127	

Example

In 'sleep' mode, an MS measures PCCH blocks

$$C_{\text{block } n} = SS_{\text{block } n} + Pb$$

where SS is as measured; Pb is 3 dB and SPLIT_PG_CYCLE value is 71 for PCCCH.
With results $C_{\text{block } n} = -82 + 3 = -79$ dBm.
Processed, this becomes

$$C_n = (1 - a)C_{n-1} + aC_{\text{block } n} \qquad C_0 = 0$$
$$= (1 - \{1/\min(n, \ \max(5, \ T_{\text{AVG\_W}}/T_{\text{DRX}}))\}C_{n-1}$$
$$+ \{1/\min(n, \ \max(5, \ T_{\text{AVG\_W}}/T_{\text{DRX}}))\}\ C_{\text{block } n}$$

and with suitable parameters for $T_{(\text{xx})}$:

$$= (1 - \{1/\min(2, \ \max(5, \ 0.67/0.9))\}(-78) + \{1/\max(2, \ \max(5, \ 0.67/0.9))\}(-82))$$
$$= (1 - 1/2)(-78) + 1/2(-82)$$
$$= -39 - 41$$
$$= -80$$

Substituting this value of C into Equation (A4.1), and using $k = 4$, $C_{n-1} = -78$ dBm,
$\Gamma_0 = 39$ dBm; $\Gamma_{\text{ch}} = 4$ dB and $\alpha = 0.2$, gives:

$$P_{\text{ch}} = \min(39 - 4 - 0.2(-80 + 48), \ 25)$$
$$P_{\text{ch}} = \min(28.6, \ 25)$$

In this case, the MS transmits GPRS_MS_TXPWR_CCCH.

A4.2 Packet transfer mode

In this mode, the MS measures the BCCH carrier and C_n is processed as

$$C_n = (1 - b)C_{n-1} + (b \times SS_n)$$

where b is $(6 \times T_{\text{av\_T}}) - 1$ and $T_{\text{av\_T}}$ is $2^{k/2}/6$, $k = 0, 1, 2, \ldots, 64$.
If the parameter PC_MEAS_CHAN tells the MS to measure on the PDCH, then

$$C_{\text{block } n} = SS_{\text{block } n} + Pb$$

and $C_{\text{block } n}$ values are filtered with

$$C_n = (1 - c)C_{n-1} + cC_{\text{block } n}$$

where c is $(12 \times T_{\text{av\_T}})^{-1}$.

A4.3   Abbreviations used in this appendix

BCCH   Broadcast control channel
CCCH   Common control channel
DL   Downlink
DRX   Discontinuous reception
GSM   Global system for mobile communication
MAC   Medium access control
MS   Mobile station
PBCCH   Packet broadcast control channel
PCCCH   Packet common control channel
PDCH   Packet data channel
PRACH   Packet random access channel
RLC   Radio link control
SS   Signal strength
UL   Uplink

Appendix 5: A Possible Problem with UL TBFs When the Application Layer is Using TCP/IP

When TCP/IP is in use by the MS application (network) layer, the worst-case problem illustrated in Figure A5.1 may occur.

The top diagram shows a TCP/IP packet delivered over a landline connection. When the far end receives the packet, it is forced to send an acknowledgment. A retransmission timer is running at the sending end and this requires reception of an acknowledgment before its expiry. In this case the acknowledgment is received before the timer expires and all is well. Had it not been received before the timer expired then the TCP/IP packet would be resent.

The lower diagram shows the same TCP/IP packet sent over the GSM air interface.The outline procedures are shown. In particular, in this worst case illustration, the radio resources are released when the TCP/IP packet is transferred across the air interface as a TBF. Now when the Internet end sends an acknowledgment, the mobile station must be paged (it is assumed in this worst-case that the non-DRX timer has expired). In the worst case, if split paging is not in use, it can take up to 15 seconds to page the mobile station. By the time the mobile station is paged and the downlink TBF is completed, the TCP retransmisssion timer has expired and the TCP layer sends the mobile station the same data PDU for retransmission!

To overcome this potential problem, network operators may keep the mobile station connected by keeping the radio resources in place until the acknowledgment PDU is received. (This could be done, for example, by the network not sending *final packet uplink ack* to the mobile station until a DL PDU arrives for it.)

GPRS in Practice: A Companion to the Specifications   Peter McGuigann
© 2004 John Wiley & Sons, Ltd   ISBN: 0-470-09507-5

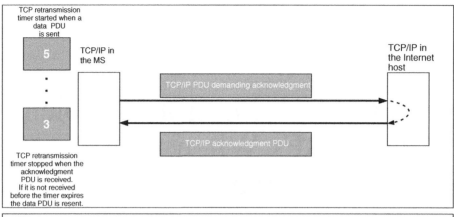

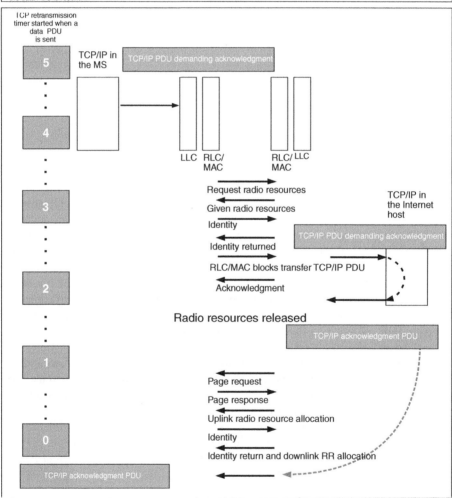

A5.1 Abbreviations used in this appendix

DL	Downlink
DRX	Discontinuous reception
GSM	Global system for mobile communication
LLC	Logical link control
MAC	Medium access control
MS	Mobile station
PDU	Protocol data unit
RLC	Radio link control
RR	Radio resources
TBF	Temporary block flow
TCP/IP	Transmission control protocol/Internet protocol
UL	Uplink

Index

Note: The page numbers in italics refer to figures or heads.

GPRS in Practice: A Companion to the Specifications   Peter McGuiggan
© 2004 John Wiley & Sons, Ltd   ISBN: 0-470-09507-5

Printed and bound in the UK by
CPI Antony Rowe, Eastbourne

Printed and bound by CPI Group (UK) Ltd, Croydon, CR0 4YY

16/04/2025

14658385-0002